T0318579

Metal Fatigue
Analysis Handbook

Metal Fatigue Analysis Handbook

Practical Problem-Solving Techniques for Computer-Aided Engineering

Yung-Li Lee

Mark E. Barkey

Hong-Tae Kang

ELSEVIER

AMSTERDAM • BOSTON • HEIDELBERG • LONDON
NEW YORK • OXFORD • PARIS • SAN DIEGO
SAN FRANCISCO • SINGAPORE • SYDNEY • TOKYO

Butterworth-Heinemann is an imprint of Elsevier

Butterworth-Heinemann is an imprint of Elsevier
225 Wyman Street, Waltham, MA 02451, USA
The Boulevard, Langford Lane, Kidlington, Oxford, OX5 1GB, UK

Notices
Knowledge and best practice in this field are constantly changing. As new research and
experience broaden our understanding, changes in research methods, professional practices,
or medical treatment may become necessary.

Practitioners and researchers must always rely on their own experience and knowledge in
evaluating and using any information, methods, compounds, or experiments described herein. In
using such information or methods they should be mindful of their own safety and the safety of
others, including parties for whom they have a professional responsibility.

To the fullest extent of the law, neither the Publisher nor the authors, contributors, or editors,
assume any liability for any injury and/or damage to persons or property as a matter of
products liability, negligence or otherwise, or from any use or operation of any methods,
products, instructions, or ideas contained in the material herein.

Library of Congress Cataloging-in-Publication Data
Lee, Yung-Li.
 Metal fatigue analysis handbook : practical problem-solving techniques for
computer-aided engineering / Yung-Li Lee, Mark E. Barkey, Hong-Tae Kang.
     p. cm.
 ISBN 978-0-12-385204-5 (hardback)
 1. Metals–Fatigue. 2. Computer-aided engineering. I. Barkey, Mark E.
II. Kang, Hong-Tae. III. Title.
 TA460.L416 2011
 620.1'66–dc23   2011020149

British Library Cataloguing-in-Publication Data
A catalogue record for this book is available from the British Library.

For information on all Butterworth-Heinemann publications
visit our website: *www.elsevierdirect.com*

Printed in the United States
Transferred to Digital Printing, 2011

Thank you to our families for their inspiration, encouragement, and patience!

To my friend, my love, my wife, Pai-Jen, and our children, Timothy and Christine

Yung-Li Lee

To my wife, Tammy, and our daughters, Lauren, Anna, and Colleen

Mark E. Barkey

To my wife, Jung-Me, and our son, Anthony

Hong-Tae Kang

Contents

Chapter 12: Design and Analysis of Metric Bolted Joints—VDI Guideline and Finite Element Analysis 461

Yung-Li Lee, Hsin-Chung Ho

Preface

Fatigue analysis of metals has been developing for more than 150 years from a simple approach involving elementary stress calculation based on the theory of elasticity, to a sophisticated treatment demanding a thorough knowledge of complex multiaxial loading and material behavior. Present engineering practice for comprehensive fatigue analysis requires computer-aided engineering (CAE) tools that use knowledge of material properties for cyclic stress–strain and fatigue behavior, structural kinematics for load simulations, finite element analysis for stress–strain calculations, and fatigue damage assessment for crack initiation and propagation life predictions.

Since the state-of-the-art technologies' load, stress, and fatigue analyses and their applications to engineering design for durability have been commonly adopted in numerous commercial analysis products, this publication strives to present, in a logical manner, the theoretical background needed for explaining and interpreting the analysis requirement and outputs. Ultimately, this book is intended to serve the reader as a theoretical manual or an analysis handbook. Beginning with coverage of background material, including references to pertinent research, the development of the formulas or theories applied in these CAE tools is followed by a number of examples to illustrate the process of designing structures to prevent fatigue failures in detail.

Considerable emphasis has been placed on including for the advanced student, as well as the practicing engineer, the present techniques used in CAE for road load simulations, pseudo stress calculations, estimates of material properties, incremental plasticity theories, and multiaxial fatigue life predictions. The understanding of these is essential to properly applying mechanical designs for durability. The specific illustrative examples are generally treated in separate sections within the chapters so that the reader can easily grasp the concepts.

The following list details the information provided in each chapter.

- **Chapter One** covers the three road-load analysis techniques to predict vehicle component loads for mechanical design for durability. They are the generic load case analysis for extreme loads; the semianalytical load analysis with the input of the acquired spindle forces, displacements, and accelerations; and the vehicle dynamics analysis with a tire model and the input of three-dimensional digitized terrain profiles. Commonly used commercial tire models are also introduced in this chapter.

- **Chapter Two** details the three pseudo stress analysis techniques: fixed reactive method, inertia relief method, and modal transient response analysis method. The pseudo stresses are the stresses calculated from a linear elastic finite element analysis. These stresses are named because they are different from the true stresses as plasticity occurs. The pseudo stresses can be directly employed for stress-based fatigue assessment in the high cycle fatigue (HCF) regime or for strain-based fatigue damage analysis in the low cycle fatigue (LCF) regime in conjunction with the multiaxial notch analysis to estimate the local true stresses and strains.

- **Chapter Three** overviews the historical rainflow cycle counting techniques for uniaxial load time history and the latest rainflow reversal extraction techniques for multiaxial load time histories. For multiaxial fatigue analyses, the uniaxial rainflow cycle counting method and the multiaxial rainflow reversal extraction method have been widely employed in the critical plane search approach and the equivalent stress–strain approach, respectively.

- **Chapter Four** provides an in-depth presentation on the stress-based uniaxial fatigue analysis, which focuses on the techniques provided by FKM-Guideline (*Analytical Strength Assessment of Components in Mechanical Engineering*) that are used to generate the synthetic nominal stress life and the local pseudo stress-life curves of a component, based on a given material's ultimate tensile strength, and to account for the mean stress effect with the mean stress sensitivity factor for various materials.

- **Chapter Five** introduces the concept of proportional and nonproportional loading, or stressing in the state of multiaxial stresses, and the nonproportional loading effect on fatigue strength; it also presents the popular stress-based theories (empirical formulas, equivalent stresses, critical plane approach, and

Dang Van multiscale method) for assessing fatigue damage under the state of multiaxial stresses.

- **Chapter Six** reviews the strain-based fatigue analysis techniques for a material under the state of uniaxial stress, which demands the familiar Ramberg–Osgood and the Masing equations for hysteresis loop simulations, the modified Neuber method or Molsky–Glinka's energy density method for the notch analysis, and finally the modified Morrow or Smith–Watson–Topper (SWT) mean stress corrected strain-life equation for fatigue behavior. The incorporation of residual stress and surface finish factor in the notch stress analysis is presented.

- **Chapter Seven** details the presentations of theories of plasticity, including the introduction of associated flow rule, consistency condition, and kinematic hardening rules, for simulating the material hysteresis behaviors under the state of multiaxial stresses, and of the multiaxial notch analysis techniques for estimating local material responses based on the input of pseudo stresses.

- **Chapter Eight** extends the discussion of multiaxial stress–strain relation and multiaxial notch analysis methods of Chapter Seven, and focuses on the presentation of strain-based fatigue damage assessment techniques for a material under multiaxial stress state. Particularly, the critical plane approach with the damage parameter such as the Fatemi–Socie, SWT, or Brown–Miller method has been introduced and discussed for the pros and cons.

- **Chapter Nine** presents an analytical solution to assess fatigue damage severity for various vibration test specifications. This chapter also reviews the fundamentals of sinusoidal and random vibration test methods and introduces the fatigue damage spectrum (FDS) calculation technique for each test method.

- **Chapter Ten** introduces some fatigue analysis techniques for seam-welded joints using linear elastic finite element analysis results. Especially, structural stress approaches developed by Dong and Femer were described in detail with examples.

- **Chapter Eleven** presents a summary of the primary factors affecting fatigue life and strength of resistance spot-welded joints and then focuses on some particular fatigue analysis techniques (the linear elastic fracture mechanics approach and the structural stress approach) and the relation of basic material specimens test to design.

- **Chapter Twelve** introduces the basics of the VDI 2230 guideline for ISO metric bolted joints to prevent potential failure modes such as embedding, clamp load loss, clamped plates crushing and slipping, bolt yielding, thread stripping, and bolt fatigue failure. The assumptions and the calculation procedure to estimate the elastic compliances of the bolt and the clamped plates are also described. In addition, the last portion of this chapter presents the commonly used finite element (FE) bolt-modeling techniques for forces and stress analysis, and recommends the appropriate fatigue damage assessment method for two different threaded bolts (e.g., rolled before and after heat treatment).

This book does not cover present nonlinear damage rules, but instead it emphasizes the application of the linear damage rule (LDR). The LDR has been universally adopted due to its simplicity and can account for load sequence and hardening effect, if used with the appropriate damage parameter. This book does not address detailed analytical models for crack growth behavior such as crack growth direction, growth rates, closures, R ratio effect, small crack effects, and threshold. However, the book does present the crack initiation life prediction models for metallic components subjected to the state of uniaxial and multiaxial stresses, and the crack growth approaches for life predictions of welded joints.

The plasticity theories introduced here do not apply to rate-dependent yielding, elevated temperature effect, and anisotropic materials. This book is not intended to be a comprehensive review of all published research in each respective subject area, but instead it presents the theories and problem-solving techniques commonly employed in automotive engineering.

Acknowledgments

Special thanks are due to Dr. Xiaobo Yang of Oshkosh Corporation, Dr. Peijun Xu of Ebco Inc., Dr. Mingchao (Mike) Guo of Chrysler Group LLC, Dr. Tana Tjhung of Chrysler Group LLC, and Mr. Hsin-Chung (Simon) Ho of Chrysler Group LLC for contributing to some chapters. Also, the authors are greatly indebted to Professor Jwo Pan of The University of Michigan–Ann Arbor, Professor Norman Dowling of Virginia Polytechnic Institute and State University, Professor Steve Tipton of the University of Tulsa, and Dr. Peter Heyes of HBM-nCode International for their valuable suggestions and encouragement.

About the Authors

Yung-Li Lee is a Technical Fellow in Science Labs & Proving Grounds at Chrysler Group LLC where he started his career as a product development engineer in 1988. He was elected as a Fellow of the Society of Automotive Engineers (SAE) and received the SAE Forest R. McFarland Award in 2006. His expertise includes durability/reliability testing and simulation.

Since 1997, he has been an Adjunct Professor of Mechanical Engineering at Oakland University in Rochester, Michigan, where he teaches graduate school courses on fundamentals of metal fatigue, accelerated test method development, reliability demonstration test planning, vibration test method development, frequency-based fatigue analysis, multiaxial fatigue, and fatigue of bolted and welded joints.

He received his B.S. in Civil Engineering from the National Central University in Taiwan in 1979, and M.S. and Ph.D. in Structural Engineering from the University of Wisconsin–Madison in 1984 and 1988, respectively. He is the primary author of the book, *Fatigue Testing and Analysis: Theory and Practice*, published by Elsevier in 2005.

Mark E. Barkey is a Professor in the Aerospace Engineering and Mechanics Department at the University of Alabama, where he joined the faculty in 1995. During this time, he has directed student theses and research in the area of fatigue testing and analysis of materials and structures. He has received support for his research from Ford, General Motors, Chrysler, Sandia National Laboratories, and NASA. He received a B.S. in Engineering Mechanics with a Minor in Mathematics from the University of Missouri-Rolla in 1989, and an M.S. and Ph.D. from the University of Illinois in Theoretical and Applied Mechanics in 1991 and 1993, respectively. He served as a Visiting Assistant Professor in the Mechanical and Industrial Engineering Department at the University of Illinois in 1994.

Dr. Barkey is a member of the Society of Automotive Engineering (SAE), the American Society of Mechanical Engineers (ASME), and the Society for Experimental Mechanics (SEM). He has made several presentations to the SAE Fatigue Design and Evaluation Committee, and was selected to receive the SAE Arch T. Colwell Merit Award for an SAE technical paper in 2005.

Hong-Tae Kang is an Associate Professor in the Mechanical Engineering Department at the University of Michigan-Dearborn where he joined the faculty in 2003. He has been conducting researches on fatigue life prediction methods using finite element analysis results for spot-welded joints and seam-welded joints. He received his B.S. in Earth Science Education from the Seoul National University in Korea in 1992, and M.S. and Ph.D. in Engineering Mechanics from the University of Alabama in 1997 and 1999, respectively.

Since graduation, he has worked for automotive companies in Detriot, Michigan, as a CAE analyst and project engineer for almost three years. He is a member of the Society of Automotive Engineering (SAE) and the American Society of Mechanical Engineers (ASME).

Nomenclature

Chapter 1

a_{cx} = engine/transmission CG x direction translational acceleration

a_{cy} = engine/transmission CG y direction translational acceleration

a_{cz} = engine/transmission CG z direction translational acceleration

$\vec{a_i}$ = i-th acceleration vector of the particles in the dynamics system

$\vec{a_E}$ = engine/transmission translational acceleration

β_X = engine/transmission angular acceleration about x axis

β_Y = engine/transmission angular acceleration about y axis

β_Z = engine/transmission angular acceleration about z axis

$\vec{\beta_E}$ = engine/transmission angular acceleration vector

CG = center of gravity

$\vec{C_i}$ = i-th constraint force vector

δW = virtual work

$\delta \vec{r_i}$ = virtual displacements of the i-th system, consistent with the constraints

F_x = spindle longitudinal force

F_y − spindle lateral force

F_z = spindle vertical force

F_{XLF} = left front tire-ground longitudinal force

F_{XLR} = left rear tire-ground longitudinal force

F_{XRF} = right front tire-ground longitudinal force

F_{XRR} = right rear tire-ground longitudinal force

F_{YLF} = left front tire-ground lateral force

F_{YLR} = left rear tire-ground lateral force

F_{YRF} = right front tire-ground lateral force

F_{YRR} = right rear tire-ground lateral force

F_{ZLF} = left front tire-ground vertical force

F_{ZLR} = left rear tire-ground vertical force

F_{ZRF} = right front tire-ground vertical force

F_{ZRR} = right rear tire-ground vertical force

$\vec{F}$ = vector forces

$\vec{F}_i$ = i-th applied forces vector

$\vec{F}_{EI}$ = engine/transmission inertial forces

$\vec{F}_{EMi}$ = engine/transmission mount forces

J = mass moment of inertia of a body

J_E = engine/transmission mass moment of inertia

LCA = low control arm

M_b = bending moment

m = mass

m_i = masses of the particles in the system

Ω = state space of a dynamic system

$[R]_{9\times6}$ = powertrain geometric rigid body transformation matrix

$\vec{r}_c$ = position vector of the centre of mass of the body

$\ddot{\vec{r}}_c$ = acceleration vector of the centre of mass of the body

$\vec{T}$ = vector moments or torques

$\vec{T}_{EI}$ = engine/transmission inertial torque

$\vec{T}_{of}$ = transfer case front output torque

$\vec{T}_{or}$ = transfer case rear output torque

$\vec{T}_{EMi}$ = engine/transmission mount moments

$\ddot{\theta}$ = angular acceleration

WFT = wheel force transducer

$X(t)$ = dynamic system state vector

$x_1(t)$ = first state of dynamic system at time t

$x_n(t)$ = n-th state of dynamic system at time t

Chapter 2

α = Rayleigh damping proportionality factor with the system mass matrix

$\{a\}$ = normal or natural modal vector

β = Rayleigh damping proportionality factor with the system stiffness matrix

C_n = generalized damping coefficient to the n-th mode shape

$[C]$ = damping matrix of a structure

Δt = time increment

$[\Phi]$ = modal matrix obtained in the solution of the undamped free vibration system

$F_n(t)$ = generalized force value to the n-th mode shape

$\{F(t)\}$ = dynamic forcing vector

γ_n = constant to normalize the modal vector

$[K]$ = stiffness matrix of a structure

K_n = generalized stiffness value to the n-th mode shape

$L_k(t)$ = k-th load magnitude at a time t

M = applied static moment at a node

$[M]$ = mass matrix of a structure

M_n = generalized mass to the n-th mode shape

ω = natural frequency of a structure

ω_n = natural frequency to the n-th mode shape

P = applied static force at a node

$\{P\}$ = applied static load vector

$\{P_0\}_{6\times1}$ = resultant force vector at the origin $(0,0,0)$

$\{P_i\}_{6\times1}$ = applied loading vector at a nodal point i

$\{\varphi\}_{6n\times1}$ = eigen-solution rigid body modal vector

$\{\varphi\}_n$ = n-th normalized modal vector

$\{\varphi\}_n^T$ = transpose of the n-th normalized modal vector

$[R_{i,0}]_{6\times6}$ = geometric rigid transformation matrix from a nodal point i to the reference point $(0,0,0)$

$\sigma_{ij}(t)$ = stress tensor at a time t

$\sigma_{ij,k}$ = stress tensor influence due to a k-th unit load source

t_i = i-th time instant

θ = nodal rotation

$\ddot{\theta}$ = nodal rotational acceleration

$\{u\}$ = flexible deformation vector of a structure

$\{u(0)\}$ = initial displacement vector

$\{u_r\}$ = rigid body motion vector with respect to the CG of a structure

$\{u_{r,0}\}_{6\times1}$ = rigid body motions from the reference point $(0,0,0)$

$\{u_{r,i}\}_{6\times1}$ = rigid body motions at a nodal point i

$\{u_t\}$ = total deformation vector of a structure

$\{\dot{u}(0)\}$ = initial velocity vector

$\{\ddot{u}\}$ = flexible acceleration vector of a structure

$\{\ddot{u}_r\}$ = rigid body acceleration vector with respect to the CG of a structure

$\{\ddot{u}_t\}$ = total acceleration vector of a structure

$\{\ddot{u}_{r,0}\}_{6\times1}$ = rigid body acceleration vector at the origin $(0,0,0)$

$\{\ddot{u}_{r,i}\}_{6\times1}$ = rigid body acceleration vector at a nodal point i

v = nodal displacement

$\ddot{v}$ = nodal translational acceleration

ξ_n = damping ratio to the n-th mode shape

$z_{n,i}$ = generalized displacement to the n-th mode shape at time t_i

$\{z\}$ = modal participation coefficients matrix

$\dot{z}_{n,i}$ = generalized velocity to the n-th mode shape at time t_i

$\ddot{z}_{n,i}$ = generalized acceleration to n-th mode shape at time t_i

Chapter 4

$2h_T$ = height of a rectangular section

$2N$ = number of reversals to a specific crack initiation length

A = fatigue parameter

a_d = constant in the size correction formula

a_G = material constant in the K_t/K_f ratio

a_M = material parameter in determining the mean stress sensitivity factor

a_N = Neuber's material constant

a_p = Peterson's material constant

a_R = roughness constant

a_{SS} = Siebel and Stieler material parameter

B = width of a plate

b = slope (height-to-base ratio) of an S-N curve in the HCF regime

b_G = material constant in the K_t/K_f ratio

b_M = material parameter in determining the mean stress sensitivity factor

b_{nw} = net width of a plate

b_W = width of a rectangular section

$C_{b,L}$ = load correction factor in bending

C_D = size correction factor

$C_{E,T}$ = temperature correction factor for the endurance limit

C_R = reliability correction factor

C_S = surface treatment factor

C_σ = stress correction factor in normal stress

$C_{\sigma,E}$ = endurance limit factor for normal stress

$C_{\sigma,R}$ = roughness correction factor for normal stress

$C_{t,L}$ = load correction factor in torsion

C_τ = shear strength correction factor

C_τ = stress correction factor in shear stress

$C_{\tau,R}$ = roughness correction factor for shear stress

$C_{u,T}$ = temperature correction factor

COV_S = coefficient of variations

D = diameter of a shaft

D_{PM} = critical damage value in the linear damage rule

d = net diameter of a notched shaft

d_{eff} = effective diameter of a cross section

$d_{eff,min}$ = minimum effective diameter of a cross section

$\Phi(-)$ = standard normal density function

G = stress gradient along a local x axis

$\overline{G}$ = relative stress gradient

$\overline{G}_\sigma()$ = relative normal stress gradient

$\overline{G}_\tau()$ = relative shear stress gradient

γ_W = mean stress fitting parameter in Walker's mean stress formula

HB = Brinell hardness

$K_{b,f}$ = fatigue notch factor for a shaft under bending

K_f = fatigue notch factor or the fatigue strength reduction factor

$K_{i,f}$ = fatigue notch factor for a superimposed notch

$K_{s,f}$ = elastic stress concentration factor for a plate under shear stress

$K_{s,f}$ = fatigue notch factor for a shaft or plate under shear

K_t = elastic stress concentration factor

$K_{t,f}$ = fatigue notch factor for a shaft under torsion

$K_{x,f}$ = fatigue notch factor for a pate under normal stress in x axis

$K_{x,t}$ = elastic stress concentration factor for a pate under normal stress in x axis

$K_{y,f}$ = fatigue notch factor for a plate under normal stress in y axis

$K_{y,t}$ = elastic stress concentration factor for a plate under normal stress in y axis

$K_{ax,f}$ = fatigue notch factor for a shaft under axial loading

k = slope factor (negative base-to-height ratio) of an S-N curve in the HCF regime

M_i = initial yielding moment

M_o = fully plastic yielding

M_σ = mean stress sensitivity factor in normal stress

N = number of cycles to a specific crack initiation length

N_E = endurance cycle limit

$N_{f,i}$ = number of cycles to failure at the specific stress event

n_i = number of stress cycles

$n_K = K_t/K_f$ ratio or the supporting factor

$n_{K,\sigma}()= K_t/K_f$ ratio for a plate under normal stress

$n_{K,\sigma,x}()= K_t/K_f$ ratio for a plate under normal stress in x axis

$n_{K,\sigma,y}()= K_t/K_f$ ratio for a plate under normal stress in y axis

$n_{K,\tau}()= K_t/K_f$ ratio for a plate under shear stress

O = surface area of the section of a component

$\varphi = 1/(4\sqrt{t/r}+2)$ = parameter to calculate relative stress gradient

q = notch sensitivity factor

R = stress ratio = ratio of minimum stress to maximum stress

R_r = reliability value

R_Z = average roughness value of the surface

r = notch root radius

r_{max} = larger one of the superimposed notch radii

S = nominal stress

S_a = stress amplitude

S_C = nominal stress of a notched component

S_E = endurance limit at 10^6 cycles

S_m = mean stress

$S_{N,E}$ = nominal endurance limit of a notched component

$S_{S,ax,E}$ = endurance limit of a notched, rod-shaped component under fully reversed loading in axial

$S_{S,ax,u}$ = ultimate strength of a notched, rod-shaped component in axial loading

$S_{S,b,E}$ = endurance limit of a notched, rod-shaped component under fully reversed loading in bending

$S_{S,b,u}$ = ultimate strength of a notched, rod-shaped component in bending

$S_{S,E,Notched}$ = nominal endurance limit of a notched component at 10^6 cycles

$S_{S,E,Smooth}$ = nominal endurance limit of a smooth component at 10^6 cycles

$S_{S,s,E}$ = endurance limit of a notched, rod-shaped component under fully reversed loading in shear

$S_{S,s,u}$ = ultimate strength of a notched, rod-shaped component in shear

$S_{S,\sigma,E}$ = endurance limit of a smooth, polish component under fully reversed tension

$S_{S,t,E}$ = endurance limit of a notched, rod-shaped component under fully reversed loading in torsion

$S_{S,t,u}$ = ultimate strength of a notched, rod-shaped component in torsion

$S_{S,\tau,E}$ = endurance limit of a smooth, polish component under fully reversed shear stress

$S_{S,\tau,u}$ = ultimate strength values of a notched, shell-shaped component for shear stress

$S_{S,x,E}$ = endurance limit of a notched, shell-shaped component under fully reversed normal stresses in x axis

$S_{S,x,u}$ = ultimate strength of a notched, shell-shaped component for normal stresses in x axis

$S_{S,y,E}$ = endurance limit of a notched, shell-shaped component under fully reversed normal stresses in y axis

$S_{S,y,u}$ = ultimate strength of a notched, shell-shaped component for normal stresses in y axis

$S_{\sigma,a}$ = normal stress amplitude in a stress cycle

$S_{\sigma,ar}$ = equivalent fully reversed normal stress amplitude

$S_{\sigma,E}$ = endurance limit for normal stress at 10^6 cycles

$S_{\sigma,FL}$ = fatigue limit in normal stress = normal stress amplitude at 10^8 cycles

$S_{\sigma,m}$ = mean normal stress in a stress cycle

$S_{\sigma,max}$ = maximum normal stresses in a stress cycle

$S_{\sigma,min}$ = minimum normal stresses in a stress cycle

$S_{t,u}$ = ultimate tensile strength with R97.5

$S_{t,u,min}$ = minimum ultimate tensile strength

$S_{t,u,std}$ = mean ultimate tensile strength of a standard material test specimen

$S_{t,y}$ = tensile yield strength with R97.5

$S_{t,y,max}$ = maximum tensile yield strength

$S_{\tau,E}$ = endurance limit for shear stress at 10^6 cycles

$S_{\tau,FL}$ = fatigue limit in shear = shear stress amplitude at 10^8 cycles

S_{max} = maximum stress

S_{min} = minimum stress

S'_f = fatigue strength coefficient

$S'_{\sigma,f}$ = fatigue strength coefficient in normal stress

σ^e = fictitious or pseudo stress

$\sigma^e(x)$ = pseudo stress distribution along x

σ^e_E = pseudo endurance limit

σ^e_{max} = maximum pseudo stress at x = 0

T = temperature in degrees Celsius

t_c = coating layer thickness in μm

V = volume of the section of a component

Chapter 5

α_S = sensitivity shear-to-normal stress parameter

α_{DV} = hydrostatic stress sensitivity

α_{NP} = nonproportional hardening coefficient for material dependence

α_{oct} = hydrostatic stress sensitivity factor

α_{VM} = mean stress sensitivity factor in the von Mises failure criterion

$\underline{\alpha}^*$ = center of the smallest von Mises yield surface

C = constant to make f_{NP} unity under $90°$ out-of-phase loading

$\Delta\varepsilon$ = strain range

E_{meso} = Young's modulus in mesoscopic level

Φ = phase angle between two loadings

$\underline{\varepsilon}^e(t)$ = macroscopic elastic strain tensor at a time instant t

$\underline{\varepsilon}^e_{meso}(t)$ = mesoscopic elastic strain tensor at a time instant t

$\underline{\varepsilon}^p_{meso}(t)$ = mesoscopic plastic strain tensor at a time instant t

η = material constant

f_{GD} = scaled normal stress factor

f_{NP} = nonproportional loading path factor for the severity of loading paths

G = factor to account the stress gradient effect

$k = \sigma_{E,R=-1}/\tau_E$

$K_{b,t}$ = elastic stress concentration factor due to bending

k_F = normal stress sensitivity factor

k_o = monotonic strength coefficient

$K_{t,t}$ = elastic stress concentration factor due to torsion

k' = cyclic strength coefficient

n_o = monotonic strain hardening exponent

n' = cyclic strain hardening exponent

φ = inclination angle between x' and z axis

φ^* = interference plane angle with respect to the x-y plane

$\underline{\rho}^*$ = residual stress tensor in the mesoscopic scale

$s_{meso,1}(t)$ = largest mesoscopic deviatoric principal stress at a time instant t

$s_{meso,3}(t)$ = smallest mesoscopic deviatoric principal stress at a time instant t

$\underline{S}(t)$ = macroscopic deviatoric stress tensor at a time instant t

$\underline{s}_{meso}(t)$ = mecroscopic deviatoric tensor at a time instant t

$SF(t)$ = safety factor at a time instant t

$[\sigma]_{xyz}$ = stress matrix relative to a global xyz coordinate system

$[\sigma]_{x'y'z'}$ = stress matrix relative to a local x'y'z' coordinate system

σ_1 = maximum principal stress

$\sigma_{1,a}$ = maximum principal stress amplitude

$\sigma_{3,a}$ = minimum principal stress amplitude

σ_a = applied in-phase normal stress amplitude

$\sigma_{E,R=-1}$ = fully reversed fatigue limit for normal stress

σ_h = hydrostatic stress

$\sigma_{eq,m}$ = equivalent mean stress

$\sigma_{meso,h}(t)$ = mesoscopic hydrostatic stress at a time instant t

$\sigma_{n,max}$ = maximum normal stress on a critical plane

$\sigma_{PS,a}$ = the maximum principal stress amplitude $(= \sigma_{1,a})$

$\sigma_{t,u}$ = ultimate tensile strength

$\sigma_{t,y}$ = yield strength in tension

$\sigma_{VM,a}$ = von Mises stress amplitude

$\sigma_{VM,a}(\Phi = 90°)$ = 90° out-of-phase von Mises stress amplitude

$\sigma_{VM,a}(\Phi = 0°)$ = in-phase von Mises stress amplitude

$\sigma_{VM,m}$ = von Mises mean stress

$\sigma_{VM,a,NP}$ = equivalent nonproportional stress amplitude

σ_x = normal stress in a local x-y coordinate

σ_x^e = pseudo normal stress in x axis

σ_y^e = pseudo normal stress in y axis

σ_f' = fatigue strength coefficient

$\vec{\sigma}_{1,max}(t)$ = maximum principal stresses at a time t

$|\vec{\sigma}_{1,max}(t)|$ = maximum absolute value of the principal stress at a time instant t

$\vec{\sigma}_{1,max}^{ref}$ = largest absolute principal stress

$|\vec{\sigma}_1(t)|$ = magnitude of the maximum principal stress at a time instant t

$|\vec{\sigma}_3(t)|$ = magnitude of the minimum principal stress at a time instant t

$\underline{\sigma}(t)$ = macroscopic stress tensor at a time instant t

$\underline{\sigma}_{meso}(t)$ = mesoscopic stress tensor at a time instant t

T = time for a cycle

$[T]$ = coordinate transformation matrix

θ = interference plane angle with respect to the x-z plane or the inclination angle between the x and x' projection vector on x-y plane

θ_1 = angle between the maximum principal stress and the local x axis

θ^* = interference plane angle with respect to the y-z plane

$\tau_{A,B,E}$ = shear fatigue limit for Case A or B crack

τ_a = applied in-phase shear stress amplitude

τ_E = fully reversed fatigue limit for shear stress

$\tau_n(\theta)$ = shear stress on an interference plane with an inclination angle (θ) to a local x axis

$\tau_{\psi^*,\varphi}$ = resultant shear stress due to $\tau_{x'y'}$ and $\tau_{x'z'}$ along a critical plane

$\tau_{MS,a}$ = maximum shear stress amplitude $(= \sigma_{1,a} - \sigma_{3,a})$

τ_{xy} = shear stress in a local x-y coordinate

$\tau_{x'y'}$ = shear stress component along a critical plane

$\tau_{x'z'}$ = shear stress component along a critical plane

τ_{xy}^e = pseudo shear stress in a local x-y plane

$\tau_{oct,a}$ = alternating octahedral shear stress

τ_{arith} = effective shear stress defined by Sonsino

$\tau_{Findley,a}$ = Findley's stress amplitude

$\tau_{McDiarmid,a}$ = McDiarmid's stress amplitude

V = ratio of the minimum to maximum principal stresses

ν = Poisson's ratio

Chapter 6

$2N_f$ = fatigue life to failure in reversals

$2N_T$ = transition fatigue life in reversals

$A_{SS} = K_t/K_f$ ratio

a_R = roughness constant

b = fatigue strength exponent

C_S = surface treatment factor

C_{ss} = material constant dependent on yield strength (σ_y) for the K_t/K_f ratio

$C_{\sigma,R}$ = roughness correction factor

c = fatigue ductility exponent

D = total damage

Δe = nominal strain range

$\Delta\varepsilon$ = true strain range

$\Delta\varepsilon^e$ = elastic strain range

$\Delta\varepsilon^p$ = plastic strain range

ΔS = nominal stress range

$\Delta\sigma$ = true stress range

$\Delta\sigma^e$ = pseudo stress range

E = modulus of elasticity

e = nominal strain

e_1 = nominal strain for initial loading

e^M = modified version of a nominal strain

ε = true strain

ε_1 = true strain for initial loading

ε_a = true strain amplitude

$\varepsilon_{a,rev}$ = fully reversed strain amplitude

ε^e = elastic strain

ε_a^e = elastic strain amplitude

ε^p = plastic strain

ε_a^p = plastic strain amplitude

ε_f' = fatigue ductility coefficient

ε_r = residual strain

$\overline{G}$ = relative stress gradient

HB = Brinell hardness

K_ε = true strain concentration factor

K_f = fatigue notch factor

k_m = mean correction factor for σ_f'

K_p = limit load factor or plastic notch factor

K_σ = true stress concentration factor

K_t = elastic stress concentration factor

K^M = modified elastic stress concentration associated with S^M

K' = cyclic strength coefficient

kn = total number of the stress blocks

L_p = load producing gross yielding of a net section

L_y = load producing first yielding of a net section

$N_{f,i}$ = number of cycles to fatigue failure

n_i = number of applied cycles to a constant stress amplitude

n' = cyclic strain hardening exponent

ψ = constant to estimate ε_f' in the uniform material law

R_Z = average roughness value of the surface in μm

r = notch radius

S = nominal stress

S_1 = nominal stress for initial loading

$S_{t,min,u}$ = minimum ultimate tensile strength

$S_{t,u}$ = ultimate tensile strength

S^M = modified version of a nominal stress

σ = true stress

σ_1 = true stress for initial loading

σ_1^e = pseudo stress for the initial loading

σ_a = true stress amplitude

$\sigma_{a,rev}$ = fully reversed stress amplitude

σ_e = pseudo stress

$\sigma^e(x)$ = theoretically calculated pseudo stress distribution near a notch root

σ_m = mean stress

σ_{max} = maximum stress

σ_{max}^e = maximum local pseudo stress

σ_r = residual stress

σ_f' = fatigue strength coefficient

σ_y' = cyclic yield stress

W_e = strain energy density at the notch root

W_S = energy density due to nominal stress and strain

x = normal distance from the notch root

Chapter 7

A = nonproportional parameter in Tanaka's model

A_1 = constant to the addition of an elastic stress increment

$\underline{A}$ = translational direction of the yield surface

$\underline{\alpha}$ = back stress tensor

$\underline{\alpha}^{(i)}$ = part of the total back stress

α_{NP} = nonproportional hardening coefficient

$\underline{B}$ = translational direction of the yield surface

b = material parameter determining the rate at which yield saturation is reached

b_r = material constant in the Zhang and Jiang model

b_{NP} = material parameter

C_c = material constant

$\underline{C}$ = internal state variable describing the internal dislocation structure

c = material constant in the Zhang and Jiang model

D = normalized difference between d and d_{max}

d = distance between the loading and the limit surfaces

d_1 = distance between a loading stress point and its conjugate point

d_{max} = maximum distance between the loading and the limit surfaces

d^m = discrete memory variable to account for the memory effect

$d\underline{e}^e$ = elastic deviatoric strain increment tensor

$d\underline{\varepsilon}$ = total strain increment tensor

$d\underline{\varepsilon}^e$ = elastic strain increment tensor

$d\underline{\varepsilon}^p$ = plastic strain increment tensor

$d\lambda_1$ = multiplier based on Levy's flow rule

$d\lambda_2$ = multiplier based on Prandtl and Reuss' flow rule

$d\lambda_3$ = multiplier based on Mises' flow rule

$d\lambda_4$ = multiplier based on Mroz's flow rule

$d\lambda_5$ = scalar associated with the back stress

$d\lambda_6$ = scalar associated with the back stress

$d\mu_4$ = scalar based on Mroz's flow rule

dp = equivalent plastic strain increment

$d\underline{S}$ = deviatoric stress increment tensor

$d\underline{\sigma}$ = stress increment tensor

$\Delta\underline{e}$ = deviatoric pseudo strain tensor

$\Delta\underline{e}^e$ = deviatoric pseudo strain increment tensor

$\Delta\underline{S}$ = deviatoric stress increment tensor

$\Delta\underline{S}^e$ = deviatoric pseudo stress increment tensor

δ_{ij} = unit tensor

E = Young's modulus of elasticity

$\underline{e}$ = deviatoric strain tensor

$\underline{e}^e$ = deviatoric pseudo strain tensor

$\underline{\varepsilon}$ = symmetric, second order strain tensor

$\underline{\varepsilon}^e$ = symmetric, second order pseudo strain tensor

ε_{eq} = equivalent (von Mises) strain

ε_{eq}^e = elastic von Mises strain

ε_{ij} = symmetric, second order strain tensor

f = func() = yield surface function or criterion

$f(\underline{S})$ = yield surface function or yield criterion depending on deviatoric stress tensor

f^L = next inactive yield or limit surface

G = elastic shear modulus

H = hardening function

h = plastic modulus, the tangent modulus of a uniaxial stress-plastic strain curve

I_{ij} = unit tensor

$\underline{I}$ = unit tensor

J_2 = second invariant of the deviatoric stress tensor

k = yield strength in shear $(= \tau_y)$

K' = cyclic strength coefficient for a true stress-strain curve

K^* = cycle strength coefficient for a pseudo stress-strain curve

M = number of back stress parts

m_1 = material constant 1 in the Zhang and Jiang model

m_2 = material constant 2 in the Zhang and Jiang model

μ_1 = constant 1 for the Armstrong-Frederick back stress rule

μ_2 = constant 2 for the Armstrong-Frederick back stress rule

$\mu_3^{(i)}$ = constant for the Armstrong-Frederick i-th back stress rule

$\mu_4^{(i)}$ = constant for the Armstrong-Frederick i-th back stress rule

n' = cyclic hardening exponent for a true stress-strain curve

n^* = cyclic hardening exponent for a pseudo stress-strain curve

$\underline{n}$ = outward normal to an active surface at $\underline{S}$

$p = \int dp$ = accumulated plastic strain

φ = phase angle between two loadings

$Q()$ = plastic potential function

q = size of the memory size in the Zhang and Jiang model

q_N = target value for nonproportional hardening

q_P = target values for proportional hardening

$R()$ = isotropic strain-hardening function

R_{NP} = saturated value with increasing plastic strain due to nonproportional loading

R^L = saturated value with increasing plastic strain

r = evolution parameter in the Zhang and Jiang model

$S_i (i = 1,2,3)$ = deviatoric principal stress component

$\underline{S}$ = deviatoric stress tensor

$\underline{S}^e$ = deviatoric pseudo stress tensor

$\underline{S}^L$ = conjugate stress point on the next inactive yield or limit surface

$\underline{\sigma}$ = symmetric, second order stress tensor

$|\underline{\sigma}| = \sqrt{\underline{\sigma}:\underline{\sigma}}$ = norm of a tensor

$\sigma_{eq} = \sqrt{3J_2}$ = von Mises or equivalent stress

σ_{eq} = equivalent (von Mises) stress

$\underline{\sigma}^e$ = symmetric, second order pseudo stress tensor

σ_{eq}^e = elastic von Mises stress

$\underline{\sigma}_h$ = hydrostatic stress tensor

σ_i = principal stress component

σ_{ij} = symmetric, second-order stress tensor

$\sigma_{VM,a}(\varphi = 90°) = 90°$ out-of-phase von Mises stress amplitude

$\sigma_{VM,a}(\varphi = 0°) =$ in-phase von Mises stress amplitude

σ_y = yield stress which is a function of accumulated plastic strain p

σ_y^L = limit or saturated yield stress

σ_{yo} = initial yield stress

τ_y = yield strength in shear (= k)

v = Poisson's ratio

$W^p = \int \sigma_{ij} d\varepsilon_{ij}^p =$ plastic work

$W^{(i)}$ = function of back stress range

X = material constant in the Zhang and Jiang model

Chapter 8

$2N$ = reversals to failure

α = nonproportional hardening coefficient

α^* = plastic work exponent

b, c = fatigue strength, ductility exponents

$d\varepsilon_{ij}^p$ = plastic strain increment tensor

$\Delta\varepsilon_I$ = maximum principle strain range

$\Delta\varepsilon_{NP}$ = nonproportional strain range

$\dfrac{\Delta\varepsilon}{2}$ = strain amplitude

$\dfrac{\Delta\varepsilon_1}{2}$ = principal strain amplitude

$\dfrac{\Delta\bar{\varepsilon}_p}{2}$ = equivalent plastic strain amplitude

$\dfrac{\Delta\gamma_{max}}{2}$ = shear strain amplitude

$\dfrac{\Delta\sigma}{2}$ = stress amplitude

$\Delta W_e, \Delta W_p, \Delta W_{total}$ = elastic, plastic, and total work per cycle

E = modulus of elasticity

$\varepsilon_1(t), \varepsilon_3(t)$ = extreme values of principle strain at a given time

ε_a = strain amplitude

$\varepsilon_a, \varepsilon_b, \varepsilon_c$ = corrected strain gage rosette readings

$\varepsilon_I(t)$ = maximum absolute value of principle strain

ε_{Imax} = maximum value of $\varepsilon_I(t)$ for the cycle

ε_n = normal strain on maximum shear strain plane

ε_o^p = von Mises equivalent plastic strain

$\varepsilon_x, \varepsilon_y, \gamma_{xy}$ = coordinate strains

$\varepsilon_x^p, \varepsilon_y^p, \varepsilon_z^p, \varepsilon_{xy}^p, \varepsilon_{xz}^p, \varepsilon_{yz}^p$ = plastic strain tensor components

$\varepsilon_a^*, \varepsilon_b^*, \varepsilon_c^*$ = uncorrected strain gage rosette readings

$\bar{\varepsilon}_a$ = equivalent strain amplitude

$\bar{\varepsilon}_e, \bar{\varepsilon}_p, \varepsilon_o$ = equivalent elastic, plastic, and total strains

F = nonproportionality factor

f_{NP} = Itoh's nonproportionality factor

G = shear modulus

γ_a = shear strain amplitude

γ_{xy}^p = engineering plastic shearing strain

k, k_τ = damage parameter fitting constants

K, K', n, n' = monotonic, cyclic Ramberg–Osgood strength coefficient

K_{ta}, K_{tb}, K_{tc} = strain gage rosette transverse sensitivity factors

$\lambda = \Delta\sigma_2/\Delta\sigma_1$ = biaxiality ratio computed from magnitude ordered principal stresses

m = material constant to account for hydrostatic stress (Mowbray)

n, n' = monotonic, cyclic Ramberg–Osgood strength exponent

ν_e, ν = Poisson's ratio

ν_o = Poisson's ratio of the material on which the strain gage factors and sensitivities were determined, 0.285

ν_p = "plastic" Poisson's ratio

S, C, k = damage parameter constants

$\sigma_f', \varepsilon_f'$ = fatigue strength, ductility coefficients

σ_{ij} = stress tensor

$\bar{\sigma}_{IP}$ = in-phase cyclic stress

σ_{max} = maximum stress

$\sigma_{n,max}$ = maximum normal stress on shear plane

σ_o = von Mises equivalent stress

$\bar{\sigma}_{OOP}$ = out-of-phase cyclic stress

$\sigma_x, \sigma_y, \sigma_z, \tau_{xy}, \tau_{xz}, \tau_{yz}$ = stress tensor components

σ_y = yield stress (when in damage parameter equation)

τ_f', γ_f' = shear fatigue strength, ductility coefficients

$\bar{\upsilon}$ = equivalent Poisson's ratio

$\xi(t)$ = angle between principle strain and max principal strain for cycle

Chapter 9

A = proportional constant in a logarithmic sweep

A_0 = coefficient in Fourier transforms of $X(t)$

A_n = coefficient in Fourier transforms of $X(t)$

A' = constant in a linear sweep

a_W = constant in Wirsching's damage equation

α = Weibull scale parameter (characteristic life)

B = half-power bandwidth

B_n = coefficient in Fourier transforms of $X(t)$

b_W = constant in Wirsching's damage equation

β = Weibull shape parameter (Weibull slope)

C = material constant of an S-N curve

c = viscous damping coefficient of an SDF system

D = linear damage value

D_{NB} = narrow band fatigue damage value

$D_{WB,Wirsching}$ = Wirsching's wide band fatigue damage value

$D_{WB,Oritz}$ = Ortiz's wide band fatigue damage value

$D_{WB,Dirlik}$ = Dirlik's wide band damage equation

dB = logarithm of the ratio of two measurements of power

$d_i(f_n)$ = fatigue damage to a system with a resonant frequency f_n, subjected to a sinusoidal input with an excitation frequency f_i

dt_i = incremental time for $x \leq X(t) \leq x + dx$

dx = incremental random variable

$\delta[n]$ = impulse to linear time invariant system

$\delta[n - k]$ = Dirac delta function

Δf = frequency increment

Δt = time increment

$E(X(t))$ = expected value of $X(t)$

$E[N_{a+}(dt)]$ = expected number of positively-sloped crossing (up-crossing) in an infinitesimal interval

$E[0^+]$ = expected rate of zero up-crossing

$E[P]$ = expected rate of peak crossing

$E[X^2(t)]$ = mean-square value of $X(t)$

η = parameter defined as $\eta = Q/n_{fn}$

$F_{Sa}(s_a)$ = cumulative distribution function of the stress amplitude

$f_{X\dot{X}}(u, v) = $ joint probability density function of $X(t)$ and $\dot{X}(t)$

$f_X(x) = $ probability density function $= P[x \leq X(t) \leq x + dx] = $ probability that $x \leq X(t) \leq x + dx$

$K = $ spring constant defined as the ratio of stress amplitude to relative displacement

$k = $ stiffness of an SDF system

$\dot{f} = $ excitation frequency rate

$f = $ excitation frequency in cycles/second (Hz)

$f_i = $ excitation frequency in cycles/second (Hz)

$f_n = $ resonant frequency in cycles/second (Hz)

$f_{min} = $ minimum excitation frequency in cycles/second (Hz)

$f_{max} = $ maximum excitation frequency in cycles/second (Hz)

$G = $ fraction of the maximum steady state response of a system

$\Gamma(.) = $ gamma function

$\gamma = $ regularity factor

$|H_1(r_i)| = $ gain function or the modulus of the transfer function

$|H(\omega)| = $ gain function or the modulus of the transfer function

$H^*(\omega) = $ complex conjugate of $H(\omega)$

$h[n] = $ response to a linear time invariant system due to an impulse $\delta[n]$

$kn = $ total number of the stress blocks

$\lambda = $ spectral width parameter

$M_j = $ j-th moment of a one-sided power spectral density function

$m = $ slope factor of an S-N curve

$m_o = $ mass of an SDF system

$\mu_X = $ mean value of $X(t)$

$N = $ number of equally spaced time intervals in $X(t)$

$N_{f,i} = $ fatigue life as the number of cycles to failure under $S_{a,i}$

$NO = $ total sample points in T

$n_{fn} = $ number of cycles of excitation between the half-power bandwidth B

$n_i = $ number of cycles in the i-th block of constant stress amplitude $S_{a,i}$

octave $= $ doubling of frequency

$\omega_k = $ frequency of the k-th harmonic frequency

$\omega_n = $ natural frequency in radians/second

PSD $= $ power spectral denstiy

$pdf(f_i) = $ probability that the stress amplitude $S_a = s_{a,i}$ may occur

$Q = $ dynamic amplification factor

$R(\tau) =$ autocorrelation function of $X(t)$

$RMS_X =$ root mean square of $X(t)$

$r_i =$ frequency ratio defined as f_i/f_n

$S_a(f_i, \xi, f_n) =$ stress amplitude

$S_{a,RMS} =$ root mean square of this stress amplitude response to an SDF system

$S_X(\omega) =$ two-sided spectral density of a stationary random process $X(t)$ with $\mu_X = 0$

$\sum_{i=1}^{k} no_i =$ total number of the sample points between x and $x + dx$

$\sigma_X =$ standard deviation of $X(t)$

$\sigma_X^2 =$ variance of $X(t)$

$T =$ total time

$t_j =$ time instant at j-th digitized point $= j \cdot \Delta t$

$\tau =$ time lag

$v_{a+} =$ expected rate of up-crossing per time unit

$W_x(f) =$ power spectral density $=$ one-sided spectral density a stationary random process

$W_z(f_i, \xi, f_n) =$ power spectral density of the relative displacement to an SDF system

$W_{\ddot{y}}(f_i) =$ power spectral density of the base random accelerations to an SDF system

$W_{\ddot{z}}(f_i, \xi, f_n) =$ power spectral density of the relative acceleration to an SDF system

$X(\omega) =$ forward Fourier transform of $X(t)$

$X(t) =$ displacement, random process

$X[n] =$ arbitrary input a linear, discrete time, time-invariant system

$X_n^* =$ complex conjugate of X_n

$\dot{X}(t) =$ velocity random process

$x =$ random variable at a time instant

$x_o(t) =$ displacement of the mass of an SDF system

$\dot{x}_o(t) =$ velocity of $x_o(t)$

$\ddot{x}_o(t) =$ acceleration of $x_o(t)$

$\ddot{x}_{RMS} =$ root mean square of the absolute acceleration response to an SDF system

$\xi =$ damping ratio

$Y[n] =$ output with a weighted sum of time-shift impulse responses

$y_o(t)$ = base input displacement of an SDF system

$\dot{y}_o(t)$ = velocity of $y_o(t)$

$\ddot{y}_o(t)$ = acceleration of $y_o(t)$

$|\ddot{y}_o(f_i)|$ = base input sine vibration with an excitation frequency of f_i

z = normalized variable for a normal distribution

$z_o(t) = x_o(t) - y_o(t)$ = relative displacement

z_{RMS} = root mean square of the relative displacement response to a SDF system

$\dot{z}_o(t)$ = velocity of $z_o(t)$

$\ddot{z}_o(t)$ = acceleration of $z_o(t)$

$|\ddot{z}_o(f_i, \xi, f_n)|$ = steady-state relative response to a SDOF system with a resonant frequency f_n, subjected to a base input sine vibration with an excitation frequency of f_i

ζ_O = constant in Ortiz's damage equation

ζ_W = rainflow correction factor based on Wirsching's study

Chapter 10

a = effective weld throat

α_{NP} = nonproportional hardening coefficient for the material dependence

α_S = sensitivity shear-to-normal stress parameter

b = fatigue strength exponent

C = fitting material coefficient

C_{NP} = constant chosen to make f_{NP} unity under 90° out-of-phase loading

D_σ = damage value due to normal stress

D_τ = damage values due to shear stress

d = partial penetration depth

ΔS_s = equivalent structural stress range

F_i = nodal force at nodes i in a local coordinate system

$F_{x1}^{(i)}(y)$ and $F_{x2}^{(i)}(y)$ = grid forces in an element $(E^{(i)})$

$f(x')$ = linear weld force distribution force as a function of a distance x' from a reference node

$f_1()$ and $f_2()$ = functional expressions

f_b = bending compliance function

f_i = unit nodal weld force at node i in a local coordinate system

$f_i^{(k)}$ = unit nodal weld force at node i in region k

f_m = membrane compliance function

$f_x^{(i)}(y)$ = unit line force in an element $(E^{(i)})$

$f_{x1}^{(i)}(y)$ and $f_{x2}^{(i)}(y)$ = grid weldline forces in an element $(E^{(i)})$

f_y = unit weldline in-plane force on the crack propagation plane

f_{NP} = nonproportional loading path factor for the severity of loading paths

$g_1^{(i)}$ and $g_2^{(i)}$ = grid points in an element $(E^{(i)})$

$I(r)$ = dimensionless function of bending stress ratio

$l_y^{(i)}$ = element edge length between the two grid points

K_n = stress intensity factor for an edge crack under structural (far-field) stresses

$K_{a/t \leq 0.1}$ = stress intensity factor dominated by the local notch stresses

$K_{a/t > 0.1}$ = stress intensity factor controlled by the structural stresses

$\Delta K_{a/t \leq 0.1}$ = stress intensity factor range in the short regime

$\Delta K_{a/t > 0.1}$ = stress intensity factor range in long crack growth regime

K_{ta} = elastic stress concentration factors under axial load

K_{tb} = elastic stress concentration factors under bending load

k = slope factor for an S-N curve

L = weld leg length

l = length of a plate or shell element

l_i = length of a plate or shell element i

M_i = nodal moment at node i in a local coordinate system

M_{kn} = stress intensity magnification factor

$M_{ms,1}$ and $M_{ms,2}$ = mean stress sensitivity factors defined in Haigh's diagram

$m_y^{(i)}(y)$ = unit line moment in an element $(E^{(i)})$

$M_{y1}^{(i)}(y)$ and $M_{y2}^{(i)}(y)$ = grid moments 1 and 2 in an element $(E^{(i)})$

$m_{y1}^{(i)}(y)$ and $m_{y2}^{(i)}(y)$ = grid line moments 1 and 2 in an element $(E^{(i)})$

m = crack growth rate exponent for the long crack growth regime

m_i = unit nodal weld moment at node i in a local coordinate system

m_x = unit weldline in-plane torsion on the crack propagation plane

n = crack growth rate exponent for the first stage of the crack growth

R = load ratio

r = bending ratio = σ_b / σ_s

ρ = weld toe radius

$\sigma_1, \sigma_2,$ and σ_3 = local stresses points 1, 2 and 3, respectively, to describe the bi-linear notch stress distribution

σ_a^e = median pseudo endurance limit at 2×10^6 cycles

σ_b = bending stresses

$\sigma_b^{(i)}(y)$ = bending stress along a weldline direction y in an element $(E^{(i)})$

$\sigma_{b,i}$ = bending stress at node i

$\sigma_b^{(k)}$ = bending stress in region k due to the bi-linear notch stress distribution

$\sigma_i^{(k)}$ = local stress at node i in region k

σ_m = membrane stress

$\sigma_{m,i}$ = membrane stress at node i

$\sigma_m^{(i)}(y)$ = membrane stress along a weldline direction y in an element $(E^{(i)})$

$\sigma_m^{(k)}$ = membrane stress in region k due to the bi-linear notch stress distribution

σ_s = structural stress normal to the crack surface

$\sigma_{s,a}$ = maximum structural stress amplitude

$\sigma_{s,m}$ = mean structural stress

$\sigma_{s,a,R=-1}$ = equivalent fully reversed stress amplitude

$\sigma_s^{(i)}(y)$ = structural stress normal to the crack surface along a weldline direction y in an element

$\sigma_{s,i}$ = structural stress at node i

$\vec{\sigma}_{1,max}^{\,ref}$ = largest absolute principal stress

$\vec{\sigma}_{1,max}(t)$ = maximum principal stresses

$\sigma_{eq,m}$ = equivalent mean stress

σ_f' = fatigue strength coefficient

$\sigma_{VM,a}(\Phi = 90°)$ = equivalent stress amplitude due to 90° out-of-phase loading

$\sigma_{VM,a}(\Phi = 0°)$ = equivalent stress amplitude due to in-phase loading

$\sigma_{VM,a}$ and $\sigma_{VM,a,NP}$ = equivalent proportional and nonproportional stress amplitudes

$\sigma_{x,a}, \sigma_{y,a}, \tau_{xy,a}$ = plane stress amplitude components

$\sigma_{x,m}$ and $\sigma_{y,m}$ = mean stress values in x and y axes, respectively

T = time period for a cycle

t = member thickness

t_1, t_2 = plate thickness

t_c = characteristic depth in bilinear notch stress distribution

θ = angle of $\vec{\sigma}_{1,max}(t)$ with respect to the x axis

θ_t = weld throat angle

$\xi(t)$ = angle between $\vec{\sigma}_{1,max}^{\,ref}$ and $\vec{\sigma}_{1,max}(t)$

Chapter 11

A = constant

A_1 = constant

a = radius of the spot weld

α = exponent for the shape of the failure surface

b = half width of the coupon

b_o = load ratio exponent

β_1 = material constant

β_2 = material parameter to correlate K_{III} mode fatigue data to K_I mode fatigue data

d = nugget diameter

ΔF = remote load range

ΔF_N = out-of-plane normal load range

$\Delta F_{N,N_f}$ = fatigue strength range in out-of-plane normal loading in lbs

ΔF_S = in-plane shear load range

$\Delta F_{S,N_f}$ = fatigue strength range in in-plane shear loading in lbs

ΔM_{ij}^* = bending moment ranges

ΔP_i = axial load range in the weld nugget

ΔQ_{ij} = membrane load ranges

ΔS_{max} = maximum structural stress ranges

F_x and F_y = in-plane interface forces

G = geometrical correction factor

h = constant

k_1 = parameter that depends on the ratio of the nugget radius and specimen span

k_2 = parameter that depends on the ratio of the nugget radius and specimen span

k_3 = material dependent geometry factor

K_{eq} = equivalent stress intensity

K_I = stress intensity factor for Mode I

K_i = fatigue damage parameter

$K_{I_{eq}}$ = equivalent stress intensity factor of Mode I

$K_{I,eq,max}$ = equivalent stress intensity factor of Mode I at the maximum applied load

K_{II} = stress intensity factor for Mode II

M = mean stress sensitivity factor

M_x and M_y = in-plane interface moments

$M_{x,y}$ = applied moment in the local x or y direction

M_z = out-of-plane interface moment

m = constant

N_f = fatigue life in cycles

N_I = number of cycles for initiation and early growth

N_{pt} = number of cycles for crack propagation through the thickness

N_{pw} = number of cycles for crack propagation through the specimen width

ω = effective specimen width ($= \pi d/3$)

P = normal component of the applied load

R = load ratio

r = nugget radius

σ_b = bending stress

$\sigma_{eq,0}$ = equivalent stress amplitude at R = 0

$\sigma_{eq,a}$ = equivalent stress amplitude

$\sigma_{eq,m}$ = mean stress

σ_n = normal stress

$\sigma_{r,max}$ = maximum radial stress

σ_{ui}, σ_{uo}, σ_{li}, and σ_{lo} = normal stresses

t = sheet thickness

τ_{max} = maximum shear stress

τ_{qu} and τ_{ql} = transverse shear stresses

τ_{ui} and τ_{li} = circumferential stresses

W = specimen width

Chapter 12

A = cross sectional area

A_D = sealing area

A_{Pmin} = minimum bolt head or nut bearing area

A_i = cross sectional area of an individual cylindrical bolt

A_N = nominal cross sectional area of the bolt

A_{d3} = cross sectional area of the minor diameter of the bolt threads

A_S = effective tensile stress area

A_{SGS} = critical shear area for the length of external thread engagement

A_{SGM} = critical shear area for the length of internal thread engagement

α = thread (flank) angle

α_A = tightening factor

$\beta_L = l_K / d_W$

C_1 = shear strength reduction factor due to internal thread dilation

C_2 = shear stress area reduction factor due to external thread bending

C_3 = shear strength reduction factor due to internal thread bending

D = major diameter of internal threads (nut) in mm

D_1 = minor diameter of internal threads in mm

D_2 = pitch diameter of internal threads in mm

D_K = projected or limiting diameter of a deformation cone at the interface

d = major (nominal) diameter of external threads (bolt) in mm

d_2 = pitch diameter of external threads in mm = $d - 0.649519P$

d_3 = minor diameter of external threads in mm = $d - 1.226869P$

d_S = effective stress diameter

$d\delta_P$ = elastic compliance of an infinite small plate thickness

ΔT = temperature difference

δ = elastic compliance, resilience, flexibility

δ_1, δ_2 = compliance of the bolt sections

δ_3 = compliance of unengaged threads

δ_G = compliance of the engaged threads

δ_K = compliance of the bolt head

δ_M = compliance of the nut or tapped hole

δ_P = elastic compliance of the clamped plates

δ_S = elastic compliance of the bolt

δ'_P = effective elastic compliance of the clamped plates ($\delta'_P = n\delta_P$)

δ'_S = effective elastic compliance of the bolt ($\delta'_S = (1-n)\delta_P + \delta_S$)

E = Young's modulus

E_M = Young's modulus of the nut

E_p = Young's modulus of the clamped material

E_S = Young's modulus of the bolt

ε_o = initial strain to a beam model due to bolt pretension

Φ = load distribution factor, percentage of the applied load to the bolt

$F_{0.2min}$ = minimum yield force

F_A = axial force on the bolted axis

F_{Kerf} = required minimum clamp force to prevent plate slippage

F_{KP} = special force required for sealing

F_{KQ} = frictional grip to transmit a transverse load (F_Q) and a torque about the bolt axis (M_T)

F_{KR} = residual clamp force

F_{KRmin} = minimum residual clamp load

F_M = bolt preload load due to M_A

F_{mGM} = ultimate shear force that will fracture the internal (nut) threads

F_{mGS} = ultimate shear force that will fracture the external (bolt) threads

F_{Mmax} = maximum bolt preload force

F_{Mmin} = minimum preload force

F_{Mzul} = permissible preload

F_{PA} = working load on the clamped plates

F_{PAmax} = portion of the working axial load, which unloads the clamped plates

F_Q = transverse force normal to the bolt axis

F_{SA} = working load on the bolt

F_{Sm} = mean nominal stress level

$F_{t,1}$ = first tangential force component on the thread surface due to the preload

$F_{t,2}$ = second tangential force component on the thread surface due to the preload

F_Z = preload loss due to embedment

f = deformation due to a force F

f_{PM} = compression of the clamped plates

f_{SM} = elongation of the bolt

f_Z = plastic deformation (embedding)

H = height of fundamental triangle in mm

I_P = polar moment of inertia

i = number of bolts (i) in the flange

k = elastic stiffness

k_τ = shear stress reduction factor used in the von Mises equivalent stress

L = component length

l_3 = length of unengaged threads

l_i = length of an individual cylindrical bolt

l_K = total clamping length

M_A = applied assembly torque

M_B = bending moment at the bolting point

M_G = thread torque

M_K = under-head torque

M_T = torque (twist moment) at the bolt position at the interface

m_{eff} = length of thread engagement

μ_G = coefficient of friction in the thread

μ_K = coefficient of friction in the bolt or nut-bearing area

μ_T = coefficient of friction between the clamp plate interfaces

N_f = fatigue life in cycle

n = load introduction factor

n_1, n_2 = percentages of the clamped plate length

P = pitch in mm

p_{Bmax} = induced surface pressure

p_G = permissible surface pressure of the clamped material

p_{imax} = maximum internal pressure to be sealed

p_{Mmax} = maximum surface pressure due to bolted joint assembly

φ = thread helix angle (lead angle)

q_F = number of slippage planes

$R_{P0.2min}$ = minimum 0.2% yield strength of the external threads in N/mm^2

$$R_S = \frac{\tau_{BM} A_{SGM}}{\tau_{BS} A_{SGS}}$$

R_Z = average surface roughness value

R_{mM} = ultimate tensile strength of the internal threads (nut) in N/mm^2

R_{mS} = ultimate tensile strength of the external threads (bolt) in N/mm^2

r_a = torque radius from the bolt axis due to M_T

S_D = safety factor against fatigue

S_F = safety factor against bolt yielding

S_G = safety factor against slipping

S_P = safety factor against clamped plates crushing

s = width across flats for nuts

σ_a = nominal stress amplitude calculated on the tensile stress area

σ_{AS} = fatigue limit in amplitude at 2×10^6 cycles

σ_{ASG} = fatigue limit of rolled threads after heat treatment

σ_{ASV} = fatigue limit of rolled threads before heat treatment

σ_{AZSG} = fatigue strength of rolled threads after heat treatment

σ_{AZSV} = fatigue strength of rolled threads before heat treatment

σ_M = bolt tensile stress on tensile stress area

σ_{Mzul} = permissible assembly stress

σ_{redB} = von Mises equivalent stress

σ_{redM} = equivalent von Mises stress for a bolt

t_e = thickness of an external thread at the critical shear plane

t_i = thickness of an internal thread at the critical shear plane

τ_{BM} = ultimate shear strength of the internal threads in N/mm^2

τ_{BS} = ultimate shear strength of the external threads in N/mm^2

τ_M = bolt torsional stress on tensile stress area

v = utilization of the initial or the gross yield stress during tightening

W_P = polar moment of resistance

w = 1 for a through-bolted joint (DSV); 2 for a tapped thread joint (ESV)

$y = D_A/d_W$

CHAPTER 1

Road Load Analysis Techniques in Automotive Engineering

Xiaobo Yang
Oshkosh Corporation

Peijun Xu
Ebco Inc.

Chapter Outline

Introduction

The concept of design-by-analysis and validation-by-testing has been proven to be the most efficient and effective way to design a vehicle structure to meet its functional objectives, and universally adopted by most of the engineering communities.

The vehicle functional objectives include the federal regulations, such as emission and safety, and the mandatory requirements set up by each manufacturer, such as durability, reliability, vehicle dynamics, ride and comfort, noise, vibration and harshness (NVH), aerothermal and electromagnetic capability (EMC), among other things. The new concept can offer a chance to optimize a vehicle structure for these multifunctional objectives in a virtual engineering domain and to validate the structure by testing it in a physical world.

The validation-by-testing concept consists of development of accelerated test methods and reliability demonstration test planning strategies. Accelerated test method development is used to develop a pass or failure criterion for durability testing by using the damage equivalence theory. Examples could be vehicle proving grounds testing to represent the extreme customer usage profiles for the life of a design vehicle (so-called duty cycle data), real-time simulation testing on systems—that is, road test simulators (RTS) and multiaxial simulation table (MAST)—or simple life testing for components (e.g., constant amplitude loading or block cycle loading test).

Once the success criterion is established, numerous reliability demonstration test planning strategies have been proposed to demonstrate the reliability and confidence level of the designed products by testing them with a limited sample size. Depending on the failure criteria, the reliability demonstration test methods are recommended for components and the repairable systems, while the reliability growth model approaches are employed only for the repairable systems. Detailed discussion of these methods is beyond the scope of this chapter, and can be found elsewhere (Lee, Pan, Hathaway, & Barkey, 2005).

The design-by-analysis concept requires reliable virtual analytical tools for analyses. The quality of these tools relies on the accuracy of the mathematical model, material characterization, boundary conditions, and load determination. For durability analysis, the three important factors are:

- Loading
- Geometry
- Material

When external forces are applied to a multibody system, these forces are transferred through that system from one component to the next, where a component is defined as an element within that system. The fatigue life of a component is governed by the loading environment to which it is subject, the distribution of stresses and strains arising from that environment, and the response of the material from which it is manufactured. As a result, the major inputs to any fatigue analysis are loading, component geometry, and cyclic material properties.

A moving vehicle is a complex dynamic system primarily subjected to various static and dynamic external loads from tire/road interaction, aerodynamics, gravity, and payload, which yield overall vehicle motion in space and relative motions among various vehicle components. The relative motions of vehicle components are always constrained by joints and compliant elements (such as springs, shock absorbers, bushings/mounts, and jounce/rebound bumpers), and would induce internal forces and stresses that will possibly result in fatigue failures. Thus, it is crucial to predict these internal responses of vehicle components and systems for any failure prevention.

Loading information can be obtained using a number of different methods. Local or nominal strains can be measured by means of strain gages. Nominal loads can be measured through the use of load cells or, more recently, they can be derived externally by means of analysis. Since early methodologies relied on measurement from physical components, the application of fatigue analysis methods has been confined to the analysis of service failures or, at best, to the later stages of the design cycle where components and systems first become available.

The ability to predict component loads analytically means that physical components are no longer a prerequisite for durability analysis and so analysis can proceed much earlier in the design cycle. It is important to note that, in this context, loading environment is defined as the set of phase-related loading sequences (time histories) that uniquely map the cyclic loads to each external input location on the component.

Many virtual analysis tools for multibody dynamics (Gipser, Hofer, & Lugner, 1997; Tampi & Yang, 2005; Bäcker et al., 2007; Abd El-Gawwad, Crolla, Soliman, & El-Sayed, 1999a, 1999b; Stadterman, Connon, Choi, Freeman, & Peltz, 2003; Berzeri et al., 2004; Haga, 2006, 2007) have been developed to accurately calculate loads for components and systems. In addition to computer memory and speed, the efficiency of their engineering applications hinges on the availability of input data sources and modeling techniques. These tools have been widely adapted by the automotive engineering industry to predict vehicle road loads for fatigue damage assessments.

The objective of this chapter is to present the virtual analysis methods employed to characterize vehicle dynamic loads for one of the functional objectives—design for durability. More specifically, this chapter will cover the road load analysis techniques to predict vehicle component loads induced by irregular road surface profiles and driver's maneuvers (steering, braking/accelerating).

Fundamentals of Multibody Dynamics

A multibody system is used to model the dynamic behavior of interconnected rigid or flexible bodies, each of which may undergo large translational and rotational displacements. The vehicle suspension is a typical example of a multibody dynamic system. Multibody systems can be analyzed using the system dynamics method.

System dynamics (Randers, 1980) is an approach used to understand the behavior of complex systems over time. Generally, a dynamic system consists of three parts. The first part is the state of a system, which is a representation of all the information about the system at some particular moment in time. For example, the state of a simple two-degrees-of-freedom (DOF) quarter-car model for vehicle suspension ride analysis, as illustrated in Figure 1.1, can be summarized

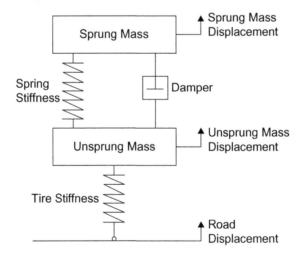

Figure 1.1
Two-DOF quarter-car model for vehicle suspension ride analysis.

by the vertical displacement and velocity of sprung and unsprung masses. In general, the symbol $X(t) = [x_1(t), \ldots, x_n(t)]$ will be used to denote the state of a system at time t.

The second part is the state space of a system. This is a set that contains all the possible states to which a system can be assigned. The state space of the two-DOF quarter-car model is the 2^n ensemble containing all the possible configurations for the n-element sprung and unsprung mass vertical motions within a given timeframe. The symbol Ω is commonly used to denote the state space of a dynamic system, and $X(t) \in \Omega$.

The third part is the state-transition function that is used to update and change the state from one moment to another. For example, the state-transition function of the two-DOF quarter-car model is defined by the governing state equation that changes the sprung and unsprung motion state at one step $X(t)$ to the next step $X(t+1)$.

The objective of dynamic systems analysis is thus to understand or predict all possible state transitions due to the state-transition function. In other words, the dynamic system analysis for the two-DOF quarter-car model is to predict the motions (displacements and velocities) of sprung and unsprung masses with given road displacement input within a given time frame. It can be seen that during the displacements and velocities are solved, the loads associated with tire stiffness, spring stiffness, and damper also can be resolved.

Depending on the differences of the state space, the state-transition function, and the excitation of a dynamic system, the dynamic response of the system may demonstrate different behaviors such as nonlinearity and hysteresis. For example, when the excitation of the road input is small, the spring stiffness, damper force-velocity characteristics, and tire stiffness may be of linear characteristics, thus the state-transition function may be expressed by a linear state equation. Then the state response of the two-DOF quarter car model may demonstrate linear behavior.

On the other hand, if the road excitation is significant enough, the nonlinear characteristics of the tire, spring stiffness, and damper force-velocity function may not be negligible. In this case, a nonlinear state-transition function is required to present the dynamic behavior of this system, thus the nonlinearity and hysteresis of the state responses may be yielded.

For a simple linear mechanical dynamics system, the state-transition function or governing state equation can be easily established based on a certain rule of the system, such as D'Alembert's principle or Newton's second law. The iterative solution for the state-transition function in a time domain is also not very difficult, as compared to a full multibody vehicle model that has large degree of complexity.

For a nonlinear multibody dynamic system, different numerical integration techniques may be required to solve for the ordinary differential equation (ODE). The linear explicit numerical integration methods with a constant time step are well applicable to most of the ODEs, but perform poorly for a class of "stiff" systems where the rates of change of the various solution components differ significantly. Consider, for example, the motion solution of a stiff suspension system when the system is being driven at a low oscillation frequency and then run into a deep pothole.

In principle, the stability region of a stiff system must include the eigenvalues of the system to be stable. Consequently, the linear explicit methods have a penalty of requiring an extremely small time step to be stable, causing unacceptable increases in the number of integration steps, integration times, and accumulated errors. On the other hand, the implicit methods with variable time steps are often recommended for stiff systems because of the better stability properties in the numerical integration process. Thus, depending on the nature of a system, stiff or nonstiff integrators may be applied to solve the dynamics equations (Newmark, 1959; Hilber, Hughes, & Taylor, 1977).

The dynamic behavior results from the equilibrium of applied forces and the rate of change in the momentum. Nowadays, the term multibody system is related to a large number of engineering fields of research, especially in vehicle dynamics. As an important feature, multibody system formalisms usually offer an algorithmic, computer-aided way to model, analyze, simulate, and optimize the arbitrary motion of possibly thousands of interconnected bodies.

Conditions of Equilibrium

The equilibrium condition of an object exists when Newton's first law is valid. An object is in equilibrium in a reference coordinate system when all external forces (including moments) acting on it are balanced. This means that the net result of all the external forces and moments acting on this object is zero. According to Newton's first law, under the equilibrium condition, an object that is at rest will stay at rest or an object that is in motion will not change its velocity.

Static equilibrium and dynamic equilibrium are termed when the object is at rest and moving in a constant velocity in a reference coordinate system, respectively. If the net result of all the external forces (including moments) acting on an object is not zero, Newton's second law applies. In this case, the object of a mass is not in equilibrium state, and will undergo an acceleration that has the same direction as the resultant force or moment. The velocity change yields an inertial force, which can be quantified by the product of mass and translational acceleration, or mass moment of inertia and angular acceleration.

D'Alembert's Principle

D'Alembert's principle, also known as the Lagrange–d'Alembert principle (Lanczos, 1970), is a statement of the fundamental classic laws of motion. The principle states that the sum of the differences among the forces acting on a system and the inertial forces along any virtual displacement consistent with the constraints of the system is zero. Thus d'Alembert's principle can be expressed as

$$\sum_i (\vec{F_i} - m_i \vec{a_i}) \cdot \delta \vec{r_i} = 0 \qquad (1.1)$$

where

$\vec{F_i}$ = applied forces
$\delta \vec{r_i}$ = the virtual displacements of the system, consistent with the constraints
m_i = the masses of the particles in the system
$\vec{a_i}$ = the accelerations of the particles in the system
$m_i \vec{a_i}$ = the time derivatives of the system momenta, or the inertial forces
i (subscript) = an integer used to indicate a variable corresponding to a particular particle

Considering Newton's law for a system of particles, the total forces on each particle are

$$\vec{F_i}^{(T)} = m_i \vec{a_i} \qquad (1.2)$$

where

$\vec{F}_i^{(T)}$ = the total forces acting on the particles of the system

$m_i \vec{a}_i$ = the inertial forces resulting from the total forces

Moving the inertial forces to the left side of the equation gives an expression that can be considered to represent quasistatic equilibrium, which is just a simple algebraic manipulation of Newton's law:

$$\vec{F}_i^{(T)} - m_i \vec{a}_i = \vec{0}. \tag{1.3}$$

Considering the virtual work, δW, done by the total and external/inertial forces together through an arbitrary virtual displacement, $\delta \vec{r}_i$ of the system leads to a zero identity, since the forces involved sum to zero for each particle:

$$\delta W = \sum_i \vec{F}_i^{(T)} \cdot \delta \vec{r}_i - \sum_i m_i \vec{a}_i \cdot \delta \vec{r}_i = 0. \tag{1.4}$$

At this point, it should be noted that the original vector equation could be recovered by recognizing that the work expression must include arbitrary displacements. Separating the total forces into external forces, $\vec{F}_i$ and constraint forces, $\vec{C}_i$, yields

$$\delta W = \sum_i \vec{F}_i \cdot \delta \vec{r}_i + \sum_i \vec{C}_i \cdot \delta \vec{r}_i - \sum_i m_i \vec{a}_i \cdot \delta \vec{r}_i = 0. \tag{1.5}$$

If arbitrary virtual displacements are assumed to be in directions that are orthogonal to the constraint forces, the constraint forces don't do work. Such displacements are said to be consistent with the constraints. This leads to the formulation of d'Alembert's principle, which states that the difference between applied forces and inertial forces for a dynamic system does not do virtual work:

$$\delta W = \sum_i (F_i - m_i a_i) \cdot \delta r_i = 0. \tag{1.6}$$

There is also a corresponding principle for static systems called the principle of virtual work for applied forces.

D'Alembert shows that we can transform an accelerating rigid body into an equivalent static system by adding the so-called inertial force and inertial torque or moment. The inertial force must act through the center of mass and the inertial torque can act anywhere. The system can then be analyzed exactly as a static system subject to inertial force and moment and the external forces. The advantage is that, in the equivalent static system, we can take moments about any point rather than only the center of mass.

This often leads to simpler calculations because any force in turn can be eliminated from the moment equations by choosing the appropriate point about which to apply the moment equation with the summation of moments equaling zero. Even in the courses of Fundamentals of Dynamics and Kinematics of Machines, this principle helps analyze the forces that act on a moving link of a mechanism. In textbooks of engineering dynamics, this is sometimes referred to as d'Alembert's principle.

A rigid body moving in a plane is subject to forces and torques; the inertial force is

$$\vec{F}_i = -m\ddot{\vec{r}}_c \qquad (1.7)$$

where

$\vec{r}_c$ = the position vector of the center of mass of the body
m = the mass of the body

The inertial torque (or moment) is

$$\vec{T}_i = -J\ddot{\vec{\theta}} \qquad (1.8)$$

where J is mass moment of inertia of the body and θ is the angular displacement of the body. In addition to the external forces and torques acting on the body, the inertia forces acting through the center of mass and the inertial torques need to be added. The system is equivalent to one in static equilibrium with following equations:

$$\sum F_x = 0$$
$$\sum F_y = 0 \qquad (1.9)$$
$$\sum T = 0$$

where

$\sum T$ = the sum of torques or moments, including the inertial moment and the moment of the inertial force taken about the axis of any point
$\sum F_x$ and $\sum F_y$ = the summation of forces along the two perpendicular axes X and Y, respectively

Multibody Dynamics Systems

During the past three decades, there has been a rapid expansion of research on multibody dynamics (Andrews & Kesavan, 1975; Andrews, Richard, & Anderson, 1988; Bainum & Kumar, 1982; Agrawal & Shabana, 1985; Allen, Oppenheim, Parker, & Bielak, 1986; Angeles, 1986a, 1986b; Amirouche, 1988). A multibody system may be defined as a collection of bodies with a given connection configuration. The system may have as few as two bodies, but multibody analyses are generally directed toward systems that have an unlimited number of bodies.

Figure 1.2 depicts a short-long-arm suspension, which is a typical multibody dynamic system. In general, a multibody system may contain a mix of rigid and flexible bodies with specified motion. Many physical systems of interest are effectively modeled by a system of pin-connected rigid bodies or compliant elements such as bushings or mounts. The interest in multibody systems stems from the ability to accurately model physical systems, such as robots, mechanisms, chains, cables, biosystems, structures, flexible beams, and vehicles. The analysis of the

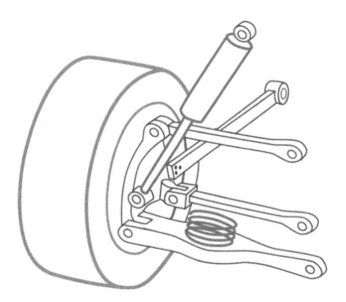

Figure 1.2
A multibody system.

corresponding multibody system has become practical with advances in computer technology and the development of supporting computational methods.

In spite of many notable advances in multibody dynamics analyses, there is no common view about which method is the best for specific applications, computational efficiency, or governing dynamic equation acquisition. There are adherents of Lagrangian methods, Newton–Euler methods, virtual work methods, Gibbs–Appell equations, and Kane's equations. Many advocate the use of pseudo-inverse methods to reduce the governing differential/algebraic equations to a consistent set of differential equations in a form suitable for numerical integration.

Others prefer null/tangent space methods, singular value decomposition, and orthogonal complement arrays. There are arguments about the best method to incorporate flexibility effects into the analyses, and the best way to take advantage of advances in finite element methods and modal methods in hybrid analyses.

The principle tasks in computational multibody dynamics are (Huston, 1991):

- To develop adequate models of interesting physical systems

- To efficiently generate the governing dynamic and constraint equations

- To accurately and efficiently solve the governing equations

Some analysts advocate lumped parameter models where the system is modeled by rigid bodies connected by springs and dampers that simulate the flexibility effects. Others, using the principles of elasticity, modal analysis, and finite element analysis, incorporate the flexibility effects directly into the multibody system.

The relative advantages and disadvantages of these approaches, especially on their accuracy, efficiency, and ease of use, are still being debated. It seems that the physical system being studied may determine which approach is optimal. The lumped parameter models appear to be the best for systems with low frequency, large-displacement movements. The innate flexible systems are better suited for systems with rapid but relatively small displacements.

The most commonly used software to generate loads is called ADAMS (Advanced Dynamic Analysis of Mechanical Systems) by MSC software (2007). It is general purpose multibody dynamic analysis software, based on the Lagrangian formulation of dynamic equations of motion. It formulates and solves a set of simultaneous differential algebraic equations (DAEs) for the dynamic and the constraint equations. This formulation leads to a stiff system of equations, which can be integrated without introducing stability problems by integrators that use backward difference formulas.

The software also utilizes symbolic derivatives and sparse matrix formulation to improve the efficiency of the solution process. The solution process includes a predictor step that utilizes a given number of solutions over previous time steps and a corrector step that is essentially a Newton–Raphson procedure for the solution of a set of nonlinear algebraic equations. The software has a built-in library of constraints such as prismatic, revolute, spherical, and such, and an extensive set of force elements such as bushings, beams, springs, and dampers, as well as generic modeling entities.

The main advantage of the software is that it can accurately handle complex nonlinear mechanical system dynamics that consist of large translational and angular displacements. It simultaneously provides the constraint forces between different bodies in the system. Since durability events are typically in the range of 0 to 50 Hz with highly nonlinear elastometric characteristics included, a time-domain formulation is ideal to handle such problems and ADAMS is well-suited to this purpose.

Generic Load Cases

Load analysis plays a pivotal role in the vehicle design program, particularly at an early design stage where it is impossible to perform an accurate and extensive load prediction since most of the vehicle data are not available. Alternatively, some extreme load cases, commonly referred to as generic load cases, have been proposed to study the reactions of a new suspension design to these load cases as compared to the current design. The generic load cases will provide engineers a good design direction, and are useful for A to B design comparison.

The generic load case can be studied independently for the front and rear suspensions so that the design proposal may be evaluated independently with minimum necessary data. However, the generic load cases cannot replace the need for vehicle proving grounds data acquisition. Once available, measured proving ground loads on every suspension component and system should be employed for detailed fatigue assessment.

Generic Load Events

The generic load events currently used by most automotive manufacturers are primarily developed based on historical test data, special cases/events studies, and special durability requirements for various structure/suspension types. They represent severe loading conditions in each primary mode of the component/ system. The fundamental objective of the generic load is to predict the most possibly severe load cases that the vehicle may experience. Meanwhile, it should be understood that the peak load for different components may not occur in one single severe load case, thus multiple load cases are needed to investigate for the actual peak load predictions for every suspension component.

Some generic load events and their input forces are listed in Tables 1.1 and 1.2, respectively. In Table 1.2, F_{XLF}, F_{XRF}, F_{XLR}, and F_{XRR} are the longitudinal

Table 1.1: List of Sample Generic Load Events

No.	Name	Static/Dynamic	Suspension	Major Functions
1	1G jounce	Static	Front and rear	Inertial load check and static load distribution
2	3G jounce	Static and dynamic	Front and rear	Vertical suspension loads and travel
3	5G jounce	Static and dynamic	Front and rear	Vertical suspension loads and travel
4	2G roll	Static	Front and rear	Roll motion and antiroll bar loads
5	Cornering	Static	Front and rear	Lateral suspension loads
6	Braking over bump	Dynamic	Front and rear	Vertical and longitudinal suspension loads
7	Left-wheel bump	Dynamic	Front and rear	Vertical suspension and antiroll bar loads
8	Curb push-off	Static	Front	Steering system loads

Note: 5G jounce events may be applied only for light-duty vehicles, such as a car.

Table 1.2: Input Loads under Various Generic Load Events

No.	Name	F_{XLF}	F_{YLF}	F_{ZLF}	F_{XRF}	F_{YRF}	F_{ZRF}	F_{XLR}	F_{YLR}	F_{ZLR}	F_{XRR}	F_{YRR}	F_{ZRR}
1	1G jounce	0	0	1Gs	0	0	1Gs	0	0	1Gs	0	0	1Gs
2	3G jounce	0	0	1Gs+ 2Gp	0	0	1Gs+ 2Gp	0	0	1Gs+ 2Gp	0	0	1Gs+ 2Gp
3	5G jounce	0	0	1Gs+ 4Gp	0	0	1Gs+ 4Gp	0	0	1Gs+ 4Gp	0	0	1Gs+ 4Gp
4	2G roll	0	0	2Gs	0	0	0	0	0	0	0	0	2Gs
5	Cornering	0	1Gs	1Gs	0	1Gs	1Gs	0	1Gs	1Gs	0	1Gs	1Gs
6	Braking over bump	2Gp	0	1Gs+ 2Gp	2Gp	0	1Gs+ 2Gp	2Gp	0	1Gs+ 2Gp	2Gp	0	1Gs+ 2Gp
7	Left-wheel bump	0	0	1Gs+ 2Gp	0	0	1Gs+ 2Gp	0	0	1Gs+ 2Gp	0	0	1Gs
8	Curb push-off	0	μGs	1Gs	0	μGs	1Gs	0	0	0	0	0	0

Note: Gs for corner GVW or GVWT weight, Gp for corner GVW or GVWR weight pulse, μ is the tire patch friction coefficient (default value as 0.88). The curb push-off lateral force is equal to the summation of left and right tire patch lateral frict on forces.

forces acting on the tire patches at four corners (LF = left front, RF = right front, LR = left rear, and RR = right rear), respectively, and the positive direction of the longitudinal forces is toward the back of the vehicle. F_{YLF}, F_{YRF}, F_{YLR}, and F_{YRR} are the lateral forces acting on the tire patches at four corners, respectively, and the positive direction of the lateral forces is toward the right side of the vehicle. F_{ZLF}, F_{ZRF}, F_{ZLR}, and F_{ZRR} are the vertical forces acting on the wheel centers at four corners, respectively, and the positive direction of the vertical forces is upward. The major assumptions in the generic load cases are as follows:

- The vehicle sprung mass is much heavier than the unsprung masses such that the motion of the sprung mass is neglected. Such an assumption may yield an over- or underprediction of the interaction forces between the sprung and unsprung masses, particular for dynamic load cases.

- During cornering, the vehicle is in steady-state at a constant forward speed, thus the inertia forces, longitudinal tire forces, and shock forces are neglected.

- The maximum possible jounce bumper forces are assumed to happen at 3G jounce static load condition, where G represents the vehicle corner weight at the gross vehicle weight (GVW) or gross vehicle weight rating (GVWR) condition.

- The input forces pulse when braking over bumps and left-wheel bump load cases are the same as that in 3G jounce dynamic load case.

- The curb push-off (CP) happens at zero vehicle speed with extremely slow steering action, thus the inertia forces, longitudinal tire forces, and shock forces are neglected.

- The vehicle rollover motion happens at extremely slow speed with a steady-state condition, thus the inertia forces, longitudinal and lateral tire forces, and shock forces are neglected.

It should be noted that the curb push-off load case is specially designed to validate the design of steering system components, thus it should not be applied to the rear suspension. Furthermore, to obtain representative loads on the frame, we need to combine the appropriate loads from the front and rear load cases.

The definition of 1G depends on the engineer. In most cases, it is equal to half the front or rear GAWR (Gross Axle Weight Rating).

However, if both front and rear suspensions are analyzed at their respective GAWRs and the loads are combined to obtain the frame loads, it should be realized that the frame is being overloaded. Alternatively, we may appropriately scale down the loads such that the frame will never experience initial static loads higher than GVWR.

Under all events except for curb push-off, all the lateral and longitudinal forces are applied at the corresponding tire patch, and all vertical forces are applied at the corresponding wheel center, as shown in Figure 1.3(a). For the curb push-off event, as shown in Figure 1.3(b), the left ($\mathbf{F}_{YLF}$) and right ($\mathbf{F}_{YRF}$) tire patch friction forces being against the curb push-off force are applied at the corresponding tire patch.

The vertical loads ($\mathbf{F}_{ZLF}$ and $\mathbf{F}_{ZRF}$) are applied at the corresponding wheel center, but the lateral curb push-off force ($\mathbf{F}_{YRF\_CP}$ or $\mathbf{F}_{YLF\_CP}$, depending on the corner of the curb push-off) is applied at a certain height (e.g., 8 inches) above the tire patch and several inches ahead or behind the wheel center, depending on the tire dimensions (see Figure 1.4). It should be noted that the configuration of the steering system directly affects the patterns of curb push-off input force and steering system motion direction, as illustrated in Figure 1.5.

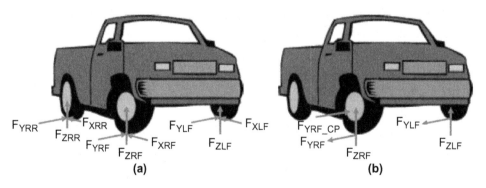

Figure 1.3
Generic loads input forces under various events.

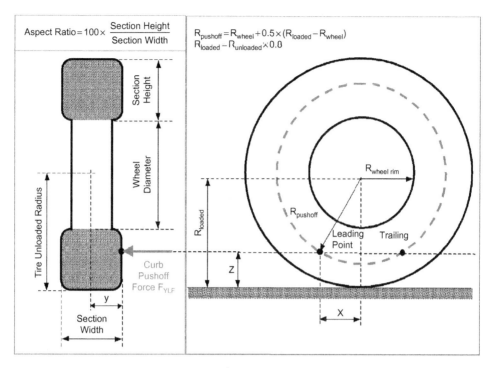

Figure 1.4
Determination of curb push-off input lateral force action point location (x, y, and z).

In generic load cases, two kinds of analyses are included, static and dynamic. Static analysis means that the simulation under a specific generic load event is based on quasistatic or static equilibrium using any multibody simulation solver, thus only the displacement-dependent and gravity forces are involved. Dynamic analysis indicates that the simulation is based on an integration of the dynamics equations using any multibody simulation solver, thus the inertial and velocity-dependent forces are also counted.

The duration (see Figure 1.6) of input force pulse has no effect on the loads from the static analysis, but affects the load calculations from dynamic analysis. The duration of an input pulse can be adjusted under the 3G jounce dynamic load case until the maximum jounce bumper load in the dynamic analysis approximately equals that in the static analysis. Such an adjustment may involve an iterative process.

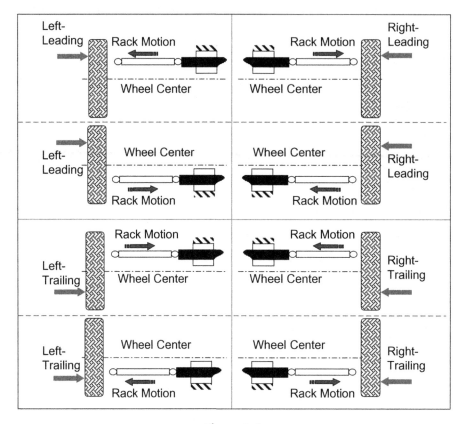

Figure 1.5
Rack motion and steering system configurations under curb push-off event.

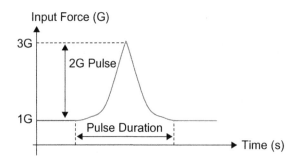

Figure 1.6
Input force pulse.

Analysis Procedure

The following procedures are applied for a typical generic load case analysis (3G jounce static load):

1. Prepare suspension and steering modeling data, including suspension hard points, properties of spring, shock, jounce bumper, and bushing.

2. Use commercial software for multibody system modeling to build the suspension model at the design position with the vehicle body fixed to the ground.

3. Adjust all the parameters to the required design values from default settings. Those parameters include geometry, mass, and inertia, spring and bushing properties, shock and jounce bumper force characteristics, and others.

4. Establish component local reference coordinates (also called markers, in multibody dynamics system) for potential output request.

5. Create force requests for all required components, including jounce bumper, frame, cradle, control arms, links, track bar, knuckles, stabilizer bar drop links, tie rods, and more.

6. Perform kinematics and compliance (K&C) analysis. If the model cannot achieve the required K&C properties, you should contact the responsible engineers to verify the geometry, spring, and bushing properties.

7. Once the 3G jounce static load is chosen, the required GVW at the front or rear axle is adjusted from the default value and all other adjustable parameters are kept in default values.

8. Run the 3G jounce static analysis and plot the calculated time histories of left and right jounce bumper forces. These jounce bumper forces are used for adjusting the pulse duration of tire patch input forces.

9. Adjust the pulse duration of tire patch input forces and run the 3G jounce dynamic analysis, until the maximum jounce bumper forces are approximately equal to those from 3G static analysis.

10. Apply the well-adjusted pulse duration parameters to all other considered generic load cases.

11. Calculate the peak values for all required loads and achieve simultaneous peak load files for all generic load cases.

For clarity, these procedures are illustrated in a flowchart shown in Figure 1.7.

Results and Report

After the static and dynamic analyses, the time histories of all the loads on each component are obtained in a coordinate system that is fixed in that component. This facilitates an easy finite element (FE) analysis using the loads. It should be noted that the inertial loads are not reported and the loads from the dynamic analysis cannot be used to perform a static FE analysis. An inertia-relief analysis must be performed instead. The results of generic load cases can be processed in two ways:

- The maximum magnitude of each load and its corresponding time are summarized. This is useful only to get a quick estimation of the maximum loads. The loads from the summary table may not be simultaneous and cannot be used to perform an FE analysis for the component.

- For each load, the maximum magnitude is given along with the rest of the loads occurring simultaneously on that component. These loads can be used to perform FE analyses. It should be noted that there may be multiple possible peak slices for a given analysis and that an FE analyst should carefully choose the slice that may cause the highest stresses in the structure.

It is important to check the results by reviewing the time histories and the animation graphic files before publishing the results. Be particularly careful about some force spikes that may result from an incorrect solution. It is a good practice to gain the simulation confidence by comparing the results with those calculated from a similar vehicle or from the same vehicle at a previous design stage. The simultaneous peak loads on a particular component can be simply examined with the summation of all attachment forces in x, y, and z directions, respectively.

If the summation in each direction is zero or close to zero, the simultaneous peak loads for the component may be considered as correct outputs. Otherwise, some attachment forces acting on that component may have not been included to extract the simultaneous peak loads. Once the results are examined and approved

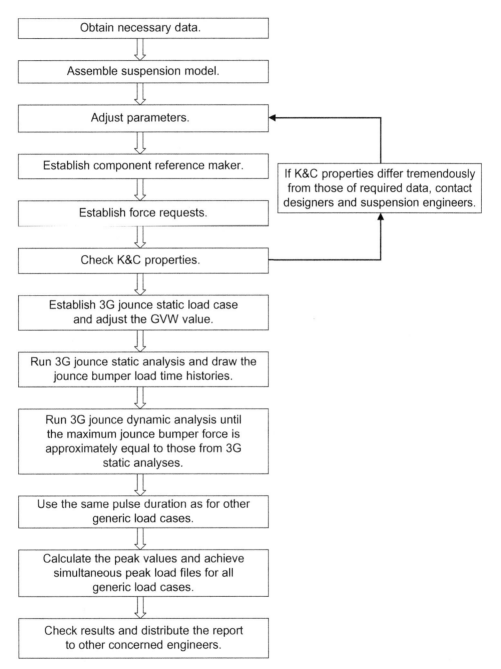

Figure 1.7
A flowchart of generic load process.

for correctness, those results can be distributed in the form of a standard report for generic loads to the design and analysis engineers for further analyses.

Semianalytical Analysis

With rapid development of advanced computer technologies, more and more virtual road load analyses or simulations have been applied by automotive manufacturers to replace the field measurement. At different stages of vehicle development, the availability of vehicle modeling data varies. Thus different simulation strategies may be applied to meet the vehicle development requirements in road loads analysis.

The semianalytical method is introduced to solve for the second-order ordinary differential motion equations by the static analysis. It is assumed that the vehicle or system of interest will meet the dynamic equilibrium, provided that all the external forces including moments and the inertia forces at every time instant are given analytically or experimentally. These analysis techniques are applied to a powertrain mount and an independent suspension system, and are presented in this section.

Powertrain Mount Load Analyses

In the process of developing a powertrain mount system, the calculation of powertrain mount loads (PML) is one of the critical inputs required for fatigue analyses of the powertrain structure and mounts. As reported by Yang, Muthukrishnan, Seo, and Medepalli (2004), the powertrain mount loads on a given powertrain architecture for durability depends on the following factors:

- Customer's usage/operating conditions

- Powertrain mount stiffness and damping characteristics

- Powertrain mount locations and orientations

- Powertrain mass and mass moments of inertia

- Powertrain reaction torques

- Rigid body vibration modes from subsystems such as engine, transmission, accessory, and exhaust systems

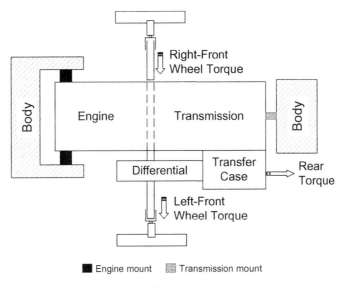

Figure 1.8
Powertrain mount system diagram (Yang et al., 2004).

To predict the powertrain mount loads for durability, an analytical model should be developed to account for all the factors just addressed.

An analytical model to calculate the mount loads at a three-mount powertrain system for an all-wheel-drive (AWD) vehicle will be presented herein. The three-mount system is schematically illustrated in Figure 1.8, where two mounts at the engine side and one mount at the transmission side are attached to the vehicle body or cradle, and the transfer case splits the input torque from the transmission into two output torques: one to the front axles and the other to the rear axles.

Based on the assumption of rigid body vibration modes and negligible damping force of the powertrain system, the free body diagram of the powertrain mount system is shown in Figure 1.9, where $\vec{F}_{EMi}$ and $\vec{T}_{EMi}$ are the force and torque vectors acting on the i-th powertrain mount; $\vec{F}_{EI}$ and $\vec{T}_{EI}$ are the powertrain inertia force and torque vectors; and $\vec{T}_{of}$ and $\vec{T}_{or}$ are the front and rear output torque vectors from the transfer case. The vehicle body or cradle is assumed to be fixed to the ground.

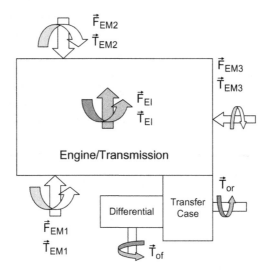

Figure 1.9
Powertrain mount loads balance.

If the two output torque vectors and the two powertrain inertia force and torque vectors can be acquired during durability testing, the remaining two unknowns ($\overrightarrow{F}_{EMi}$ and $\overrightarrow{T}_{EMi}$) can be solved from the following two static equilibrium equations:

$$\sum_{i=1}^{3} \overrightarrow{F}_{EMi} + \overrightarrow{F}_{EI} = 0 \tag{1.10}$$

$$\sum_{i=1}^{3} \overrightarrow{T}_{EMi} + \overrightarrow{T}_{of} + \overrightarrow{T}_{or} + \overrightarrow{T}_{EI} = 0. \tag{1.11}$$

$\overrightarrow{T}_{of}$ and $\overrightarrow{T}_{or}$ can be measured by placing torsional bridges on two front axle shafts and one rear propshaft, respectively. Furthermore, the inertial force and torque vectors of powertrain system ($\overrightarrow{F}_{EI}$ and $\overrightarrow{T}_{EI}$) can be calculated from the measured powertrain translational acceleration vector ($\overrightarrow{a}_{E}$) and angular acceleration vector ($\overrightarrow{\beta}_{E}$) in the following manner:

$$\overrightarrow{F}_{EI} = M_E \, \overrightarrow{a}_{E} \tag{1.12}$$

$$\overrightarrow{T}_{EI} = J_E \, \overrightarrow{\beta}_{E} \tag{1.13}$$

where M_E and J_E are the mass and mass moment of inertia of the powertrain system, respectively. $\vec{a_E}$ and $\vec{\beta_E}$ have six components and are expressed in this equation:

$$\vec{a_E} = \left\{ \begin{array}{c} a_{cx} \\ a_{cy} \\ a_{cz} \end{array} \right\} \text{ and } \vec{\beta_E} = \left\{ \begin{array}{c} \beta_{cx} \\ \beta_{cy} \\ \beta_{cz} \end{array} \right\} \qquad (1.14)$$

where

a_{cx}, a_{cy}, and a_{cz} = the powertrain CG translational accelerations along fore/ aft, lateral, and vertical directions, respectively

β_{cx}, β_{cy}, and β_{cz} = the powertrain angular acceleration in roll, pitch, and yaw directions, individually

Here the reference coordinates attached to powertrain rigid body are defined as X axis-positive toward back of the vehicle; Y axis-positive toward vehicle right-side; and Z axis-positive upward.

Since translational acceleration ($\vec{a_E}$) and angular acceleration ($\vec{\beta_E}$) of engine/ transmission have six components in total, we need at least six measured channels for calculating the engine/transmission translational and angular accelerations. In summary, Table 1.3 lists the necessary channels for calculating AWD vehicle powertrain mount loads. For the front- or rear-wheel-drive vehicle, only one channel for the powertrain output torque is needed.

It should be mentioned that the total channels needed to resolve the powertrain mount durability loads are based on the assumption that all measured accelerations are accurate and can be applied to nonsingularly resolve the powertrain CG accelerations. Furthermore, all those high frequency components due to the nonrigid body vibrations are neglected.

Table 1.3: Summary of Necessary Measured Channels

Channels Needed	Channel Number	Total Channel Number
Powertrain acceleration	6	8
Powertrain output torque (front and rear)	2	

Practically, the accelerations at a powertrain center of gravity (CG) are usually obtained by measuring the nine accelerations located at three different locations of powertrain sides. It is assumed that the i-th location of an arbitrary accelerometer can be described as (x_i, y_i, z_i) in a Cartesian coordinate system and that the origin of the Cartesian axes is located at the powertrain CG.

A geometric rigid body transformation matrix ($[R]_{9\times6}$), which relates rigid body accelerations ($\{a_{cE}\}_{6\times1}$) from the CG location to those accelerations at a nodal point i ($\{a_{iE}\}_{9\times1}$), can be defined as follows:

$$\{a_{iE}\}_{9\times1} = [R]_{9\times6}\{a_{cE}\}_{6\times1} \tag{1.15}$$

or

$$
\begin{Bmatrix}
a_{1x} \\
a_{1y} \\
a_{1z} \\
a_{2x} \\
a_{2y} \\
a_{2z} \\
a_{3x} \\
a_{3y} \\
a_{3z}
\end{Bmatrix}
=
\begin{bmatrix}
1 & 0 & 0 & 0 & z_1 & -y_1 \\
0 & 1 & 0 & -z_1 & 0 & x_1 \\
0 & 0 & 1 & y_1 & -x_1 & 0 \\
1 & 0 & 0 & 0 & z_2 & -y_2 \\
0 & 1 & 0 & -z_2 & 0 & x_2 \\
0 & 0 & 1 & y_2 & -x_2 & 0 \\
1 & 0 & 0 & 0 & z_3 & -y_3 \\
0 & 1 & 0 & -z_3 & 0 & x_3 \\
0 & 0 & 1 & y_3 & -x_3 & 0
\end{bmatrix}
\begin{Bmatrix}
a_{cx} \\
a_{cy} \\
a_{cz} \\
\beta_{cx} \\
\beta_{cy} \\
\beta_{cz}
\end{Bmatrix}. \tag{1.16}
$$

Using the D-optimal technique (Lee, Lu, & Breiner, 1997), with the given $\{a_{iE}\}_{9\times1}$ and $[R]_{9\times6}$, the best estimated $\{a_{cE}\}_{6\times1}$ can be obtained as follows:

$$\{a_{cE}\} = ([R]^T[R])^{-1}[R]^T\{a_{iE}\}. \tag{1.17}$$

Once the powertrain CG accelerations are estimated, they can be used to back-calculate accelerations at measured locations. Then the calculated accelerations can be compared with those measured. The calculated powertrain mount accelerations using nine channels were compared with those measured. The maximum error for all nine channels is below 0.1 g, which shows that the back-calculated powertrain mount accelerations are reasonably well-correlated with those measured ones.

Figure 1.10 displays the maximum error of all nine channels between calculated and measured a_{1x}. Finally, the calculated powertrain CG accelerations are illustrated in Figure 1.11.

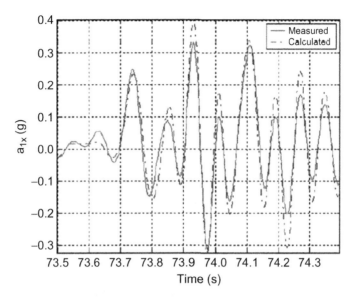

Figure 1.10

Snapshot of maximum error between calculated and measured powertrain mount accelerations a_{1x}.

It should be mentioned that the measured accelerations are first filtered by a low-pass sixth-order forward-backward Butterworth digital filter with a certain band limit (in this example, 15 Hz). The determination of the filter bandwidth is based on the highest frequency of powertrain rigid-body modes under given mount stiffness and mass/inertia properties. Table 1.4 shows the frequencies of six rigid body modes. It can be seen that the maximum frequency for the rigid body roll mode is 14 Hz.

To predict the powertrain mount loads using the developed analytical model, Equation (1.17), the input variables (the inertial loads and powertrain output torques) should be available. The inertial loads can be calculated using the resultant powertrain CG accelerations, while the powertrain output torques can be either directly measured or estimated using the measured front- and rear-wheel driving torques divided by the transmission ratios.

In the calculation, the measured front- and rear-wheel driving torques are served as the input variables. All powertrain mount translational stiffness characteristics are measured. The torsional and conical stiffness values of powertrain mounts are

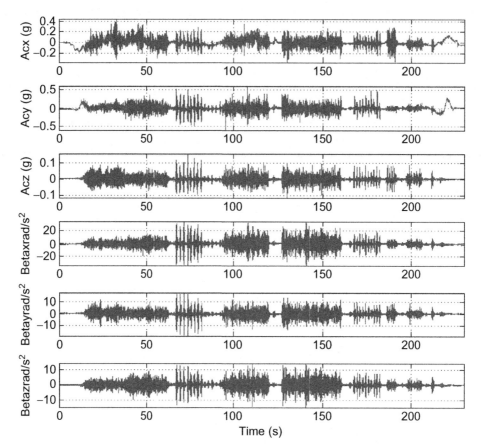

Figure 1.11
Calculated powertrain CG accelerations.

**Table 1.4: Six Rigid Body Modes for a Sample
Powertrain Mount System**

Mode No.	Frequency (Hz)	Mode
1	4.83	Lateral
2	5.72	Fore-aft
3	6.54	Yaw
4	8.04	Bounce
5	10.58	Pitch
6	13.97	Roll

estimated using the historical powertrain mount data. Equations (1.12) and (1.13) are solved using a quasistatic analysis technique to yield the powertrain mount loads.

The calculated powertrain mount loads are reasonably correlated to those measured, except the transmission mount lateral loads. Figure 1.12 shows the comparison of the measured and calculated left front mount vertical load in time history. Table 1.5 illustrates the comparison of the RMS values for calculated and measured powertrain mount loads.

Table 1.5 shows that the RMS values for both measured and calculated loads are very close except for transmission mount lateral loads, where the relative error is 63%. The results may indicate that some parameters in the current model are not accurate or the current approach may need to be refined to include some other effects, such as the interaction between the powertrain and exhaust system, assuming that the measured transmission mount lateral load is accurate.

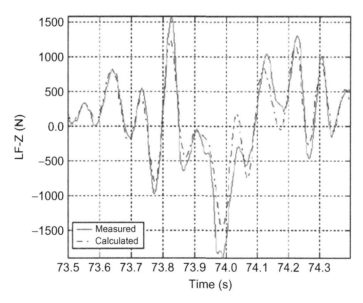

Figure 1.12
Comparison of left front mount vertical loads.

Table 1.5: Comparison of RMS Values for PML

Location/Direction		Calculated	Measured	Relative Error (%)
LF (N)	F_x	113.4	100.0	13.4
	F_y	132.7	133.9	−0.9
	F_z	300.7	358.0	−16.0
RF (N)	F_x	105.0	85.2	23.3
	F_y	137.7	134.7	2.2
	F_z	293.8	369.8	−20.6
TM (N)	F_x	70.7	69.8	1.3
	F_y	155.4	95.2	63.4
	F_z	333.0	293.1	13.6

It should be noted that the proposed method only uses the nominal mount stiffness characteristics and doesn't consider the change of those properties due to manufacture tolerance, operating temperature variation, and so on. The mass and inertia values used in the model include only the engine and transmission, but the powertrain is actually rigidly attached to the exhaust system.

Furthermore, the contributions due to the vibrations of exhaust and accessories attached to the powertrain are entirely ignored in this example. The front half-shaft and rear prop shaft torques are estimated using the measured wheel force driving torque, which may not be capable of capturing the dynamic torque as well as transmission efficiency loss, since a constant torque ratio is used in the calculation.

Suspension Component Load Analysis

A semianalytical method is presented in this section to calculate suspension component loads with minimum instrumentation effort and data acquisition channels, as compared with the conventional road load data acquisition approach where all the component loads are measured with strain gauges or load cells. In this method, the vehicle body is fixed to the ground, and the reaction forces acting on suspension components are calculated by the quasistatic analysis technique, based on the measured input data from wheel force transducers, steering angle, shock force, and inertia forces calculated from acquired accelerations.

As an example of a McPherson strut front suspension (see Figure 1.13) of a vehicle, two load prediction techniques are introduced. The first one, called the

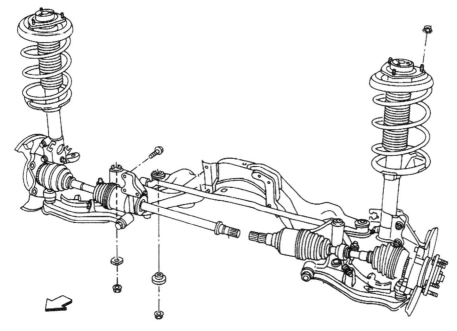

Figure 1.13
A McPherson strut front suspension.

vertical displacement input method (Sommerfeld & Meyer, 1999; Tatsuya, 2001), requires the inputs of six forces from every wheel force transducer (WFT), except that the vertical force in each WFT is replaced by the vertical displacement calculated from the measured lower control arm (LCA) angle. In the vertical displacement method, the measured or calculated jounce bumper force is also used as an input.

The other technique, the vertical WFT force input method, uses all six forces from every wheel force transducer as inputs, where the jounce bumper force is calculated by using the jounce bumper force-deflection curve. If the inertia force, measured shock absorber force, and measured WFT forces are accurate, we can calculate the induced forces acting on every suspension component.

Vertical Displacement Input Method

In this method (Sommerfeld & Meyer, 1999; Tatsuya, 2001), the input channels to a multibody dynamics model first are summarized in Table 1.6, and then discussed.

Table 1.6: Input Channels for Vertical Displacement Input Method

Channels Used in the Simulation	No. of Channels	Input to ADAMS Model
WFT force (left, right)	4	Longitudinal and lateral force
WFT moment (left, right)	6	Overturning, driving, and aligning moment
Spindle acceleration (left, right)	6	Inertia force of spindle, rotor and caliper, and WFT
LCA angle (left, right)	2	Vertical displacement of WFT position
Steering angle	1	Steering angle
Shock absorber force (left, right)	2	Shock absorber force
Jounce bumper force (left, right)	2	Jounce bumper force
Braking signal	1	Braking signal
Total channels	**24**	

WFT Forces Including Moments

The WFT forces are measured between tire and spindle axis. The measured WFT forces are not actual tire patch forces because of the inertia effect of the tire. However, the WFT forces are actual external forces applied on the spindle axis. The measured WFT forces are applied at the wheel force transducer center position, instead of the wheel center position, because the WFT forces are measured at that point. The WFT central plane is typically 4 inches from the wheel central plane. The WFT forces and moments are applied with respect to the wheel-fixed reference frame.

Inertia Force

The inertia forces of spindle, rotor, caliper, and WFT are calculated by using the measured spindle acceleration and the mass of each component. The rotational inertia force is not considered. The inertia forces of the other components whose mass is negligible are ignored. The tire mass is not considered because the WFT forces are measured between spindle and tire.

The inertia force should be applied at the CG of each component in the analysis. However, the CG of a bulk shape component such as a spindle is usually inside the component. It is almost impossible to measure the acceleration of CG with only one accelerometer because we cannot cut the knuckle to measure the acceleration.

In the case of a McPherson strut type suspension, it can be assumed that the motion of the spindle is usually translational and the angular acceleration is small. Even when the spindle is steered, the angular acceleration may not be high. Thus the same acceleration can be used to calculate the inertia force of spindle, rotor, caliper, and WFT. In the test, three translational accelerations are measured at one location on the spindle. The calculated inertia forces are applied at the CG of each component with respect to the spindle-fixed reference frame.

Vertical Displacement of WFT Position

The vertical displacement of WFT is derived from the measured lower control arm angle. The lower control arm angle is a relative angle between vehicle body and LCA. Thus the calculated vertical displacement is the relative displacement between vehicle body and spindle. A special simulation technique is required to ensure the wheel moves vertically, by creating a kinematics curve between the vertical displacement of WFT center and LCA angle, as shown in Figure 1.14. Figure 1.15 shows the comparison between the calculated and measured LCA angles for a diagonal trench road event.

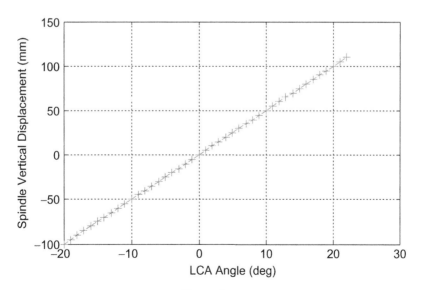

Figure 1.14
Kinematics curve between the spindle vertical displacement and LCA angle.

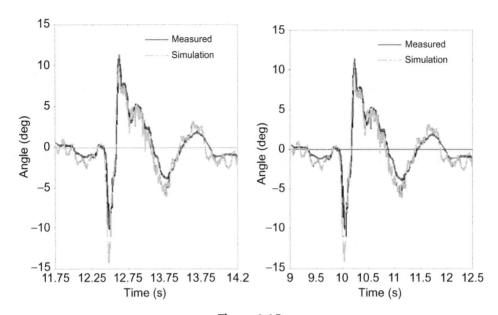

Figure 1.15
Comparison between the calculated (*left*) and measured (*right*) LCA angles
in a diagonal trench event.

Note that the spindle acceleration can also be used to calculate the vertical displacement by the double integration approach. However, the integrated vertical displacement is not a relative displacement but the absolute displacement with respect to the fixed ground. The displacement should be relative to the vehicle body because the vehicle body is fixed to the ground in this quasistatic simulation method.

The absolute displacement of the spindle can be used under the assumption that the vertical movement of vehicle body is negligible, but the vehicle body should not be fixed to the ground to use the absolute displacement. In that case, the independent analysis of front and rear suspension is impossible and we need accurate mass properties of vehicle body.

Shock Absorber Force

The measured shock absorber force is the preferable force input for better accuracy. However, in the absence of the measured shock absorber force, an accurate shock force-velocity curve is needed to estimate the shock force using the calculated shock axial velocity by differentiating its axial displacement estimated from the kinematics relationship between the measured LCA angle and LCA ball joint.

Jounce Bumper Force

The measured jounce bumper force is the preferable external force input. The jounce bumper force can be calculated from a jounce bumper force versus deflection curve. It should be noted that the calculated jounce bumper force is very sensitive to the initial static position of the analysis. A small displacement error in an initial position will yield large jounce bumper force discrepancy due to its highly nonlinear force-deflection characteristics.

Brake Signal

The brake signal is very important to identify the brake or driving torque (Ty) from the WFT moment measurement. For the front-wheel drive suspension illustrated in Figure 1.16, the drive torque is not reacted by the suspension, thus

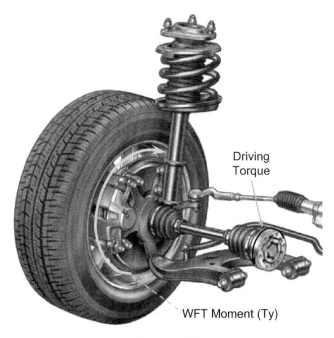

Figure 1.16
Illustration of front-wheel drive suspension.

the application of Ty depends on whether braking is on or off. If braking is on, the caliper holds the brake disk and the braking torque acts on the suspension. If braking is off, then Ty should not be applied at all on the suspension. This principle can be applied to any other type of torque-driven independent suspensions.

Vertical WFT Force Input Method

In this method, the vertical WFT force instead of the vertical displacement of the WFT is used. The input channels to a multibody dynamics model are summarized in Table 1.7. The jounce bumper force is determined by the measured inertia force and WFT vertical force according to the D'Alembert principle. Thus, the accurate measurement of shock force and acceleration is required to obtain accurate load prediction. In general, the calculated jounce bumper force in this method is not sensitive to the jounce bumper force-deflection curve as well as the initial suspension static position.

The suspension load analysis procedures with the WFT force input method are summarized as follows:

1. Check polarity, calibration, unit, and quality of the measured data for accuracy. If necessary, noise and outliers should be removed prior to any engineering usage.

2. Build the suspension model using any commercially available software.

Table 1.7: Input Channels for Vertical Force Input Method

Channels Used in the Simulation	No. of Channels	Input to ADAMS Model
WFT force (left, right)	6	Longitudinal, lateral, and vertical force
WFT moment (left, right)	6	Overturning, driving, and aligning moment
Spindle acceleration (left, right)	6	Inertia force of spindle, rotor and caliper, and WFT
Steering angle	1	Steering angle
Shock absorber force (left, right)	2	Shock absorber force
Braking signal	1	Braking signal
Total channels	**22**	

3. Calculate the suspension kinematics and compliance (K&C) and compare them with the targeted data. The K&C in this section include the suspension ride and roll rates as well as the steering ratio.

4. Estimate the WFT inboard mass for inertial load compensation. The inboard mass includes any unsprung mass inboard of the WFT interface with the rim and moving together with the knuckle, which can be the WFT, knuckle, brake disc, brake caliper, and the like.

5. Apply the time histories of WFT loads, spindle accelerations, shock forces, and steering wheel angle to the suspension model, and perform the quasistatic analysis to resolve the suspension component loads.

The correlations of calculated and measured suspension load data for a McPherson strut front suspension system are presented in Figures 1.17 through 1.21.

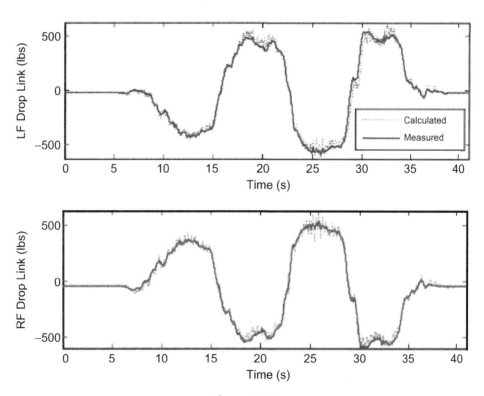

Figure 1.17
Antiroll bar drop link axial loads in a figure-eight event.

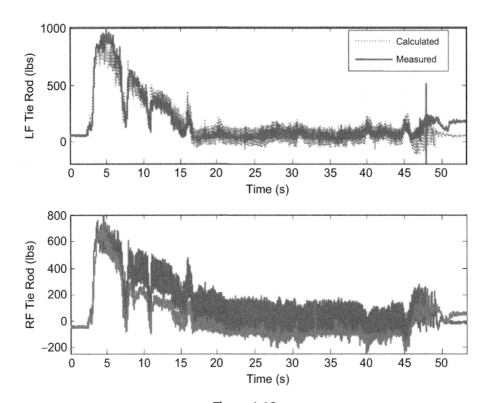

Figure 1.18
Steer tie rod axial loads in a wide-open throttle event.

Some discrepancies between the calculated and experimental strut forces in longitudinal and lateral directions are observed in Figures 1.19 and 1.20, which are believed to be attributable to the compliance of the strut subjected to high forces, resulting in larger elastic deflection such that the two signals acquired from the transducers on the strut tower would be distorted from its original local coordinate system fixed to a rigid vehicle body. By considering the contributions due to the flexibility of some critical components, such as body, frame, subframe, suspension control arms, strut tube, tie rod, and so on, the accuracy of the road load analysis can be greatly improved.

Incorporating the structural flexibility of various components into time domain dynamic analyses of multibody systems have been studied by various authors in the literature (Griffis & Duffy, 1993; Kang, Yun, Lee, & Tak, 1997; Knapczyk & Dzierźek, 1999; Medepalli & Rao, 2000; Ambrósio & Gonçalves, 2001; Kang,

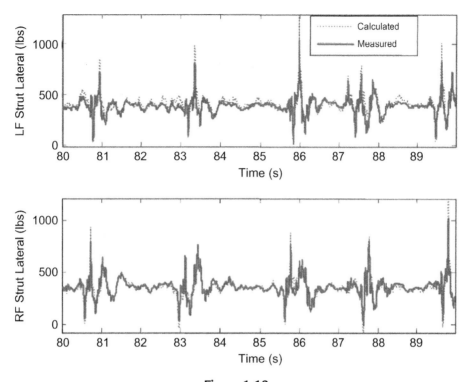

Figure 1.19
Strut lateral loads in a very rough road event.

Bae, Lee, & Tak, 2003; Davidson & Hunt, 2004; Knapczyk & Maniowski, 2006; Lee & Han, 2009; Hong, You, Kim, & Park, 2010). Discussion of these approaches is beyond the scope of this chapter.

Vehicle Load Analysis Using 3D Digitized Road Profiles

As mentioned in the previous section, the suspension component loads can be efficiently predicted by using the measured spindle loads and some other channels as well as an accurate multibody dynamics suspension model. This is possible when the prototype vehicle for spindle loads measurement is available. However, during a vehicle development process, the design changes after the prototype vehicle cannot be avoided. These changes may be due to the suspension tuning, powertrain mount system optimization, body structure optimization, and even the vehicle weight target setting.

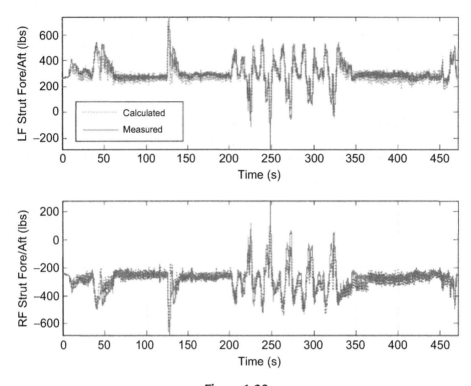

Figure 1.20
Strut fore-aft loads in a severe braking event.

In the current severely competitive automotive market, it is impossible to build the prototype vehicles for road loads data acquisitions along with all these changes in different development stages. Thus it is critical to effectively and accurately predict the spindle loads change due to the design changes from the prototype vehicle by utilizing the measured data acquired from the previous prototype vehicle. To ensure the accuracy of the spindle loads prediction with full vehicle model, the following three components are important:

- Road profile representation
- Vehicle and tire model representation
- Operating condition representation

The road profiles in a proving ground for durability assessment typically include various types, such as concrete flat surface, deterministic transverse and

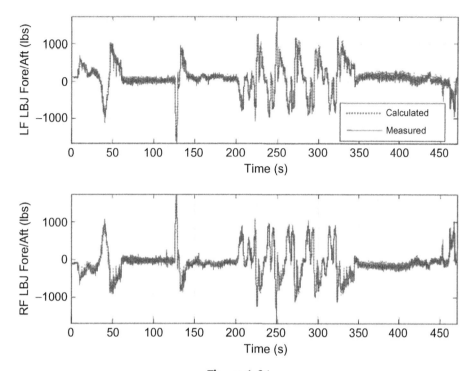

Figure 1.21
Lower ball joint fore-aft loads in a severe braking event.

diagonal trenches, random nature pothole, cobblestone, and Belgian block with different material properties for the road. For full vehicle simulation on these road profiles, the geometrical representation of the road profile is essential since the tire–road interaction heavily depends on the local contact area normal and shear forces representation. The representations of these forces are directly related to the road profile geometry.

The emergence of a high-fidelity 3D terrain measurement technique and durability tire model evolution has made virtual road load analysis and data acquisition possible. A high-speed inertial profiler developed by General Motors Research in the 1960s (Spangler & Kelly, 1966) may be the first simple device to measure 2D road roughness. As computer power increased and signal processing evolved, 3D terrain measurement systems were developed (Wagner & Ferris, 2009), which incorporated a scanning laser that is rigidly mounted to the body of a host vehicle. This vehicle traverses the terrain while simultaneously acquiring terrain measurement.

To obtain accurate terrain measurements, the motion of the vehicle must be accurately measured so that it can be removed from the laser measurement. Modern systems use an Inertial Navigation System (INS) to measure the vehicle motion (Kennedy, Hamilton, & Martell, 2006). The accuracy of the INS depends on the alignment of the Inertial Measurement Unit (IMU) to the laser and satellite coverage of the Global Positioning System (GPS).

Discussion about the emerging 3D terrain measurement capabilities can be found elsewhere (Chemistruck, Binns, & Ferris, 2011; Smith & Ferris, 2010a, 2010b; Detweiler & Ferris, 2010), and is beyond the scope of this chapter. However, it is the objective of this section to present the vehicle suspension load analysis based on measured 3D terrain profiles.

The operating condition representation usually refers to the driver's control and maneuver when driving the vehicle according to the proving ground durability schedules, such as the braking/accelerating and cornering. These maneuvers are required for different road profiles of the proving ground according to the durability schedules. It is thus important to simulate the same driver's control maneuvers as required by the durability schedule, to have the vehicle yield the same spindle loads in longitudinal, lateral, and vertical directions as those measured. Besides the road profile and the operating condition representations, the vehicle and tire models are also critical for the spindle loads prediction.

Vehicle Model Description

The accuracy of the vehicle load analysis hinges on the high-fidelity of vehicle and tire models. A typical vehicle model with meshed body representations is illustrated in Figure 1.22, where the front suspension is a typical short-long arm (SLA) with a rack-pinion steering system. The rear suspension is a multilink type with solid axle and coil springs. The vehicle body is modeled as a rigid part and its mass and mass moments of inertia are derived from the finite element model.

As a result of the fact that a suspension travel and bushing deformations are large in proving grounds road events, the force at every bushing follows a non-linear function of its deflection and a linear function of its velocity in all three

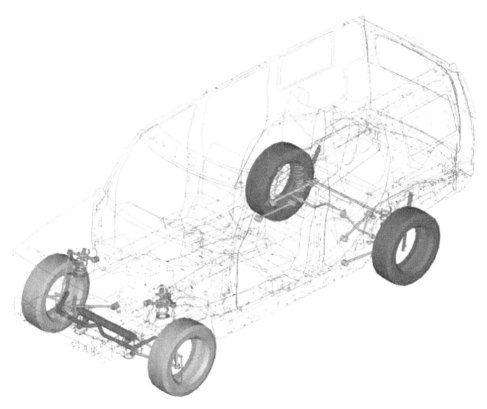

Figure 1.22
Full vehicle model with flexible body graphical representation.

translational and three rotational directions. All suspension components are represented with rigid bodies. The vehicle model needs to be well correlated with its target kinematics and compliance properties before being used for full vehicle simulations.

Tire Model Description

The tire model is the most critical and difficult module in a full vehicle system model due to its complexity of nonlinear characteristics and interaction with terrain profiles. For practical applications, a tire model must prove its quantitative quality and its possibility for adaptation to different driving conditions (Yang & Medepalli, 2009).

Over the years, many tire models have been developed for vehicle dynamics and road loads simulations, among which the most commonly applied tire models are RMOD-K (Oertel & Fandre, 1999, 2001; Pacejka, 2006), FTire (Gipser, 2007), CDTire (Gallrein, De Cuyper, Dehandschutter, & Bäcker, 2005), and SWIFT (Schmeitz, Besselink, & Jansen, 2007; Jansen, Verhoeff, Cremers, Schmeitz, & Besselink, 2005). The four tire models have different approaches and levels of complexity, resulting in differences in computational effort and accuracy.

The agreement with experimental data may be significantly different, depending on the type of application. All the models aim at similar motion input ranges and application types. These include steady-state (combined) slip, transient, and higher frequency responses, covering at least the rigid body modes of vibration of the belt. The models are also designed to roll over three-dimensional terrain unevenness, typically exhibiting the enveloping properties of the tire. The four tire models are introduced as follows.

RMOD-K is a detailed finite element model (FEM) for the actual tire structure. It features a flexible belt that is connected to the rim with a simplified sidewall model with pressurized air. The belt is modeled by one or more FE layers that interact with each other. Terrain contact is activated through an additional sensor layer. In each sensor point, the normal and frictional forces are calculated. The contact area (with possible gaps) and pressure distribution are determined from the rolling and compressed FEM.

Depending on the need of an application, the complexity of the model may be reduced from a fully FEM to a hybrid, or a discrete structure representation. Three-dimensional uneven terrain surfaces can be dealt very well with this sophisticated model. However, the computational effort remains relatively high. Friction functions are also included to allow generation of both adhesion and sliding areas with various friction levels dependent on temperature and contact pressure.

SWIFT (Short Wavelength Intermediate Frequency Tire) is a relatively simple model in representing the actual physical structure of a tire. However, this model relies heavily on experimental data concerning the tire-terrain slip properties. The belt is represented by a rigid ring with numerous residual springs that connect the ring with the contact patch. This simplification limits the frequency of the model application to 100 Hz.

Contact dynamic behavior is represented by a bushing model with nonlinear frequency response characteristics. The wavelength of horizontal tire motion is limited to, but not less than, 10 cm. Rolling over terrain unevenness is accomplished by the so-called effective road plane defined by the vertical positions of a 2D tandem or 3D multiple set of oval cams that travel over a real sharp-edged terrain surface. A scaling factor can be used to account for any change that may occur in a frictional or service condition. Moreover, full-scale tire slip measurements, straightforward rolling experiments over oblique cleats, and some special tests are required to determine the parameters pertinent to tire dynamic rolling behavior.

Alternatively, the MF (Magic Formula) quasi-steady-state or transient tire model was introduced by approximating some parameters in the SWIFT model. This model is applicable for any low frequency response to about 10 Hz, relatively large motion wavelengths, and smooth road surfaces. The MF tire model is a simple variant of the SWIFT model. Both produce identical steady-state responses.

More information on MF-Swift can be found in Jansen et al. (2005) and Schmeitz et al. (2007). Typical applications are vehicle handling and stability, vehicle ride and comfort analysis, suspension vibration analysis, and the development of vehicle control systems such as antilock brake systems (ABS) and electronic stability programs (ESP).

FTire features a flexible belt provided with a large number of friction elements on tread blocks. The flexible belt is modeled by 80 to 200 segments, each of which possesses five degrees of freedom including twisting and bending about the circumferential axis. The segments are connected with respect to the wheel rim by nonlinear spring-damper elements. One thousand to 10,000 friction elements are attached to each of the segments through five to 50 tread blocks. Through these elements, normal and frictional forces are generated. Friction functions are used to make distinction between sticking and sliding friction. The tread block may represent a simple tread pattern design.

A thermal model can also account for the change in the temperature of structure and contact surface, thereby varying friction, inflation pressure, and stiffness properties. In addition, the model features a tread wear model based on the concept of friction power. The contact patch contour and pressure distribution are determined by the model's flexible properties. The model can roll over arbitrary uneven, possibly sharp-edged terrain surfaces.

The frequency bandwidth of the model is limited to 150 Hz (up to first-order bending modes), and the horizontal motion wavelength should be constrained to 5 to 15 cm. Special measurements are required to obtain the geometry, inertia, stiffness, damping, friction, and material properties of the tire of interest. Parameterization may be conducted either by physical oblique cleat tests or virtual finite element simulations.

CDTire is a high frequency FEM tire model. The CDTire family offers three packages with variable characteristics, which include

1. CDTire-20 with a rigid ring and long wavelength surfaces

2. CDTire-30 with a single flexible ring, short wavelength surfaces, and constant lateral height profile

3. CDTire-40 with multiple flexible rings

The unique features of these CDTire models are the smaller local enveloping surfaces on the tire patch as well as the ability to present tire-specific vibrations up to 40 Hz and higher. It is worth mentioning that CDTire-40 is suitable for irregular terrain profiles such as Belgian block, cobblestone, cleats with variable heights and arbitrary positions, and so on.

Model Validation Process

This section discusses the validation process from tire model parameterization to full vehicle model simulation.

Tire Model Validation

The CDTire model is chosen here as an example because this model has been proven (Cuyper, Furmann, Kading, & Gubitosa, 2007) to be suitable for durability road load simulations. The CDTire model is expressed as a GFORCE statement in an ADAMS® environment. Many laboratory bench tests for tire model parameterization are performed according to the published SAE J-documents (SAE 2704, 2005; SAE J2705, 2005; SAE J2707, 2005; SAE J2710, 2005; SAE J2717, 2006; SAE J2718, 2006; SAE J2730, 2006). As a result, the parameters used in the

CDTire model are identified by optimizing the tire model with the experimental data from these bench tests.

The dynamic drum cleat test (SAE J2730, 2006) is the bench test selected to illustrate the validity of analytical predictions, as compared with the tested results. In the test, the tire spindle is fixed to a rigid structure and the tire can freely rotate around the spindle. The tire is mounted with a preload on a 1.7 m diameter drum, which is driven. The drum diameter of 1.7 m is chosen such that the tire oscillations due to cleat impact during the test decay away prior to the second encounter with the cleat at speeds of 64.4 km/h or less. A WFT is installed to measure the spindle loads of the tested tire. One steel cleat with a cross-section dimension of 15 mm × 15 mm is rigidly attached on the drum at 90° or 45°. The drum rotates at multiple speeds to yield the different dynamic responses of the parameterized tire.

A virtual dynamic drum cleat testing with the parameterized CDTire model is simulated. The spindle force comparison for the 45° cleat at 40 km/h is illustrated in Figure 1.23, where the CDTire-40 model is applied for simulation. The CDTire-40 model can predict the longitudinal (F_x) and vertical load (F_z) very well for the 45° cleat test as shown in the figure. The lateral force (F_y) comparison shows apparent discrepancies. It should be argued that the spindle longitudinal and lateral forces in the 45° cleat test are expected to demonstrate similar trends, while the measured longitudinal and lateral spindle forces between 0.1 and 0.15 second show extremely different patterns. This requires further investigation.

Vehicle Model Validation

Vehicle model validation involves the full vehicle dynamics analysis with the input of 3D digitized terrain profiles as well as the vehicle performance simulation on a flat road surface. There are two types of terrain profiles applied for any full vehicle simulation at a constant speed, including deterministic and random types. The deterministic terrains includes 10" × 2" wood plank on concrete road, transverse and diagonal trenches in concrete. The random terrains include Belgian blocks, random potholes, and cobblestones. Furthermore, the vehicle performance simulations include figure-eight cornering and straight-line acceleration and braking.

During analysis, the terrain profile interface with the tire is difficult to characterize in terms of its friction. The friction between the tire and road surface at

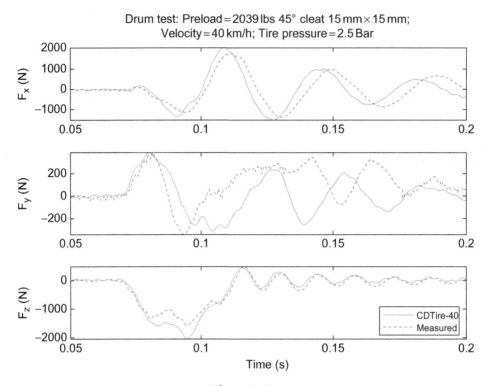

Figure 1.23
Spindle longitudinal, lateral, and vertical force over a 45° cleat.

various locations is not the same, and it is hard to model the friction coefficient as a function of road surface location, tire pressure, tire load, and tire material. In this validation, the friction coefficient for all road surfaces applied is assumed to be 0.9.

Deterministic Road and Performance Event Simulations

The validation study is based on the time histories of measured and predicted spindle forces from the deterministic terrain profiles and the performance type simulations.

For a deterministic road simulation, Figure 1.24 shows the comparison of left-rear spindle force-time histories for a simulated vehicle over the diagonal trenches at 20 mph. It can be seen that the simulated spindle loads match reasonably well with the measured. It can also be seen that the phase between the simulated and measured spindle loads are not negligible.

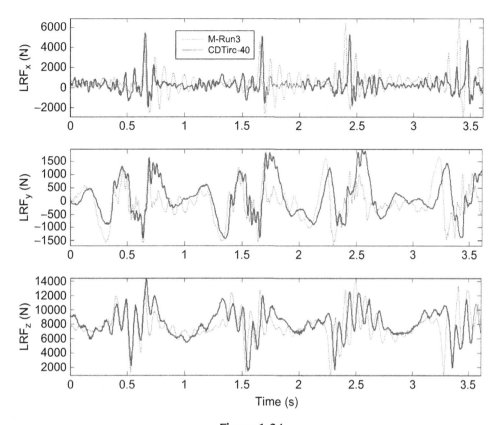

Figure 1.24
Left-rear spindle longitudinal, lateral, and vertical force responses under
diagonal trenches event at 20 mph.

The reason behind this is that during the measurement, it is very hard for the driver to maintain the constant forward speed, while in simulation the virtual driver model is "smart" enough to keep the vehicle as constant speed. Since the spindle loads are very sensitive to the driving speed in this kind of event, the spindle load's difference between the measured and simulated are not completely due to the tire/vehicle model accuracy. This fact may be one of the many challenges for the validation of the virtual simulation model.

For a vehicle performance simulation on braking and acceleration, the measured driving and/or braking torques at front and rear axles are applied as input. The predicted spindle longitudinal and vertical spindle loads, as well as the vehicle forward velocity are compared with the measured to illustrate

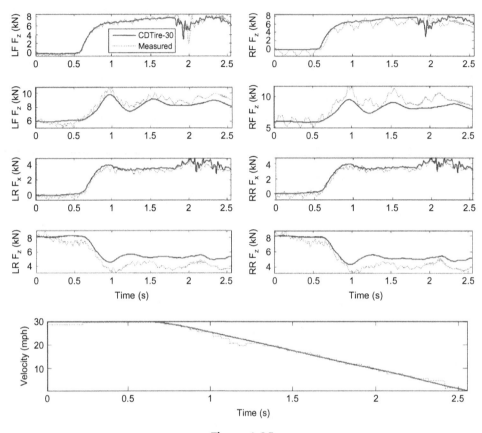

Figure 1.25
Spindle longitudinal and vertical forces and vehicle velocity under hard-stop
braking event (start speed is 30 mph).

the validity of the tire model. Figures 1.25 and 1.26 illustrate the comparison
of the predicted and measured spindle longitudinal force, vertical force, and
the vehicle velocity under a hard-stop and a heavily accelerating event,
respectively.

It can be seen that the predicted spindle loads (especially those of the longitudinal forces) and vehicle velocity, in general, match very well with the measured
data. This indicates that the CDTire-30 tire model can be applied to accurately
predict the relationship among the tire driving/braking torque and longitudinal
and vertical forces on such a performance simulation. It should also be noted
that the vertical spindle forces in front and rear axles underpredict the load

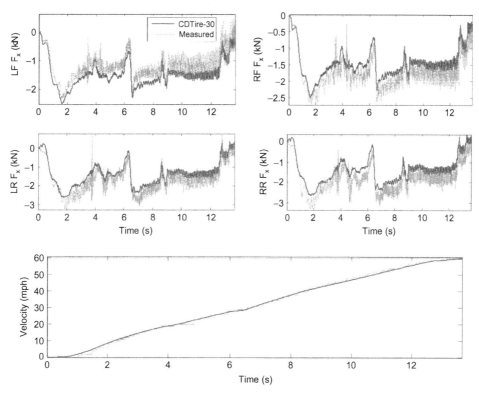

Figure 1.26
Spindle longitudinal forces and vehicle velocity under acceleration
event (start speed is 0 mph).

transfer due to deceleration, which may indicate some modeling errors of the vehicle body mass, CG location, or inertia properties.

For a vehicle performance simulation subjected to the figure-eight event, the driver's behavior significantly affects vehicle dynamic responses and spindle loads. The validation is not to validate a closed-loop driver-vehicle system model; it focuses on the predicted longitudinal, lateral, and vertical spindle loads under the given steering wheel angle and driving/braking torque at front and rear axles.

Figure 1.27 illustrates the dynamic spindle loads at four corners for the predicted and measured spindle loads. The results clearly indicate that the CDTire-20 model can accurately predict the dynamic spindle loads for the figure-eight event.

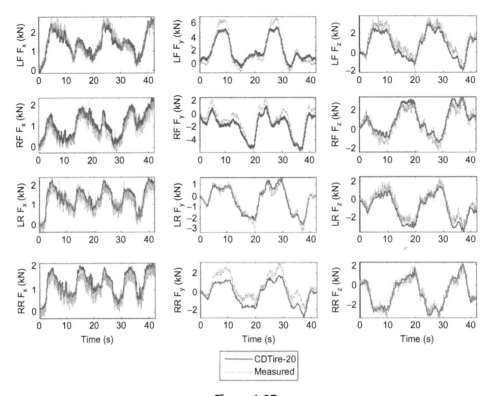

Figure 1.27
Spindle longitudinal, lateral, and vertical forces under the figure-eight event.

Figure 1.28 further illustrates the comparison of the vehicle velocity between the predicted and measured, as well as the predicted vehicle center of gravity (CG) trajectory when negotiating the figure-eight event. It can be seen that the predicted vehicle forward velocity is somewhat close to the measured, except for a certain time frame (from 20–25 seconds) where the measured velocity is constant at 8 mph and the predicted velocity increases to 11.8 mph and then decreases.

A further investigation of the measured wheel torque within the time frame indicates that the vehicle should be accelerating and then braking. Thus we can conclude that the measured velocity signal may not be accurate. The predicted vehicle CG trajectory response indicates that it is hard for a driver to follow the exact same figure-eight trajectory for two figure-eight runs. This further indicates that the vehicle dynamic responses under such a severe cornering event are greatly dependent on the driver's behavior.

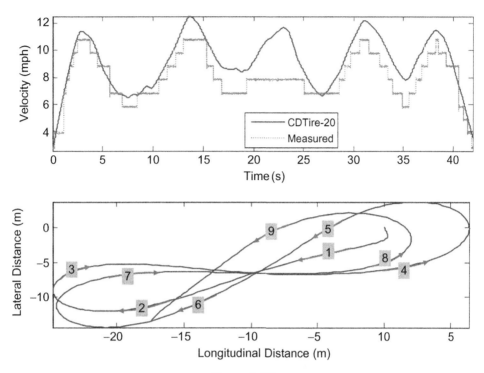

Figure 1.28
Vehicle speed and CG trajectory under the figure-eight event.

Random Road Simulations

The random type road events presented in this study include cobblestones, Belgian blocks, and combinations of random potholes, diagonal trenches, and cobblestone. For these types of roads, the cumulative exceedance plot from the rainflow cycle counting output and the pseudo-fatigue damage are used for comparison. For the pseudo-damage calculation, a fictitious material load-life curve with a slope of −0.2 and 20,000 lbs intercepts at 1 reversal is applied. The discussion of the rainflow cycle counting techniques and the linear damage theory can be found elsewhere (Lee et al., 2005) and is beyond the scope of this chapter.

CDTire-40 (CDT40) is the most comprehensive tire model in the CDTire family. It can be used to simulate vehicle longitudinal, lateral, and vertical dynamics responses on the road profiles with long and short wavelengths. However, due to its relative complexity thus a longer simulation time, it is not recommended to use

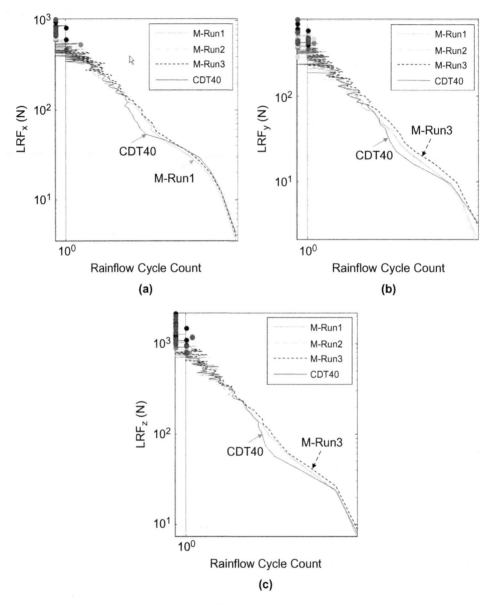

Figure 1.29
Left-rear spindle longitudinal (a), lateral (b), and vertical (c) force responses
under Belgian block event at 5 mph.

a CDTire-40 model to simulate a road profile with very large wavelength, such as flat surface, or an event with only vertical and longitudinal vehicle dynamics involved, such as transverse trench on a flat concrete surface. On the other hand, for random type roads that will stimulate all three-direction vehicle dynamics, the CDTire-40 model should be applied.

Figure 1.29(a–c) illustrates the comparison of cumulative exceedance plots for the measured and predicted longitudinal, lateral, and vertical spindle forces, respectively, with a vehicle moving over Belgian blocks at 5 mph. Figure 1.30 also provides a normalized pseudo-damage ratio with respect to the pseudo-damage-using model.

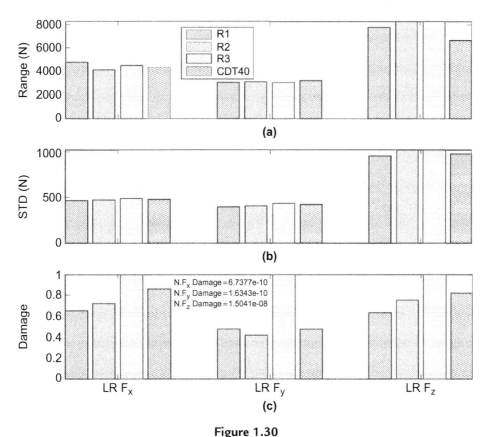

Figure 1.30

Left-rear spindle longitudinal (a), lateral (b), and vertical (c) force responses under Belgian block event at 5 mph displayed in bar graphs.

The comparisons between the three-time measured and predicted peak-to-peak range (Range) and standard deviation (STD) of the left-rear spindle loads are also illustrated in Figure 1.30. It can be seen that the spindle loads predicted with CDT40 model, in general, are well-correlated with the measured ones.

Note that a validation of the durability tire model usually focuses on the predictability of the rainflow cycle counts of spindle loads under given road events and operating conditions. If the rainflow cycle counts of the predicted spindle loads match very well with the measured data, it indicates that the predicted spindle loads will yield comparable damage due to the spindle loads, which demonstrates the validity of the tire model for durability assessment.

Based on the previous studies, the following conclusions can be drawn:

- The current LMS CDT models can be applied to predict the spindle loads under various road events with reasonable accuracy, achieving significant improvement in longitudinal, lateral, and vertical dynamics compared with the Durability 521 tire model (Yang & Medepalli, 2009).

- The validation of a tire model is strongly challenged by the measurement run-to-run variations from the tire bench test and full vehicle measurement. Simulations cannot exactly replicate physical events due to road surface and driving variability. It is highly recommended to develop a standard worldwide validation procedure to make consistent validation criteria for comparisons of different tire models and sources of validated data. Furthermore, the tire bench test data followed by the published SAE J-documents are applied in this study, which indicates that the documents cover the necessary tire data to parameterize the CDTire model for "nonmisuse" road loads simulation purpose.

Summary

In the current competitive automotive business environment, decisions on automotive product development greatly rely on accurate and fast estimations of loads input data to perform durability, fatigue analysis, and bench tests for the designed vehicle system, subsystems, and components. Conventional data acquisitions, which measure the component loads with installed strain gages, are time-consuming, expensive, and rarely reusable. These methods thus have almost been replaced with the wheel force transducers technology, a fast and reliable method, which measures only wheel spindles loads, accelerations, and a few correlation channels.

The measured spindle loads can be directly applied to perform full vehicle fatigue simulation testing with road test simulators in the laboratory. They can also be effectively applied to predict the vehicle component loads by using the semi-analytical load simulation methodology, which combines the measured data and analytical models. With the measured or calculated component loads, it is convenient to conduct the stress and fatigue analysis for components through FE model simulation.

Multibody loads simulation focuses on the rigid and flexible body combination and integration with finite element analysis. All analytical models for loads analyses may be further integrated to perform the vehicle dynamics simulation. Ultimately, the advanced virtual road load simulation technology allows fewer prototype vehicles, and less proving ground and laboratory testing. It can also perform design of experiments (DOE) and optimal design with low cost and high confidence.

Sensitivity analysis using DOE and contribution ratio calculation can help determine the necessary signal to be measured in the experiments. Signals having little effect on the loads can be removed from the measurement list to reduce the number of transducer channels, thereby speeding up instrumentation, data acquisition, and data analysis activities.

References

Abd El-Gawwad, K. A., Crolla, D. A., Soliman, A. M. A., & El-Sayed, F. M. (1999a). Off-road tire modeling I: The multi-spoke tire model modified to include the effect of straight lugs. *Journal of Terramechanics, 36*, 3–24.

Abd El-Gawwad, K. A., Crolla, D. A., Soliman, A. M. A., & El-Sayed, F. M. (1999b). Off-road tire modeling IV: Extended treatment of tire-terrain interaction for the multi-spoke model. *Journal of Terramechanics, 36*, 77–90.

ADAMS User Manual. (2007). MSC software corporation. Santa Ana, CA.

Agrawal, O. P., & Shabana, A. A. (1985). Dynamic analysis of multibody systems using component modes. *Computers and Structures, 21*(6), 1303–1312.

Allen, R. H., Oppenheim, I. J., Parker, A. R., & Bielak, J. (1986). On the dynamic response of rigid body assemblies. *Earthquake Engineering and Structural Dynamics, 14*, 861–876.

Ambrósio, J. A. C., & Gonçalves, J. P. C. (2001). Complex flexible multibody systems with application to vehicle dynamics. *Multibody System Dynamics, 6*, 163–182.

Amirouche, F. M. L. (1988). Dynamics of large constrained flexible structures. *Journal of Dynamic Systems, Measurement, and Control, 110*(1), 78–83.

Andrews, G. C., & Kesavan, K. (1975). The vector-network model: a new approach to vector dynamics. *Mechanism and Machine Theory, 10*, 57–75.

Andrews, G. C., Richard, M. J., & Anderson, R. J. (1988). A general vector-network formulation for dynamic systems with kinematic constraints. *Mechanism and Machine Theory, 23*(3), 243–256.

Angeles, J. (1986a). Automatic computation of the screw parameters of rigid-body motions, part I: Finitely separated positions. *Journal of Dynamic Systems, Measurement, and Control, 108*, 32–38.

Angeles, J. (1986b). Automatic computation of the screw parameters of rigid-body motions, part II: Infinitesimal separated positions. *Journal of Dynamic, Systems Measurement, and Control, 108*, 39–43.

Bäcker, M., Möller, R., Kienert, M., Bertan, B., Özkaynak, M. U., & Yaman, H. (2007). Load identification for CAE based fatigue life prediction of a new bus type. SAE Paper No. 2007-01-4281. Warrendale, PA: SAE International.

Bainum, P. M., & Kumar, Y. K. (1982). Dynamics of orbiting flexible beams and platforms in the horizontal orientation. *Acta Astronautica, 9*(3), 119–127.

Berzeri, M., Dhir, A., Ranganathan, R., Balendran, B., Jayakumar, P., & O'Heron, P. J. (2004). A new tire model for road loads simulation: Full vehicle validation. SAE Paper No. 2004-01-1579. Warrendale, PA: SAE International.

Chemistruck, H. M., Binns, R., & Ferris, J. B. (2011). Correcting INS drift in terrain surface measurements. *Journal of Dynamic Systems Measurement and Control, 133*(2), 021009.

Cuyper, D., Furmann, J., Kading, M., & Gubitosa, M. (2007). Vehicle dynamics with LMS Virtual.Lab Motion. *Vehicle System Dynamics, 45*(1), 199–206.

Davidson, K., & Hunt, K. (2004). *Robots and screw theory.* New York: Oxford University Press.

Detweiler, Z. R., & Ferris, J. B. (2010). Interpolation methods for high-fidelity three dimensional terrain surfaces. *Journal of Terramechanics, 47*(4), 209–217.

Gallrein, A., De Cuyper, J., Dehandschutter, W., & Bäcker, M. (2005). Parameter identification for LMS CDTire. *Vehicle System Dynamics, 43*(1), 444–456.

Gipser, M., Hofer, R., & Lugner, P. (1997). Dynamical tire forces response to road unevenness, tire models for vehicle dynamic analysis. *Supplement to Vehicle System Dynamics, 27*, 94–108.

Gipser, M. (2007). FTire—The tire simulation model for all applications related to vehicle dynamics. *Vehicle System Dynamics, 45*(S1), 139–151.

Griffis, M., & Duffy, J. (1993). Global stiffness modeling of a class of simple compliant couplings. *Mechanism and Machine Theory, 28*, 207–224.

Haga, H. (2006). Evaluation method for road load simulation using a tire model and an applied example. SAE Paper No. 2006-01-1256. Warrendale, PA: SAE International.

Haga, H. (2007). Evaluation of road load simulation on rough road using a full vehicle model-load prediction for durability using a tire model. SAE Paper No. 2007-5868. Warrendale, PA: SAE International.

Hilber, H. M., Hughes, T. J. R., & Taylor, R. L. (1977). Improved numerical dissipation for time integration algorithms in structural dynamics. *Earthquake Engineering and Structural Dynamics, 5*(3), 283–292.

Hong, E. P., You, B. J., Kim, C. H., & Park, G. J. (2010). Optimization of flexible components of multibody systems via equivalent static loads. *Structural and Multidisciplinary Optimization, 40,* 549–562.

Huston, R. L. (1991). Multibody dynamics—modeling and analysis methods. *Applied Mechanics Reviews, 44*(3), 109–117.

Jansen, S. T. H., Verhoeff, L., Cremers, R., Schmeitz, A. J. C., & Besselink, I. J. M. (2005). MF-Swift simulation study using benchmark data. *Vehicle System Dynamics, 43*(1), 92–101.

Kang, J. S., Yun, J. R., Lee, J. M., & Tak, T. O. (1997). Elastokinematic analysis and optimization of suspension compliance characteristics. SAE Paper No. 970104. Warrendale, PA: SAE International.

Kang, J. S., Bae, S., Lee, J. M., & Tak, T. O. (2003). Force equilibrium approach for linearization of constrained mechanical system dynamics. *ASME Journal of Mechanical Design, 125,* 143–149.

Kennedy, S., Hamilton, J., & Martell, H. (2006). *Architecture and system performance of SPAN-NovAtel's GPS/INS solution.* Piscataway, NJ: Institute of Electrical and Electronics Engineers Inc.

Knapczyk, J., & Dzierźek, S. (1999). Analysis of spatial compliance of the multi-link suspension system. *Advances in multibody systems and mechatronics* (pp. 243–252). Duisburg: University.

Knapczyk, J., & Maniowski, M. (2006). Elastokinematic modeling and study of five-rod suspension with subframe. *Mechanism and Machine Theory, 41,* 1031–1047.

Lanczos, C. (1970). *The variational principles of mechanics* (4th ed., p. 92). New York: Dover Publications.

Lee, D. C., & Han, C.-S. (2009). CAE (Computer Aided Engineering) driven durability model verification for the automotive structure development. *Finite Elements in Analysis and Design, 45,* 324–332.

Lee, Y. L., Lu, M. W., & Breiner, R. W. (1997). Load acquisition and damage assessment of a vehicle bracket component. *International Journal of Materials and Product Technology, 12*(4–6), 447–460.

Lee, Y. L., Pan, J., Hathaway, R., & Barkey, M. (2005). *Fatigue testing and analysis: theory and practice.* Boston: Elsevier/Butterworth-Heinemann.

Medepalli, S., & Rao, R. (2000). Prediction of road loads for fatigue design—a sensitivity study. *International Journal of Vehicle Design, 23*(1/2), 161–175.

Newmark, N. M. (1959). A method of computation for structural dynamics. *Journal of Engineering, Mechanics Division, ASCE, 85,* 67–94.

Oertel, C., & Fandre, A. (1999). Ride comfort simulations and steps towards life time calculations: RMOD-K and ADAMS. International ADAMS Users' Conference, Berlin.

Oertel, C., & Fandre, A. (2001). RMOD-K tyre model system/a contribution to the virtual vehicle. ATZ worldwide eMagazines edition: 2001–11.

Pacejka, H. B. (2006). *Tyre and vehicle dynamics* (2nd ed.). Boston: Butterworth-Heinemann.

Randers, J. (1980). *Elements of the system dynamics method*. Cambridge: MIT Press.

SAE J2704 (2005). Tire normal force/deflection and gross footprint dimension test, January 5.

SAE J2705 (2005). Tire quasi-static envelopment of triangular/step cleats test, October 1.

SAE J2707 (2005). Wear Test Procedure on Inertia Dynamometer for Brake Friction Materials, February 1.

SAE J2710 (2005). Modal testing and identification of lower order tire natural frequencies of radial tires, October 12.

SAE J2717 (2006). Test to define tire size (geometry), mass, and inertias, April 18.

SAE J2718 (2006). Test for tire quasi-static longitudinal force vs. longitudinal displacement and quasi-static lateral force vs. lateral displacement, February 22.

SAE J2730 (2006). Dynamic cleat tests with perpendicular/inclined obstacles, August 23.

Schmeitz, A. J. C., Besselink, I. J. M., & Jansen, S. T. H. (2007). TNO, MF-SWIFT. *Vehicle System Dynamics, 45*(1), 121–137.

Smith, H., & Ferris, J. B. (2010a). Calibration surface design and validation for terrain measurement systems. *Journal of Testing and Evaluation, 38*(4), 431–438.

Smith, H., & Ferris, J. B. (2010b). Techniques for averting and correcting inertial errors in high-fidelity terrain topology measurements. *Journal of Terramechanics, 47*(4), 219–225.

Sommerfeld, J., & Meyer, R. (1999). Correlation and accuracy of a wheel force transducer as developed and tested on a Flat-Trac® tire test system. SAE Paper 1999-01-0938. Warrendale, PA: SAE International.

Spangler, E. B., & Kelly, W. J. (1966). *GMR road profilometer—a method for measuring road profile* (Report No. 121, pp. 27–54) Washington, DC: Highway Research Board.

Stadterman, T. J., Connon, W., Choi, K. K., Freeman, J. S., & Peltz, A. L. (2003). Dynamic modeling and durability analysis from the ground up. *Journal of the IEST, 46*, 128–134.

Tampi, M., & Yang, X. (2005). Vehicle cradle durability design development. SAE Paper No. 2005-01-1003. Warrendale, PA: SAE International.

Tatsuya, S. (2001). Development of wheel force transducer equivalent to actual wheel. *Proceedings of JSAE Annual Congress, 107*(1), 11–14.

Wagner, S. M. & Ferris, J. B. (2009), Developing stable autoregressive models of terrain topology, *International Journal of Vehicle Systems Modelling and Testing, 4*(4), 306–317.

Yang, X., & Medepalli, S. (2009). Comfort and durability tire model validation. *Tire Science & Technology, 37*(4), 302–322.

Yang, X., Muthukrishnan, G., Seo, Y. J., & Medepalli, S. (2004). Powertrain mount loads prediction and sensitivity analyses. SAE Paper 2004-01-169. Warrendale, PA: SAE International.

CHAPTER 2

Pseudo Stress Analysis Techniques

Yung-Li Lee
Chrysler Group LLC

Mingchao Guo
Chrysler Group LLC

Chapter Outline

Introduction

To expedite the computational process, numerous notch analysis techniques (Neuber, 1961; Molsky & Glinka, 1981; Hoffmann & Seeger, 1989; Barkey, Socie, & Hsia, 1994; Lee, Chiang, & Wong, 1995; Moftakhar, Buczynski, & Glinka, 1995; Gu & Lee, 1997; Lee & Gu, 1999; Buczynski & Glinka, 2000) have been developed to estimate the local stress–strain responses based on the stress output from a linear, elastic FEA. The stress calculated from a linear, elastic FE analysis is often termed pseudo stress or fictitious stress to differentiate the true stress as plasticity occurs.

Calculation of pseudo stresses becomes a crucial step in the notch analysis to estimate the true local stress–strain responses. Static stress analysis and modal transient response analysis are the two commonly used techniques for pseudo stress analysis. In the static stress analysis, the pseudo stresses can be obtained by superimposing all the pseudo stress influences from the applied loads at every time step.

In addition, in the modal transient response analysis, the pseudo stresses can be calculated from the stress influences from the normal modes and modal coordinates. If resonant fatigue is of primary concern, the modal transient response analysis would be recommended; otherwise, the static stress analysis could be a primary choice. Both the static stress analysis and the modal transient response analysis are described in this chapter.

Static Stress Analysis

Static stress analysis often assumes the relationship between the applied load and its structural response is linear and the material follows elastic behavior. Static stress analysis herein is referred to a linear, elastic static stress analysis.

If a structure subjected to external loads experiences small deformation response and elastic material behavior, a linear elastic static analysis would be the best option to use because it has advantages in simplifying stress and fatigue calculations. One advantage is that the computational process is very efficient because the structural stiffness matrix is constant and does not require any update as load increases.

Another advantage is that the response of a structure due to an identical loading but with a different magnitude can be scaled by the load magnitude ratio. Moreover, if a structure is subjected to multiple load cases, responses of the structure can be calculated by superimposing the same response of each load case with an appropriate load magnitude ratio. This is called superposition of structure responses.

The combination of scaling and superposition of structure responses plays an important role in stress and fatigue analyses for a long, complicated multiaxial load time history. The time history of pseudo stresses of a structure due to the time history of N different load sources can be obtained by the following equation:

$$\sigma_{ij}(t) = \sum_{k=1}^{N} \sigma_{ij,k} \cdot L_k(t) \qquad (2.1)$$

where

$\sigma_{ij}(t)$ = the stress tensor at a time t
$\sigma_{ij,k}$ = the stress tensor influence due to a unit load at the k-th load source
$L_k(t)$ = the k-th load magnitude at a time t

Depending on the boundary condition of a structure, the static stress analysis has two different problem-solving techniques, the fixed reactive and inertia-relief methods for constrained and unconstrained structures, respectively.

Fixed Reactive Analysis

The fixed reactive analysis is a common and fundamental solution for a constrained structure subjected to a set of time-independent actions such as forces, moments, torques, or temperatures. If the constrained structure can be modeled by finite elements, the nodal displacements of the analytical model are usually unknown and can be solved by the following force equilibrium equation:

$$[K]\{u\} = \{P\} \qquad (2.2)$$

where

$[K]$ = a system stiffness matrix
$\{u\}$ and $\{P\}$ = the nodal displacement and force vectors, respectively

Once the nodal displacements are known, any desired output, such as element forces or strains and stresses, can be computed on an element-by-element basis.

Example 2.1

A cantilever beam with a length of 1016 mm and a box section of 50.8 × 25.4 mm is subjected to a set of loads at the free end, as shown in Figure 2.1(a), and all the degrees of freedom are fixed at the constrained end. The constrained beam is made of steel with Young's modulus of

210,000 MPa. Conduct linear static stress analyses for the following load cases:

1. A single unit force 1000 N in the vertical/z direction, $F_z = 1000$ N

2. A single unit force 1000 N in the lateral/y direction, $F_y = 1000$ N

3. A single unit force 1000 N in the longitudinal/x direction, $F_x = 1000$ N

4. A single force 2224 N in the vertical/z direction, $F_z = 2224$ N

5. A force with components of $F_z = 2224$ N, $F_y = 1500$ and $F_x = 3000$ N

Solution

The finite element model is meshed by using eight-node hexahedral solid elements. The analyses are conducted using "Static" stress analysis solution in ABAQUS® (ABAQUS, Inc.). The contours of stresses in the x direction (σ_{xx}) are shown in Figure 2.1(b) through (f) for all the load cases where the maximum compressive stresses on the same element for load cases (1), (2), and (3) ($\sigma_{xx} = -93.0$ MPa, -69.4 MPa, and -0.8 MPa) are listed.

Actually the maximum compressive stress σ_{xx} under load case (4) can be easily calculated as -207.0 MPa, by scaling the stress result of load case (1) with a load ratio of 2.224, because load case (4) has the same loading condition as load case (1) but a different magnitude of force. Similarly, by scaling and superimposing the stresses of load cases (1) to (3), the maximum compressive stress σ_{xx} under load case (5) can be calculated as -313.2 MPa ($= -93.0 \times 2.224 - 69.4 \times 1.5 - 0.8 \times 3.0$).

For displacement, since it is a vector, scaling calculation is still applicable but superimposing calculation should be conducted by the vector calculation rules. In this example, the maximum displacements in the loading directions at the loading point of load cases (1) to (3) are 24.3 mm, 6.08 mm, and 0.004 mm, respectively. The maximum displacement of load case (4) is 54.0 mm ($= 24.3$ mm $\times 2.224$).

In addition, the maximum displacement in magnitude at the loading point of load case (5) should be calculated as follows: $[(24.3 \times 2.224)^2 + (6.08 \times 1.5)^2 + (0.004 \times 3.0)^2]^{1/2} = [(54.0)^2 + (9.12)^2 + (0.012)^2]^{1/2} = 54.78$ mm. Note that it would be wrong by simply adding the three scaled displacement values for the magnitude of the maximum displacement such as $54.0 + 9.12 + 0.012 = 63.13$ mm.

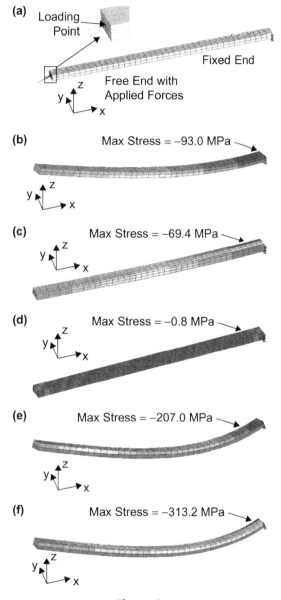

Figure 2.1

Linear static stress analyses of a cantilever beam: (a) a cantilever beam with loading and boundary conditions; (b) contour plot of σ_{xx} under $F_z = 1000$ N; (c) contour plot of σ_{xx} under $F_y = 1000$ N; (d) contour plot of σ_{xx} under $F_x = 1000$ N; (e) contour plot of σ_{xx} under $F_z = 2224$ N; and (f) contour plot of σ_{xx} under $F_z = 1000$ N, $F_y = 1500$ N, and $F_x = 3000$ N.

Inertia Relief Analysis

Solving for the responses of an unconstrained structure subjected to constant or slowly varying external loads is commonly performed throughout the automotive and aerospace industry. It is assumed that the external loads are steady-state loads applied to the structure and the structural transient responses have dampened out. Typical example problems are a rocket undergoing constant or slowly varying acceleration during lift-off, and a vehicle driven on a test track where the local responses are not critical.

These problems can be solved by using static stress analysis with the inertia relief technique. Conventional finite element static analyses cannot be performed on unconstrained structures because of the singularity of the stiffness matrix due to rigid body motions. With the inertia relief technique, it is assumed that inertial loads of rigid body motions and external loads are in balance with respect to a reference point, thus the unconstrained structure is in a state of static equilibrium. The rigid body accelerations at the reference point will be calculated and applied along with the external loads back to the finite element structure to produce a load-balanced static formulation. With the steady state equilibrium condition, the relative structure displacements, the internal forces, and the stresses can be finally determined.

Even though numerous techniques for inertia relief have been developed and published elsewhere (Barnett, Widirck, & Ludwiczak, 1995; Gaffrey & Lee, 1994; ABAQUS, 2008), a simple, concise solving technique for static analysis with inertia relief is presented in this section.

Approximation to Dynamic Equilibrium Equations

If an unconstrained structure has rigid body modes with respect to its center of gravity (CG), the static analysis with inertia relief is considered as an approximation to the dynamic equilibrium equations where the damping force is excluded because of the assumption of steady-state responses. This concept is illustrated here.

It is assumed that the total deformation vector $\{u_t\}$ of the unconstrained structure is a combination of a rigid body motion vector with respect to its

CG $\{u_r\}$ and a flexible deformation vector $\{u\}$, which can be written as follows:

$$\{u_t\} = \{u_r\} + \{u\}. \tag{2.3}$$

Therefore, the corresponding total acceleration vector is

$$\{\ddot{u}_t\} = \{\ddot{u}_r\} + \{\ddot{u}\}. \tag{2.4}$$

When a steady-state external force vector $\{P\}$ is applied to the unconstrained structure with a mass matrix $[M]$ and a stiffness matrix $[K]$, the dynamic equilibrium equation becomes

$$[M]\{\ddot{u}_r\} + [M]\{\ddot{u}\} + [K]\{u\} = \{P\}. \tag{2.5}$$

It is also assumed that the inertia force due to flexible body motions of the structure is negligible as compared to that from rigid body motions. Thus the relative acceleration term $[M]\{\ddot{u}\}$ is dropped off from the preceding equilibrium equation. Equation (2.5) can be rewritten as

$$[M]\{\ddot{u}_r\} + [K]\{u\} = \{P\}. \tag{2.6}$$

This is the fundamental equation for the static analysis with inertia relief when the CG of the structure is the center of its rigid body motions.

Basic Approximation of Finite Element Analysis

If an unconstrained structure has rigid body modes with respect to an arbitrary reference point different from its center of gravity (CG), all the applied loads and the inertia forces generated by the applied loads must be balanced at this point for force and moment equilibrium. This concept is demonstrated in this section.

It is assumed that an arbitrary nodal point i (x_i, y_i, z_i) of an unconstrained structure can be described in a Cartesian coordinate system and that the origin of the Cartesian axes is the reference point. A geometric rigid body transformation matrix $([R_{i,0}]_{6\times6})$ that relates rigid body motions $(\{u_{r,0}\}_{6\times1})$ from the reference point $(0, 0, 0)$ to those motions at a nodal point i $(\{u_{r,i}\}_{6\times1})$ can be defined as follows:

$$\{u_{r,i}\}_{6\times1} = [R_{i,0}]_{6\times6}\{u_{r,0}\}_{6\times1} \tag{2.7}$$

where

$$\{u_{r,i}\} = \begin{Bmatrix} v_{x,i} \\ v_{y,i} \\ v_{z,i} \\ \theta_{x,i} \\ \theta_{y,i} \\ \theta_{z,i} \end{Bmatrix} \quad \{u_{r,0}\} = \begin{Bmatrix} v_{x,0} \\ v_{y,0} \\ v_{z,0} \\ \theta_{x,0} \\ \theta_{y,0} \\ \theta_{z,0} \end{Bmatrix} \quad [R_{i,0}] = \begin{bmatrix} 1 & 0 & 0 & 0 & z_i & -y_i \\ 0 & 1 & 0 & -z_i & 0 & x_i \\ 0 & 0 & 1 & y_i & -x_i & 0 \\ 0 & 0 & 0 & 1 & 0 & 0 \\ 0 & 0 & 0 & 0 & 1 & 0 \\ 0 & 0 & 0 & 0 & 0 & 1 \end{bmatrix}$$

where

v and θ = the nodal displacement and rotation

x, y, and z (the subscripts) = the Cartesian axes

If a reference point different from the origin is introduced, the geometric rigid body transformation matrix will be revised accordingly. The choice of the reference point is arbitrary and will not affect the results of inertia relief.

With the application of the geometric rigid body transformation matrix, the following node-to-origin displacement and force equations apply:

$$\{\ddot{u}_{r,i}\}_{6\times1} = [R_{i,0}]_{6\times6}\{\ddot{u}_{r,0}\}_{6\times1} \tag{2.8}$$

$$[R_{i,0}]^T_{6\times6}\{P_i\}_{6\times1} = \{P_0\}_{6\times1} \tag{2.9}$$

where

$\{\ddot{u}_{r,0}\}_{6\times1}$ = the rigid body acceleration vector at the origin

$\{\ddot{u}_{r,i}\}_{6\times1}$ = the rigid body acceleration vector at a nodal point i

$\{P_0\}_{6\times1}$ = the resultant force vector at the origin

$\{P_i\}_{6\times1}$ = the applied loading vector at a nodal point i

These accelerations and forces are noted as follows:

$$\{\ddot{u}_{r,0}\} = \begin{Bmatrix} \ddot{v}_{x,0} \\ \ddot{v}_{y,0} \\ \ddot{v}_{z,0} \\ \ddot{\theta}_{x,0} \\ \ddot{\theta}_{y,0} \\ \ddot{\theta}_{z,0} \end{Bmatrix} \quad \{\ddot{u}_{r,i}\} = \begin{Bmatrix} \ddot{v}_{x,i} \\ \ddot{v}_{y,i} \\ \ddot{v}_{z,i} \\ \ddot{\theta}_{x,i} \\ \ddot{\theta}_{y,i} \\ \ddot{\theta}_{z,i} \end{Bmatrix} \quad \{P_0\}_{6\times1} = \begin{Bmatrix} P_{x,0} \\ P_{y,0} \\ P_{z,0} \\ M_{x,0} \\ M_{y,0} \\ M_{z,0} \end{Bmatrix}$$

where

$\ddot{v}$ and $\ddot{\theta}$ = the nodal translational and rotational accelerations
P and M = the applied force and moment at a node

Consider a structure modeled by a finite element model with a total of n nodal points where the external loading applies. The total resultant forces and the accelerations at the origin can be obtained in the following equations:

$$\{\ddot{u}_r\}_{6n\times1} = [R_{i,0}]_{6n\times6}\{\ddot{u}_{r,0}\}_{6\times1} \tag{2.10}$$

$$[R]^T_{6\times6n}\{P\}_{6n\times1} = \{P_0\}_{6\times1} \tag{2.11}$$

where

$$[R]_{6n\times6} = \begin{bmatrix} [R_{1,0}] \\ [R_{2,0}] \\ \cdot \\ \cdot \\ \cdot \\ [R_{n,0}] \end{bmatrix} \quad [\ddot{u}_r] = \begin{bmatrix} [\ddot{u}_{1,0}] \\ [\ddot{u}_{2,0}] \\ \cdot \\ \cdot \\ \cdot \\ [\ddot{u}_{n,0}] \end{bmatrix} \quad \{P\}_{6n\times1} = \begin{Bmatrix} \{P_1\} \\ \{P_2\} \\ \cdot \\ \cdot \\ \cdot \\ \{P_n\} \end{Bmatrix}.$$

Similarly, the inertia forces at all the nodes can be transformed into the inertia forces at the origin by taking moments about this reference point:

$$[R]^T_{6\times6n}[M]_{6n\times6n}\{\ddot{u}_r\}_{6n\times1} = [R]^T_{6\times6n}[M]_{6n\times6n}[R]_{6n\times6}\{\ddot{u}_{r,0}\}_{6\times1} \tag{2.12}$$

where

$$[M] = \begin{bmatrix} [M_1] & 0 & 0 & 0 & 0 & 0 \\ 0 & [M_2] & 0 & 0 & 0 & 0 \\ \cdot & \cdot & \cdot & \cdot & \cdot & 0 \\ \cdot & \cdot & \cdot & \cdot & \cdot & 0 \\ \cdot & \cdot & \cdot & \cdot & \cdot & 0 \\ 0 & 0 & 0 & 0 & 0 & [M_n] \end{bmatrix}.$$

Rigid body mechanical loads are balanced at the origin, meaning at which reference point the total resultant inertia forces are equal to the total resultant loads. This can be expressed as

$$[R]^T_{6\times6n}[M]_{6n\times6n}[R]_{6n\times6}\{\ddot{u}_{r,0}\}_{6\times1} = [R]^T_{6\times6n}\{P\}_{6n\times1}. \tag{2.13}$$

The rigid body accelerations at the origin can be solved:

$$\{\ddot{u}_{r,0}\}_{6\times1} = ([R]^T_{6\times6n}[M]_{6n\times6n}[R]_{6n\times6})^{-1}[R]^T_{6\times6n}\{P\}_{6n\times1}. \qquad (2.14)$$

By applying the balanced loads to the finite element structure in a linear static formulation, the relative nodal flexible displacements with respect to the origin ($\{u\}_{6n\times1}$) can be solved from the following equation:

$$[M]_{6n\times6n}[R]_{6n\times6}\{\ddot{u}_{r,0}\}_{6\times1} + [K]_{6n\times6n}\{u\}_{6n\times1} = \{P\}_{6n\times1}. \qquad (2.15)$$

Since the stiffness matrix is singular for the unconstrained structure, it requires a special technique to solve for the relative nodal displacements. The technique used in MSC-NASTRAN® (MSC Software Corporation; NASTRAN®, National Aeronautics and Space Administration) is based on the fact that the relative nodal displacements matrix is orthogonal or decoupled from eigen-solution rigid body mode shapes ($\{\varphi\}_{6n\times1}$). For low strain rigid body modes, geometric rigid body transformation vector ($[R]_{6n\times6}$) can be expressed as a linear combination of $\{\varphi\}_{6n\times1}$.

Therefore, a rigid body decoupling constraint is met by the following criterion:

$$[\varphi]^T[M]\{u\} = 0 \quad \rightarrow \quad [R]^T[M]\{u\} = 0. \qquad (2.16)$$

Adding Equations (2.12), (2.14), and (2.15) to Equation (2.16) obtains

$$\underbrace{\begin{bmatrix} [K] & [M][R] \\ [R]^T[M] & [R]^T[M][R] \end{bmatrix}}_{\text{a nonsingular matrix}} \begin{Bmatrix} \{u\} \\ \{\ddot{u}_{r,0}\} \end{Bmatrix} = \begin{Bmatrix} \{P\} \\ [R]^T\{P\} \end{Bmatrix}. \qquad (2.17)$$

Therefore, the relative nodal displacements with respect to the origin can be solved:

$$\begin{Bmatrix} \{u\} \\ \{\ddot{u}_{r,0}\} \end{Bmatrix} = \begin{bmatrix} [K] & [M][R] \\ [R]^T[M] & [R]^T[M][R] \end{bmatrix}^{-1} \begin{Bmatrix} \{P\} \\ [R]^T\{P\} \end{Bmatrix}. \qquad (2.18)$$

Example 2.2

The same beam in dimensions as in Example 2.1 is subjected to a thrust force of 2224 N in the z direction acting at one end, but does not have any constraints, as shown in Figure 2.2(a). The beam is made of steel with the mass density of $7.8e^{-9}$ Mg/mm^3 and Young's modulus of

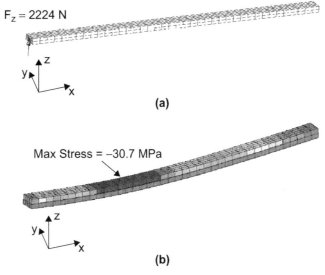

$F_z = 2224$ N

(a)

Max Stress = –30.7 MPa

(b)

Figure 2.2
A free beam analyzed using initial relief technique: (a) a finite element model of the unconstrained beam and (b) stress contour plot of σ_{xx}.

210,000 MPa. Conduct stress analyses of this unconstrained beam using inertia relief technique.

Solution
The finite element beam model is meshed by using eight-node hexahedral solid elements. It is analyzed using "Static" stress analysis solution with the "Inertia relief" option in ABAQUS. The analyzed stress contour plot in x direction σ_{xx} is shown in Figure 2.2(b) where the maximum compressive stress σ_{xx} is found to be –30.7 MPa at the marked location.

Modal Transient Response Analysis

In the cases where local dynamic responses of a structure cannot be ignored or the relative acceleration term cannot be dropped from the dynamic equilibrium equation as it does in inertial relief method, dynamic structural response analyses are needed. One of the analyses is modal transient response analysis.

Modal transient response analysis is an approach to compute the transient response of a linear structure in a modal coordinate system. The method uses the mode shapes of the structure to reduce size, uncouple equations of motion, and then perform numerical integration method. There is no need to use all the computed modes in the transient response solution. Only the few lowest ones are sufficient for the dynamic analysis. Mode truncation assumes that an accurate solution can be obtained using a reduced set of modes whose frequencies are below the cutoff frequency.

It should be noted that truncating modes in a particular frequency range may truncate a significant portion of the behavior in that frequency range. It is often necessary to evaluate the frequency contents of transient loads and to determine a frequency above which no modes are noticeable excited. This frequency is called the cutoff frequency. It is recommended to set the cutoff frequency to five times the forcing frequency and to run the analysis a second time with additional modes for a final verification.

The equations of motion describing a forced vibration of a linear discrete N degrees-of-freedom system can be written in a matrix notation as

$$[M]\{\ddot{u}\} + [C]\{\dot{u}\} + [K]\{u\} = \{F(t)\} \tag{2.19}$$

where [M], [C], and [K] are the system mass, damping, and stiffness matrices, individually. The system displacement and force vectors are given by $\{u\}$ and $\{F(t)\}$. Often the system is given some initial displacements and velocities, which are represented by $\{u(0)\}$ and $\{\dot{u}(0)\}$.

Natural Frequencies and Normal Modes

To perform a modal analysis the force vector $\{F(t)\}$ and the damping matrix [C] must be equal to zero in Equation (2.19), namely

$$[M]\{\ddot{u}\} + [K]\{u\} = \{0\}. \tag{2.20}$$

The solutions of Equation (2.20) in free vibrations of the undamped structure have the form

$$\{u\} = \{a\}\sin(\omega t - \alpha) \tag{2.21}$$

where

    $\{a\}$ = the normal or natural mode shape of vibration
    ω = the natural frequency of the system

The substitution of Equation (2.21) into (2.20) and factoring out $\sin(\omega t - \alpha)$ gives

$$[[K] - \omega^2[M]]\{a\} = \{0\}. \tag{2.22}$$

The solution for which not all $a_i = 0$ requires that the determinant of the matrix be equal to zero, meaning

$$|[K] - \omega^2[M]| = 0. \tag{2.23}$$

The eigenvalues, or the square of each natural frequency ω_i^2, can be obtained from the expansion of the determinant in Equation (2.23), resulting in a characteristic equation of degree n in ω^2. The corresponding mode shape or eigenvector at a specific eigenvalue can be solved from Equation (2.22) by the substitution of ω_i^2.

Orthonormalization of Normal Modes

It is sometimes convenient to work with normalized, or orthonormal mode shapes. In particular we want to scale each mode shape such that

$$\{\varphi\}_n^T[M]\{\varphi\}_m = 0 \quad \text{for} \quad n \neq m \tag{2.24a}$$

$$\{\varphi\}_n^T[M]\{\varphi\}_n = 1 \quad \text{for} \quad n = m \tag{2.24b}$$

where $\{\varphi\}_n$ is the n-th normalized modal vector. To accomplish this we seek some constant γ_n such that

$$\{\varphi\}_n = \gamma_n\{a\}_n. \tag{2.25}$$

The substitution of Equation (2.25) into Equation (2.24b) yields

$$\gamma_n\{a\}_n^T[M]\gamma_n\{a\}_n = \gamma_n^2\{a\}_n^T[M]\{a\}_n = 1. \tag{2.26}$$

Solving Equation (2.26) for γ_n, it has

$$\gamma_n = \frac{1}{\sqrt{\{a\}_n^T[M]\{a\}_n}} = \frac{1}{\sqrt{M_n}} \tag{2.27}$$

where M_n is the n-th generalized mass. Hence the n-th normalized mode is given by

$$\{\varphi\}_n = \frac{1}{\sqrt{M_n}}\{a\}_n. \tag{2.28}$$

The normalized modes can be conveniently arranged in the modal matrix as

$$[\Phi] = \left\{\{\varphi\}_1, \{\varphi\}_2, \dots, \{\varphi\}_n\right\} = \begin{bmatrix} \varphi_{11} & \varphi_{12} & \cdots & \varphi_{1n} \\ \varphi_{12} & \varphi_{22} & \cdots & \varphi_{2n} \\ \vdots & \vdots & \cdots & \vdots \\ \varphi_{n1} & \varphi_{n2} & \cdots & \varphi_{nn} \end{bmatrix}. \tag{2.29}$$

We can rewrite Equation (2.22) for the i-th normalized mode as

$$[K]\{\varphi\}_n = \omega_n^2[M]\{\varphi\}_n. \tag{2.30}$$

Premultiplying Equation (2.30) by $\{\varphi\}_n^T$, we can obtain the orthogonality condition

$$\{\varphi\}_n^T[K]\{\varphi\}_m = 0 \quad \text{for} \quad n \neq m; \quad \{\varphi\}_n^T[K]\{\varphi\}_n = \omega_n \quad \text{for} \quad n = m. \tag{2.31}$$

Example 2.3

A two-degree-of-freedom model with two springs and two masses is illustrated in Figure 2.3. It is given that $m_1 = 0.1\,\text{kg} = 0.1\ \text{N} \cdot \text{sec}^2/\text{m}^2$, $m_2 = 10\,\text{kg} = 10\,\text{N} \cdot \text{sec}^2/\text{m}^2$, $k_1 = 100\,\text{N/m}$, and $k_2 = 10{,}000\,\text{N/m}$. Determine the normalized normal modes and the modal matrix.

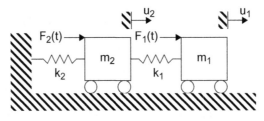

Figure 2.3
A two degree-of-freedom model.

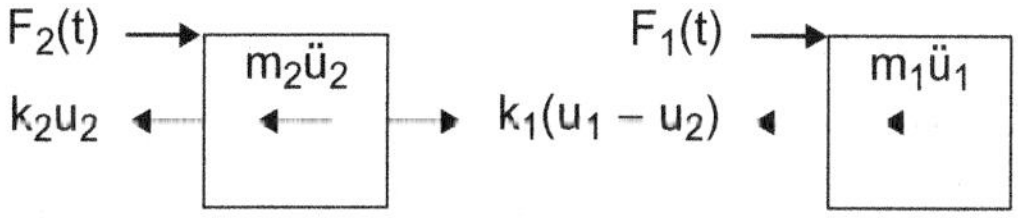

Figure 2.4
Free body diagrams and force equilibrium.

Solution

The following equations of motion are obtained from the free body diagrams as shown in Figure 2.4 by equating the sum of the forces acting on each mass to zero, respectively. Hence,

$$m_1\ddot{u}_1 + k_1(u_1 - u_2) = F_1(t)$$
$$m_2\ddot{u}_2 - k_1(u_1 - u_2) + k_2u_2 = F_2(t).$$

The preceding equations are written in a matrix form as

$$\begin{bmatrix} m_1 & 0 \\ 0 & m_2 \end{bmatrix} \begin{Bmatrix} \ddot{u}_1 \\ \ddot{u}_2 \end{Bmatrix} + \begin{bmatrix} k_1 & -k_1 \\ -k_1 & k_1 + k_2 \end{bmatrix} \begin{Bmatrix} u_1 \\ u_2 \end{Bmatrix} = \begin{Bmatrix} F_1(t) \\ F_2(t) \end{Bmatrix}.$$

The mass and stiffness matrices are given, respectively, by

$$[M] = \begin{bmatrix} m_1 & 0 \\ 0 & m_2 \end{bmatrix} = \begin{bmatrix} 0.1 & 0 \\ 0 & 10 \end{bmatrix}$$

$$[K] = \begin{bmatrix} k_1 & -k_1 \\ -k_1 & k_1 + k_2 \end{bmatrix} = \begin{bmatrix} 100 & -100 \\ -100 & 10,100 \end{bmatrix}.$$

The equation of motion for a free vibration system is expressed as

$$[[K] - \omega_i^2[M]]\{a\} = \begin{bmatrix} 100 - 0.1\omega^2 & -100 \\ -100 & 10,100 - 10\omega^2 \end{bmatrix} \begin{Bmatrix} a_1 \\ a_2 \end{Bmatrix} = \begin{Bmatrix} 0 \\ 0 \end{Bmatrix}.$$

For a nontrivial solution, the determinant of the matrix needs to be set equal to zero, that is,

$$|[K] - \omega_i^2[M]| = (100 - 0.1\omega^2)(10,100 - 10\omega^2) - (-100)^2 = 0.$$

The expansion of this equation leads to

$$\omega^4 - 2010\omega^2 + 1,000,000 = 0.$$

The roots of this quadratic, eigenvalues are

$$\omega_1^2 = 904.9$$

$$\omega_2^2 = 1105.$$

Thus, the natural frequencies of the structure are

$$\omega_1 = 30.08 \, \text{radians/sec} = 4.787 \, \text{cycles/sec}$$

$$\omega_2 = 33.24 \, \text{radians/sec} = 5.290 \, \text{cycles/sec}$$

The first normal mode or modal shape can be obtained by substituting the first natural frequency, ω_1 = 30.08 radians/sec, back to the equation of motion for free vibration, that is,

$$\begin{bmatrix} 100 - 0.1 \times 904.9 & -100 \\ -100 & 10,100 - 10 \times 904.9 \end{bmatrix} \begin{Bmatrix} a_{11} \\ a_{21} \end{Bmatrix} = \begin{Bmatrix} 0 \\ 0 \end{Bmatrix} \Rightarrow \begin{Bmatrix} a_{11} \\ a_{21} \end{Bmatrix} = \begin{Bmatrix} 1.000 \\ 0.0951 \end{Bmatrix}.$$

Similarly, substituting the second natural frequency, ω_1 = 33.24 radians/sec, into the equation of motion for free vibration, we obtain the following second normal mode,

$$\begin{Bmatrix} a_{12} \\ a_{22} \end{Bmatrix} = \begin{Bmatrix} 1.000 \\ -0.105 \end{Bmatrix}.$$

The generalized masses M_1 and M_2 corresponding to the first and second normal modes are calculated as

$$M_1 = \begin{Bmatrix} a_{11} \\ a_{21} \end{Bmatrix}^T \begin{bmatrix} m_1 & 0 \\ 0 & m_2 \end{bmatrix} \begin{Bmatrix} a_{11} \\ a_{21} \end{Bmatrix} = \begin{Bmatrix} 1.00 \\ 0.0951 \end{Bmatrix}^T \begin{bmatrix} 0.1 & 0 \\ 0 & 10 \end{bmatrix} \begin{Bmatrix} 1.00 \\ 0.0951 \end{Bmatrix} = 0.1904$$

$$M_2 = \begin{Bmatrix} a_{12} \\ a_{22} \end{Bmatrix}^T \begin{bmatrix} m_1 & 0 \\ 0 & m_2 \end{bmatrix} \begin{Bmatrix} a_{12} \\ a_{22} \end{Bmatrix} = \begin{Bmatrix} 1.00 \\ -0.105 \end{Bmatrix}^T \begin{bmatrix} 0.1 & 0 \\ 0 & 10 \end{bmatrix} \begin{Bmatrix} 1.00 \\ -0.105 \end{Bmatrix} = 0.2103.$$

and the normalized normal modes for the two natural frequencies are defined as

$$\{\varphi_1\} = \begin{Bmatrix} \varphi_{11} \\ \varphi_{21} \end{Bmatrix} = \frac{1}{\sqrt{M_1}} \begin{Bmatrix} a_{11} \\ a_{21} \end{Bmatrix} = \frac{1}{\sqrt{0.1904}} \begin{Bmatrix} 1.000 \\ 0.0951 \end{Bmatrix} = \begin{Bmatrix} 2.2917 \\ 0.2179 \end{Bmatrix}$$

$$\{\varphi_2\} = \begin{Bmatrix} \varphi_{12} \\ \varphi_{22} \end{Bmatrix} = \frac{1}{\sqrt{M_2}} \begin{Bmatrix} a_{12} \\ a_{22} \end{Bmatrix} = \frac{1}{\sqrt{0.2103}} \begin{Bmatrix} 1.000 \\ -0.105 \end{Bmatrix} = \begin{Bmatrix} 2.1806 \\ -0.2290 \end{Bmatrix}.$$

Finally, the modal matrix is

$$[\Phi] = \begin{bmatrix} 2.2917 & 2.1806 \\ 0.2179 & -0.2290 \end{bmatrix}.$$

Decoupled Modal Equations of Motion

To uncouple the equations of motion in Equation (2.19), the transformation of coordinates is introduced:

$$\{u\} = [\Phi]\{z\} \tag{2.32}$$

where

[Φ] = the modal matrix obtained in the solution of the undamped free vibration system

{z} = the modal participation coefficients matrix

The substitution of Equation (2.32) and its derivatives into Equation (2.19) leads to

$$[M][\Phi]\{\ddot{z}\} + [C][\Phi]\{\dot{z}\} + [K][\Phi]\{z\} = \{F(t)\}. \tag{2.33}$$

Premultiplying Equation (2.33) by the transpose of the n-th modal vector $\{\varphi\}_n^T$ yields

$$\{\varphi\}_n^T[M][\Phi]\{\ddot{z}\} + \{\varphi\}_n^T[C][\Phi]\{\dot{z}\} + \{\varphi\}_n^T[K][\Phi]\{z\} = \{\varphi\}_n^T\{F(t)\}. \tag{2.34}$$

The orthogonality properties of the modal shapes are

$$\{\varphi\}_n^T[M]\{\varphi\}_m = 0 \quad m \neq n \tag{2.35a}$$

$$\{\varphi\}_n^T[M]\{\varphi\}_n = M_n \quad m = n \tag{2.35b}$$

$$\{\varphi\}_n^T[K]\{\varphi\}_m = 0 \quad m \neq n \tag{2.35c}$$

$$\{\varphi\}_n^T[K]\{\varphi\}_n = K_n \quad m = n. \tag{2.35d}$$

It is assumed that a similar reduction is applied to the damping term,

$$\{\varphi\}_n^T[C]\{\varphi\}_m = 0 \quad m \neq n \tag{2.36a}$$

$$\{\varphi\}_n^T[C]\{\varphi\}_n = C_n \quad m = n. \tag{2.36b}$$

This equation of motion (Equation 2.34) can be decoupled as

$$M_n\ddot{z}_n + C_n\dot{z}_n + K_nz_n = F_n(t) \tag{2.37}$$

where

$$M_n = \{\varphi\}_n^T[M]\{\varphi\}_n = 1 \tag{2.38a}$$

$$K_n = \{\varphi\}_n^T[K]\{\varphi\}_n = \omega_n^2 M_n = \omega_n^2 \tag{2.38b}$$

$$C_n = \{\varphi\}_n^T[C]\{\varphi\}_n = 2\xi_n\omega_n M_n = 2\xi_n\omega_n \tag{2.38c}$$

$$F_n(t) = \{\varphi\}_n^T\{F(t)\} \tag{2.38d}$$

or, alternatively as

$$\ddot{z}_n + 2\xi_n\omega_n\dot{z}_n + \omega_n^2 z_n = F_n(t). \tag{2.39}$$

Here ξ_n is the damping ratio to the n-th mode shape.

The normal modal coordinator transformation, Equation (2.32), will uncouple the damping force if the damping matrix [C] follows the Rayleigh damping equation as follows:

$$[C] = \alpha[M] + \beta[K] \tag{2.40}$$

where α and β are the damping proportionality factors. This can be demonstrated by premultiplying both sides of the equation by the transpose of the n-th mode $\{\varphi\}_n^T$ and postmultiplying by the modal matrix $[\Phi]$. We obtain

$$\{\varphi\}_n^T[C][\Phi] = \alpha\{\varphi\}_n^T[M][\Phi] + \beta\{\varphi\}_n^T[K][\Phi]. \tag{2.41}$$

Equation (2.41) can be rewritten on the orthogonality conditions as follows:

$$\{\varphi\}_n^T[C]\{\varphi\}_m = 0 \quad m \neq n \tag{2.42a}$$

$$\{\varphi\}_n^T[C]\{\varphi\}_n = C_n \quad m = n \tag{2.42b}$$

where

$$C_n = \alpha + \beta \omega_n^2. \tag{2.43}$$

To determine these constants for any desired values of damping ratios ξ_ns in any specified number of modes, setting Equation (2.38c) equal to Equation (2.43) provides

$$C_n = 2\xi_n \omega_n = \alpha + \beta \omega_n^2. \tag{2.44}$$

When two damping ratios (ξ_1 and ξ_2) and their corresponding natural frequencies (ω_1 and ω_2) of a structure are given, Equation (2.44) reduces to

$$2\xi_1 \omega_1 = \alpha + \beta \omega_1^2 \tag{2.45a}$$

$$2\xi_2 \omega_2 = \alpha + \beta \omega_2^2. \tag{2.45b}$$

Thus, the Rayleigh damping factors can be obtained by solving Equations (2.45a) and (2.45b) as

$$\beta = \frac{2\xi_1 \omega_1 - 2\xi_2 \omega_2}{\omega_1^2 - \omega_2^2} \tag{2.46}$$

$$\alpha = 2\xi_1 \omega_1 - \beta \omega_1^2. \tag{2.47}$$

Example 2.4

It is assumed that a structure has a constant damping ratio of 2% at the two natural frequencies of 200 Hz and 1000 Hz. Determine the Rayleigh damping factors.

Solution

Using Equations (2.46) and (2.47) we can find α and β as

$$\beta = \frac{2\xi_1 \omega_1 - 2\xi_2 \omega_2}{\omega_1^2 - \omega_2^2} = \frac{2 \cdot 0.02 \cdot (2\pi \cdot 200) - 2 \cdot 0.02 \cdot (2\pi \cdot 1000)}{(2\pi \cdot 200)^2 - (2\pi \cdot 1000)^2} = 5.3 \times 10^{-6}$$

$$\alpha = 2\xi_1 \omega_1 - \beta \omega_1^2 = 2 \cdot 0.02 \cdot (2\pi \cdot 200) - 5.3 \cdot 10^{-6} \cdot (2\pi \cdot 200)^2 = 41.90.$$

Equation (2.44) can be rearranged as

$$\xi_n = \frac{\alpha + \beta \omega_n^2}{2\omega_n}.$$

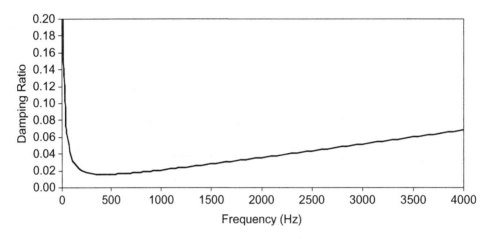

Figure 2.5
Rayleigh damping ratios versus natural frequencies.

This equation can be used to plot the Rayleigh damping ratios versus the natural frequencies as shown in Figure 2.5.

Numerical Integration Method for Equations of Motion

Explicit and implicit numerical integration methods are commonly used for solving the equations of motion. For explicit schemes the equations of motion are evaluated at the current time step t_i, and the implicit methods use the equations of motion at the new time step t_{i+1}. LS-DYNA3D® (Livermore Software Technology Corporation) and NASTRAN uses the central difference time integration. The central difference scheme is an explicit method.

Explicit Integration Scheme: The Central Difference Method

As illustrated in Figure 2.6, the central difference equations for velocity and acceleration at discrete times are

$$\dot{z}_{n,i} = \frac{1}{2 \cdot \Delta t} \left(z_{n,i+1} - z_{n,i-1} \right) \tag{2.48}$$

and

$$\ddot{z}_{n,i} = \frac{1}{\Delta t} \left(\dot{z}_{n,i+\frac{1}{2}} - \dot{z}_{n,i-\frac{1}{2}} \right) = \frac{1}{(\Delta t)^2} \left(z_{n,i+1} - 2 z_{n,i} + z_{n,i-1} \right). \tag{2.49}$$

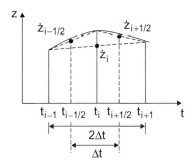

Figure 2.6
Central finite difference approximation.

The equilibrium at time t_i is

$$M_n \ddot{z}_{n,i} + C_n \dot{z}_{n,i} + K_n z_{n,i} = F_{n,i}. \qquad (2.50)$$

By substituting Equations (2.48) and (2.49) into Equation (2.50), the equations of motion can be written as

$$\left(\frac{M_n}{(\Delta t)^2}\right)(z_{n,i+1} - 2z_{n,i} + z_{n,i-1}) + \left(\frac{C_n}{2 \cdot \Delta t}\right)(z_{n,i+1} - z_{n,i-1}) + K_n z_{n,i} = F_{n,i}. \qquad (2.51)$$

Collecting terms, the equation of motion can be rearranged as

$$\left(M_n + \frac{1}{2}\Delta t C_n\right) z_{n,i+1} = F_{n,i} + \left(2M_n - (\Delta t)^2 K_n\right) z_{n,i} + \left(\frac{1}{2}\Delta t C_n - M_n\right) z_{n,i-1}. \qquad (2.52)$$

In the case where M_n, C_n, K_n, and Δt are constant throughout the analysis and they do not change with time, the solution for $z_{n,i+1}$ can be obtained from Equation (2.52) based on the given displacements and force at time t_{i-1} and t_i, such as $z_{n,i-1}, z_{n,i}$, and $F_{n,i}$. The same procedure is repeated to calculate the quantity at the next time step t_{i+2} and the process is continued to any desired final time.

Transformation of Initial Conditions

In order to specify the constants $z_{n,0}$ and $\dot{z}_{n,0}$ in Equations (2.51) and (2.52), the initial conditions must be transformed from the physical coordinates to the modal coordinates. Since the mode shapes are normalized, at time $t = 0$

$$\{u(0)\} = [\Phi]\{z(0)\} \tag{2.53}$$

$$\{\dot{u}(0)\} = [\Phi]\{\dot{z}(0)\}. \tag{2.54}$$

Premultiplying both side of Equations (2.53) and (2.54) by $[\Phi]^T[M]$ gives

$$[\Phi]^T[M]\{u(0)\} = [\Phi]^T[M][\Phi]\{z(0)\} \tag{2.55}$$

$$[\Phi]^T[M]\{\dot{u}(0)\} = [\Phi]^T[M][\Phi]\{\dot{z}(0)\}. \tag{2.56}$$

Due to the orthogonality property of the mass matrix it follows

$$\{z(0)\} = [\Phi]^T[M]\{u(0)\} \tag{2.57}$$

$$\{\dot{z}(0)\} = [\Phi]^T[M]\{\dot{u}(0)\}. \tag{2.58}$$

Initial conditions $z_{n,0}$ and $\dot{z}_{n,0}$ are set at time t_0.

Based on Equation (2.50) we can find $\ddot{z}_{n,0}$ as

$$\ddot{z}_{n,0} = \frac{1}{M_n}F_{n,0} - \frac{C_n}{M_n}\dot{z}_{n,0} - \frac{K_n}{M_n}z_{n,0}. \tag{2.59}$$

In the central difference method it is necessary to know the value of $z_{n,-1}$ to calculate $z_{n,1}$, which can be estimated as

$$z_{n,-1} = z_{n,0} - \Delta t \cdot \dot{z}_{n,0}. \tag{2.60}$$

Converting Back to Physical Coordinates

Once the N modal solutions $z_n(t)$ have been obtained the physical solutions $u_n(t)$ are obtained by employing the modal transformation as

$$\{u_n(t)\} = [\Phi]\{z_n(t)\} \tag{2.61}$$

or

$$u_i(t) = \sum_{k=1}^{N} \varphi_{i,k} z_k(t). \tag{2.62}$$

and the local stresses can be calculated as

$$\sigma_{ij}(t) = \sum_{k=1}^{N} \sigma_{ij,k}(\varphi_k) z_k(t). \tag{2.63}$$

Example 2.5

The same beam in dimensions as in Examples 2.1 and 2.2 is subjected to a dynamic thrust force with a peak of 2224 N in the z direction at one end, as shown in Figure 2.7(a). The beam is unconstrained, made of steel with mass density of $7.8e^{-9}$ Mg/mm^3 and Young's modulus of 210,000 MPa. Conduct modal transient response analyses for stresses of the beam as the force is cyclically applied with loading frequencies of (a) 125 Hz and (b) 62.5 Hz, respectively, as illustrated in Figure 2.7(b).

Solution
The finite element beam model is meshed by using eight-node hexahedral solid elements. The analyses are conducted using ABAQUS. Both options "Frequency" and "Modal dynamic" are employed in sequence as the solution methods. A damping ratio is taken to 0.05, and both the initial displacement and velocity are set to 0.0.

The normal mode analysis is carried out first to extract natural frequencies and mode shapes of the beam that are essential for the modal transient response analysis. The first four natural frequencies (129.7 Hz, 257.7 Hz, 355.9 Hz, and 693.6 Hz) of the beam and their corresponding normal modes are presented in Figure 2.7(c). It can be found that all the four modes, except for Mode 2, are bending modes in the vertical direction, which happens to be the loading direction.

In the case where the loading frequency of 125 Hz is considered, using the recommended rule for the cutoff frequency (625 Hz) being five times the loading frequency, the first four mode shapes are assumed to be sufficient for use in the modal transient response analysis. Consequently, the stress contours for σ_{xx} and the maximum compressive stress of −39.4 MPa are shown in Figure 2.7(d).

To verify the assumption and check accuracy of the stress solution, the first five and 20 modes are employed in analyses. The corresponding maximum compressive stress σ_{xx} at the same element and location are

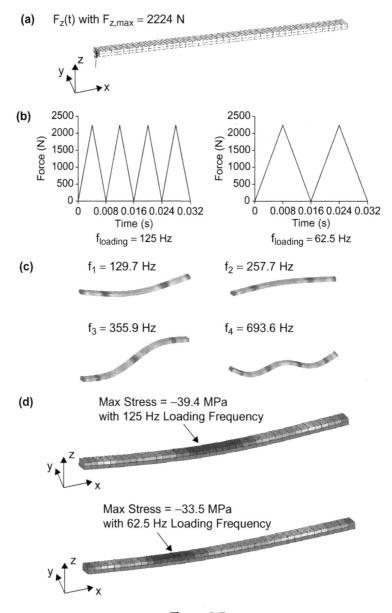

Figure 2.7
A free beam analyzed using the modal transient response analysis: (a) a finite element model of the unstrained beam with a dynamic force, (b) load time histories with two different loading frequencies, (c) the first four natural frequencies and mode shapes, and (d) stresses in the x direction.

found to be −39.4 MPa and −39.9 MPa, respectively. This indicates that the transient response solution using the first four normal modes yields very satisfactory stress results.

Since this loading frequency of 125 Hz is close to the first fundamental frequency of 129.7 Hz, the beam is excited at a frequency close to the resonant frequency, resulting in the maximum stress increase by 28%, as compared to the maximum compressive stress of −30.7 MPa from the analysis using the inertia relief technique in Example 2.1. This indicates that in this case modal transient analysis is essential to capture the resonant effect on responses of the beam.

In the case where the loading frequency of 62.5 Hz is considered, the cutoff frequency is calculated as 312.5 Hz and the first three modes are used in the modal transient response analysis. As a result, the stress contours for σ_{xx} and the maximum compressive stress of −33.4 MPa are illustrated in Figure 2.7(d). Next, the analysis is performed based on the first four and five normal modes, separately. Both analyses result in the same stress −33.5 MPa, indicating the first three modes are sufficient for use in the modal transient response analysis.

Summary

Three commonly used techniques for pseudo stress analysis such as fixed reactive analysis, inertia relief analysis, and modal transient response analysis have been presented. The pseudo stress output can be used in conjunction with a multiaxial notch analysis technique to estimate local true stresses/strains at stress concentration areas for fatigue damage assessment.

The fixed reactive stress analysis is a common, fundamental solution for a constrained structure that is subjected to a set of constant or time independent actions such as forces, moments, torque, and/or temperatures. The actions and reactions from constraint are statically balanced.

The inertial relief analysis is to solve for the responses of an unconstrained structure subjected to constant or slowly varying external loads. It is assumed that the external loads are steady-state loads applied to the structure, the structural transient responses have damped out, and local dynamic responses are ignored.

The modal transient response analysis is used in cases where local dynamic responses of a structure cannot be ignored or the relative acceleration term cannot be dropped from the dynamic equilibrium equation. Modal transient response analysis is to compute the transient response of a linear structure in a modal coordinate system. The method uses the mode shapes of the structure to reduce the size, uncouple the equations of motion, and then perform the numerical integration technique.

References

ABAQUS Analysis User's Manual. (2008). Part IV, Analysis Techniques, Version 6.8, Dassault Systems, SIMULIA.

Barnett, A. R., Widirck, T. W., & Ludwiczak, D. R. (1995). Closed-form static analysis with inertia relief and displacement-dependent loads using a MSC/NASTRAN DMAP alter. NASA Technical Memorandum 106836.

Barkey, M. E., Socie, D. F., & Hsia, K. J. A. (1994). A yield surface approach to the estimation of notch strains for proportional and nonproportional cyclic loading. *Journal of Engineering Materials and Technology, 116*, 173–180.

Buczynski, A., & Glinka, G. (2000). Multiaxial stress-strain notch analysis. In Kalluri, S., & Bonacause, P. J. (Eds.), *Multiaxial fatigue and deformation: Testing and prediction*, ASTM 1387 (pp. 82–98). West Conshohocken, PA: American Society for Testing and Materials. 2000

Gaffrey, J. P., & Lee, J. M. (1994). *MAS/NASTRAN Linear Static Analysis, User's Guide*, Vol. 68. The Macneal-Schwendler Corporation.

Gu, R., & Lee, Y. L. (1997). A new method for estimating nonproportional notch-root stresses and strains. *ASME Journal of Engineering Materials and Technology, 119*, 40–45.

Hoffmann, M., & Seeger, T. (1989). Stress-strain analysis and life predictions of a notched shaft under multiaxial loading. In Leese, G. E., & Socie, D. (Eds.), *Multiaxial fatigue: Analysis and experiments*, Chap. 6, AE-14, SAE (pp. 81–96). Warrendale, PA: SAE International.

Lee, Y. L., Chiang, Y. J., & Wong, H. H. (1995). A constitutive model for estimating multiaxial notch strains. *ASME Journal of Engineering Materials and Technology, 117*, 33–40.

Lee, Y. L., & Gu, R. (1999). Multiaxial notch stress-strain analysis and its application to component life predictions. In Cordes, T., & Lease, K. (Eds.), *Multiaxial fatigue of an inducted hardened shaft*, Chap. 8, SAE AE-28 (pp. 71–78). Warrendale, PA: SAE International.

Moftakhar, A., Buczynski, A., & Glinka, G. (1995). Calculation of elasto-plastic strains and stress in notches under multiaxial loading. *International Journal of Fracture, 70*, 357–373.

Molsky, K., & Glinka, G. (1981). A method of elastic-plastic stress and strain calculation at a notch root. *Material Science and Engineering, 50,* 93–100.

Neuber, H. (1961). Theory of stress concentration for shear-strained prismatic bodies with arbitrary nonlinear stress-strain law. *Journal of Applied Mechanics, Transactions of ASSM,* Section E, *28,* 544–550.

CHAPTER 3

Rainflow Cycle Counting Techniques

Yung-Li Lee
Chrysler Group LLC

Tana Tjhung
Chrysler Group LLC

Chapter Outline

Introduction

This chapter focuses on the process of extracting cycles from a complicated loading history, where each cycle is associated with a closed stress–strain hysteresis loop. For a uniaxial load time history, the rainflow cycle counting technique introduced in 1968 by Matsuishi and Endo (1968) was the first accepted method used to extract closed loading cycles.

The "rainflow" analogy is derived from a comparison of this method to the flow of rain falling on a pagoda and running down the edges of the roof. Due to the importance of the rainflow cycle counting method, many different algorithms have been proposed in the literature, namely, the three-point (Richards et al., 1974; Downing & Socie, 1982; Conle et al., 1997; ASTM E 1049-85, 2005) and the four-point (Amzallag et al., 1994; Dreβler et al., 1995) cycle counting techniques. These algorithms are computationally more efficient and will be reviewed in this chapter.

For a complicated multiaxial load time history, there are two broad cycle counting techniques used for fatigue damage assessment of the structure. In the first approach (the critical plane method), the material volume is segregated into candidate planes. On each candidate plane, the uniaxial cycle counting method is used and the fatigue damages are calculated. The plane that accumulates the most damage is deemed to be the critical plane and the structure life is assessed from this plane.

In the second approach (the equivalent stress or strain amplitude method), a multiaxial cycle counting technique is developed based on the assumption that the fatigue damage or life can be evaluated from the cycles identified from the complicated equivalent loading history. The first multiaxial reversal counting technique was developed by Wang and Brown (1996).

The Wang–Brown technique incorporates a novel extension of the uniaxial rainflow reversal counting technique along with the equivalent strain amplitude concept. Two variants of the Wang–Brown reversal counting technique, namely, the Lee–Tjhung–Jordan equivalent stress reversal counting method (Lee et al., 2007) and the path dependent maximum range (PDMR) reversal counting method (Dong et al., 2010) were developed later. These multiaxial reversal counting techniques are discussed in this chapter.

Uniaxial Rainflow Cycle Counting Techniques

Consider the stress and strain time histories and the corresponding stress–strain response behavior shown in Figure 3.1. Since the hysteresis loops are associated with energy dissipation and fatigue damage, most established uniaxial fatigue damage parameters are calculated from the cycles identified from the hysteresis loops. The rainflow cycle counting method specifically identifies these hysteresis loops (cycles) within a load, stress, or strain time history.

Rainflow Counting Method by Matsuishi and Endo

The rainflow counting technique introduced in 1968 by Matsuishi and Endo (1968) is the first accepted method used to extract closed loading reversals or cycles.

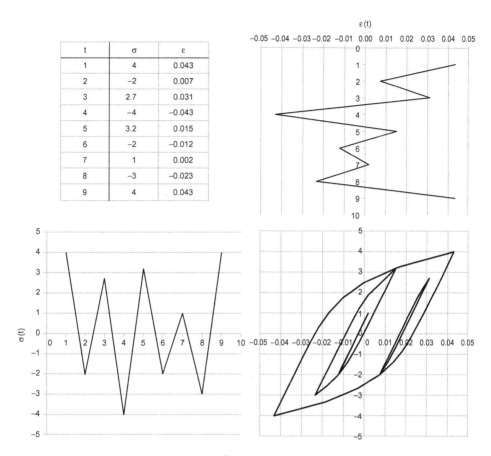

t	σ	ε
1	4	0.043
2	−2	0.007
3	2.7	0.031
4	−4	−0.043
5	3.2	0.015
6	−2	−0.012
7	1	0.002
8	−3	−0.023
9	4	0.043

Figure 3.1
Stress and strain time histories and corresponding hysteresis loops.

The "rainflow" was named from a comparison of this method to the flow of rain falling on a pagoda and running down the edges of the roof. The rainflow cycle counting algorithm is summarized as follows:

1. Rotate the loading history 90° such that the time axis is vertically downward and the load time history resembles a pagoda roof.

2. Imagine a flow of rain starting at each successive extremum point.

3. Define a loading reversal (half-cycle) by allowing each rainflow to continue to drip down these roofs until:

   a. It falls opposite a larger maximum (or smaller minimum) point.

  b. It meets a previous flow falling from above.

  c. It falls below the roof.

4. Identify each hysteresis loop (cycle) by pairing up the same counted reversals.

Example 3.1

Perform the rainflow cycle counting technique on a given service load time history as shown in Figure 3.2(a) where it has been constructed to start from the largest maximum point A and to end with the same load value at A.

Solution

1. Rotate the load time history 90° clockwise.

2. Designate A as the first extremum point, the largest peak in this load time history.

3. Identify the first largest reversal A–D as the flow of rain starts at A and falls off the second extremum point D, the smallest valley in this load time history.

4. Identify the second largest reversal D–A as the flow initiates at D and ends at the other extremum point, which happens to be the first one, A.

5. In the first largest reversal A–D,

  a. Identify a reversal B–C as the rain starts flowing at B and terminates at C because D is a larger maximum than B.

  b. Identify a reversal C–B as the rain starts flowing at C and meets a previous flow at B.

  c. Complete all the points in the first large reversal A–D.

6. In the second largest reversal D–A,

  a. Identify a reversal E–H as the rain starts flowing at E and falls off the roof at H.

  b. Identify a reversal H–E as the rain starts flowing at point H and meets a previous flow at E.

  c. Identify a reversal F–G as the rain starts flowing at F and terminates at G because H is a larger maximum than F.

  d. Identify a reversal G–F as the rain starts flowing from the successive extremum point G and meets a previous flow at F.

  e. Complete all the points in the second largest reversal D–A.

7. The rainflow cycle counting results in terms of reversals and cycles are given in Tables 3.1 and 3.2, respectively.

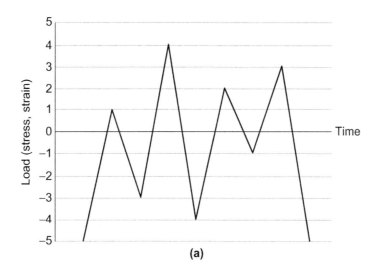

(a)

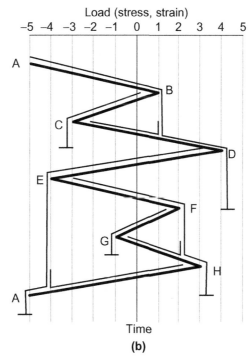

Time

(b)

Figure 3.2

Illustration of the rainflow counting technique. Identification of reversals and loading cycles: (a) a service load-time history and (b) rainflow cycle counting the loading history.

Table 3.1: Reversal Counts Based on the Rainflow Counting Technique

No. of Reversals	From	To	From	To	Range	Mean
1	A	D	−5	4	9	−0.5
1	B	A	4	−5	9	−0.5
1	B	C	1	−3	4	−1
1	C	B	−3	1	4	−1
1	E	H	−4	3	7	−0.5
1	H	E	3	−4	7	−0.5
1	F	G	4	−1	3	0.5
1	G	F	−1	2	3	0.5

Table 3.2: Cycle Counts Based on the Rainflow Counting Technique

No. of Cycles	From	To	From	To	Range	Mean
1	A	D	−5	4	9	−0.5
1	B	C	1	−3	4	−1
1	E	H	−4	3	7	−0.5
1	F	G	2	−1	3	0.5

When a loading history is periodic, the loading history needs to be rearranged to start from the largest extremum point and this extremum point is repeated at the end, in effect closing the largest hysteresis loop. All inner reversals therefore pair up to form cycles. Otherwise, for the nonperiodic loading case, where the loading history does not start and end with the largest extremum point, the rainflow technique will identify unpaired reversals, or half-cycles, in addition to full cycles.

Three-Point Counting Technique

ASTM E 1049-85 recommends a cycle counting method commonly known as the three-point method because this method repeatedly evaluates the loading history three consecutive peak/valley points at a time. The basic three-point cycle counting rule is illustrated in Figure 3.3, where a hanging cycle and a standing cycle are identified in (a) and (b), respectively.

The labels and values of the three peak/valley points are designated as P1, P2, and P3. Define the range $X = |P3-P2|$, and the previous adjacent range $Y = |P2-P1|$.

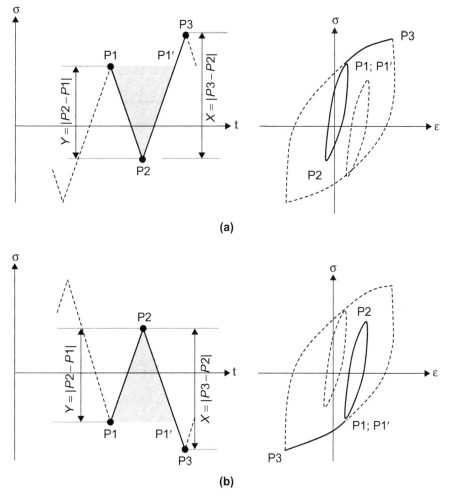

Figure 3.3
ASTM 3-point rainflow cycle counting rule. X ≥ Y:
(a) hanging cycle, (b) standing cycle.

A cycle or hysteresis loop from P1 to P2 and back to P1′ (= P1) is defined if X ≥ Y, and no cycle is counted if X < Y.

Cycle Counting for a Periodic Load Time History

When a loading history is periodic, the loading history needs to be rearranged such that it contains only peaks and valleys and starts with either the highest peak or lowest valley, whichever is greater in absolute magnitude. Then the cycle identification rule is applied to check every three consecutive points from the beginning until a closed loop is defined. The two points P1 and P2 are discarded from the loading history and the remaining points are connected together. This procedure is repeated until the remaining data are exhausted.

Example 3.2

Perform the rainflow cycle counting technique on a given service load time history as shown in Figure 3.4(a). It is assumed this loading history is periodic.

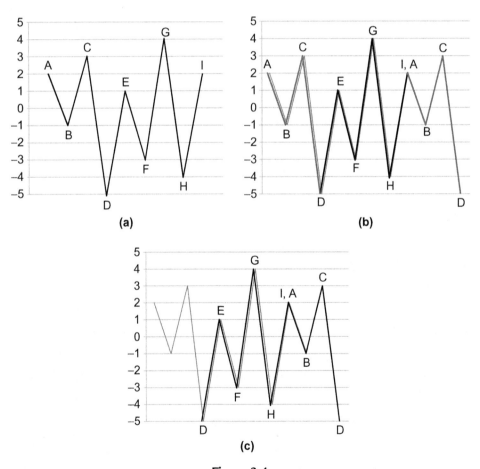

Figure 3.4
Rearrangement of a service load time history: (a) a service load time history, (b) a periodic service load time history, and (c) the rearranged load time history.

Solution

Since this loading history is periodic, the loading history needs to be rearranged such that it starts at the minimum valley D that is the largest extremum point. As shown in Figure 3.4(b), this can be accomplished by cutting off all the points before and at D (namely, A, B, C, and D) and by appending these data to the end of the original history. Please note that additional point D is included in the newly constructed load time history, as illustrated in Figure 3.4(c), to close the largest loop for conservatism.

The three-point cycle counting method is illustrated in Figure 3.5 as follows:

- Consideration of D, E, and F. $X = |F-E|$; $Y = |E-D|$ (Figure 3.5(a)). Since $X < Y$, no cycle is counted and E is designated as the new starting point.

- Consideration of E, F, and G. $X = |G-F|$; $Y = |F-E|$ (Figure 3.5(b)). Since $X \geq Y$, count a cycle from E to F, remove E and F, and connect D and G. Designate D as the new starting point.

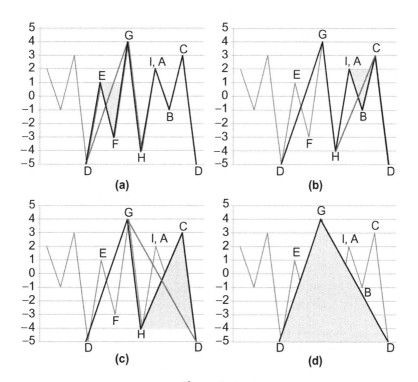

Figure 3.5

Cycles extracted by the three-point counting method: (a) cycle E–F extracted, (b) cycle A–B extracted, (c) cycle H–C extracted, and (d) cycle D–G extracted.

- Consideration of D, G, and H. X = $|H–G|$; Y = $|G–D|$ (Figure 3.5(c)). Since X < Y, no cycle is counted and G is designated as the new starting point.

- Consideration of G, H, and A. X = $|A–H|$; Y = $|H–G|$ (Figure 3.5(c)). Since X < Y, no cycle is counted and H is designated as the new starting point.

- Consideration of H, A, and B. X = $|B–A|$; Y = $|A–H|$ (Figure 3.5(c)). Since X < Y, no cycle is counted and A is designated as the new starting point.

- Consideration of A, B, and C. X = $|C–B|$; Y = $|B–A|$ (Figure 3.5(c)). Since X ≥ Y, count a cycle from A to B, remove A and B, and connect H and C. Designate G as the new starting point.

- Consideration of G, H, and C. X = $|C–H|$; Y = $|H–G|$ (Figure 3.5(c)). Since X < Y, no cycle is counted, and H is designated as the new starting point.

- Consideration of H, C, and D. X = $|D–C|$; Y = $|C–H|$ (Figure 3.5(d)). Since X ≥ Y, count a cycle from H to C, remove H and C, and connect G and D. Designate D as the new starting point.

- Consideration of D, G, and D. X = $|D–G|$; Y = $|G–D|$ (Figure 3.5(e)). Since X ≥ Y, count a cycle from D to G and conclude the counting process.

- The summary of the cycle counting result is given in Table 3.3.

Table 3.3: Cycle Counts Based on the Three-Point Counting Technique

No. of Cycles	From	To	From	To	Range	Mean
1	E	F	1	−3	4	−2
1	A	B	2	−1	3	0.5
1	H	C	−4	3	7	−0.5
1	D	G	−5	4	9	−0.5

Reversal Counting for a Nonperiodic Load Time History

For a nonperiodic loading case, where the loading history does not start and end with the largest extremum point, the rainflow technique will identify unpaired reversals, or half-cycles, in addition to full cycles. The two rules to follow are:

- If X ≥ Y and point P1 is not the starting point of the loading history, then a cycle is counted (as shown in Figure 3.3).

- If X ≥ Y and point P1 is the starting point of the loading history, then a reversal or half-cycle is counted from P1 to P2, and only point P1 is removed (as illustrated in Figure 3.6).

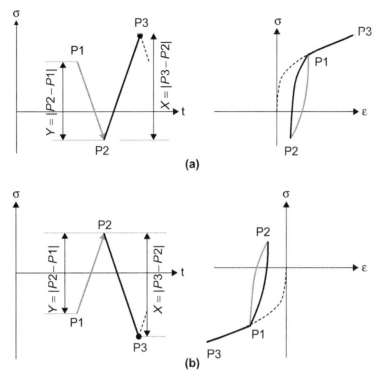

Figure 3.6
Reversal extracted at the starting point of the load time history by the three-point counting method: (a) reversal P1 to P2 and (b) reversal P1 to P2.

Example 3.3

Perform the three-point counting technique on a given service load time history as shown in Figure 3.4(a). It is assumed this loading history is nonperiodic.

Solution
The three-point counting technique to extract the reversals from this complicated history is illustrated in Figure 3.7.

- Consideration of A, B, and C. $X = |C–B|$; $Y = |B–A|$ (Figure 3.7(a)). Since $X \geq Y$ and A is the starting point of this history, count a reversal from A to B and remove A. Designate B as the new starting point.

- Consideration of B, C, and D. $X = |D–C|$; $Y = |C–B|$ (Figure 3.7(b)). Since $X \geq Y$ and B is the starting point of this history, count a reversal from B to C and remove B. Designate C as the new starting point.

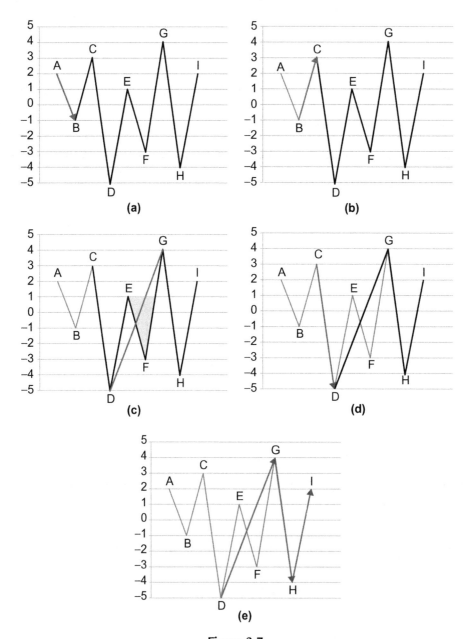

Figure 3.7
Reversal extraction for the nonperiodic load-time history by the three-point counting method: (a) reversal A–B extracted, (b) reversal B–C extracted, (c) cycle E–F extracted, (d) reversal C–D extracted, and (e) residue.

- Consideration of C, D, and E. X= |E–D|; Y= |D–C| (Figure 3.7(c)). Since X < Y, no cycle is counted and D is designated as the new starting point.

- Consideration of D, E, and F. X= |F–E|; Y= |E–D| (Figure 3.7(c)). Since X < Y, no cycle is counted and E is designated as the new starting point.

- Consideration of E, F, and G. X= |G–F|; Y= |F–E| (Figure 3.7(c)). Since X ≥ Y, count a cycle from E to F, remove E and F, and connect D and G. Designate C as the new starting point.

- Consideration of C, D, and G. X= |G–D|; Y= |D–C| (Figure 3.7(d)). Since X ≥ Y and C is the starting point of this history, count a reversal from C to D and remove C. Designate D as the new starting point.

- Consideration of residue D, G, F, and I (Figure 3.7(e)). The entire loading history has been evaluated and no further reversals or cycles can be counted by the three-point rule. The remaining points constitute the residue. By the ASTM three-point method, the remaining ranges are counted as reversals.

- The summary of the reversal counting result is given in Table 3.4.

Table 3.4: Reversal Counts Based on the Three-Point Counting Technique

No. of Cycles	From	To	From	To	Range	Mean
1	A	B	2	−1	3	0.5
1	B	C	−1	3	4	1
2	E	F	1	−3	4	−1
1	C	D	3	−5	8	−1
1	D	G	−5	4	9	−0.5
1	G	I I	4	−4	9	0
1	H	A, I	−4	2	6	−1

In contrast to the classical rainflow method described in Chapter 2, the three-point technique counts inner loops before outer loops. Computationally, this is a more efficient algorithm compared to the classic rainflow method. It also allows the three-point method to be used for real-time cycle counting applications. In the postprocessing case, ASTM recommends rearrangement and closure for periodic load histories. In the real-time case, where rearrangement cannot be performed, the ASTM three-point technique has provisions to count half-cycles in addition to full cycles.

Four-Point Counting Technique

Consider four consecutive peak/valley points P1, P2, P3, and P4, as shown in Figure 3.8. If P2 and P3 are contained within P1 and P4, then a cycle is counted from P2 to P3 (and back to P2'); otherwise no cycle is counted. One way to code this rule is given as:

- Define ranges $X = |P4–P3|$, $Y = |P3–P2|$, and $Z = |P2–P1|$.

- If $X \geq Y$ AND $Z \geq Y$ then FROM = P2 and TO = P3, end.

Similar to the three-point counting technique, the four-point counting method can be easily implemented for a real-time cycle counting acquisition for a nonperiodic loading history. But this method can only recognize the closed cycles for fatigue

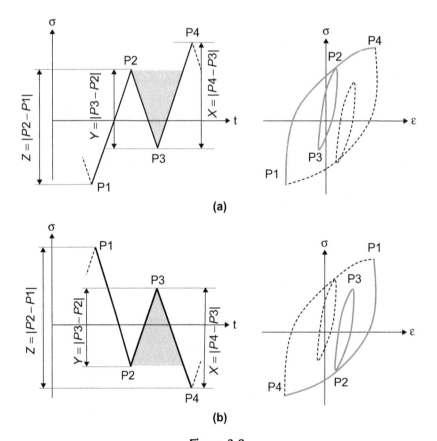

(a)

(b)

Figure 3.8
The four-point rainflow cycle counting rule: (a) hanging cycle and (b) standing cycle.

analysis and excludes the contribution from the residues (unpaired reversals), which is different from the three-point technique.

Example 3.4

Perform the four-point counting technique on a given service load time history as shown in Figure 3.4(a).

Solution

The four-point counting method, as illustrated in Figure 3.9, is as follows:

- Consideration of A, B, C, and D. $X = |D–C|$, $Y = |C–B|$, $Z = |B–A|$. $X \geq Y$ but $Z < Y$, therefore no cycle is counted. Designate B as a new starting point.

- Consideration of B, C, D, and E. $X = |E–D|$, $Y = |D–C|$, $Z = |C–B|$. $X < Y$ and $Z < Y$, therefore no cycle is counted. Designate C as a new starting point.

- Consideration of C, D, E, and F. $X = |F–E|$, $Y = |E–D|$, $Z = |D–C|$. $X < Y$ and $Z \geq Y$, therefore no cycle is counted. Designate D as a new starting point.

- Consideration of D, E, F, and G. $X = |G–F|$, $Y = |F–E|$, $Z = |E–D|$. $X \geq Y$ and $Z \geq Y$, therefore count a cycle from E to F. Remove E and F, and join D to G. Designate C as a starting point.

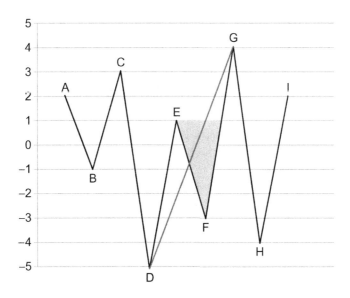

Figure 3.9
Cycle extraction of a service load time history using the four-point cycle counting rule.

Table 3.5: Cycle Counting Result Based on the Four-Point Counting Technique

No. of Cycles	From	To	From	To	Range	Mean
1	E	F	1	−3	4	−1

- Consideration of C, D, G, and H. $X = |H–G|$, $Y = |G–D|$, $Z = |D–C|$. $X < Y$ and $Z < Y$, therefore no cycle is counted. Designate D as a new starting point.

- Consideration of D, G, H, and I. $X = |I–H|$, $Y = |H–G|$, $Z = |G–D|$. $X < Y$ and $Z \geq Y$, therefore no cycle is counted. The residue (A, B, C, D, G, H, and I) are found.

- The summary of the four-point cycle counting result is given in Table 3.5.

Alternatively, the four-point counting technique (Dreβler et al., 1995) offers a procedure to obtain the identical cycle counting results from using the three-point counting method where the load time history has been rearranged to begin and end with a global extremum point. The procedure is given as follows:

- Extract the cycles and the residue based on the four-point cycle counting technique.

- Duplicate the residue to form a sequence of [residue + residue].

- Perform the four-point cycle counting technique on the sequence of [residue + residue].

- Add the newly extracted cycles to the original cycle count.

Example 3.5

Assume the service load time history in Figure 3.4(a) is periodic. Use the result from Example 3.4 using the four-point counting technique (where one extracted cycle E–F and the residue (A, B, C, D, G, H, and I) are identified) to produce a cycle counting result based on the three-point counting method.

Solution

- Duplication of residue to form a sequence of [residue + residue] is shown in Figure 3.10(a).

- Consideration of G, H, A, and B (Figure 3.10(a)). $X = |B–A|$, $Y = |A–H|$, $Z = |H–G|$. $X < Y$ and $Z \geq Y$, therefore no cycle is counted. Designate H as a new starting point.

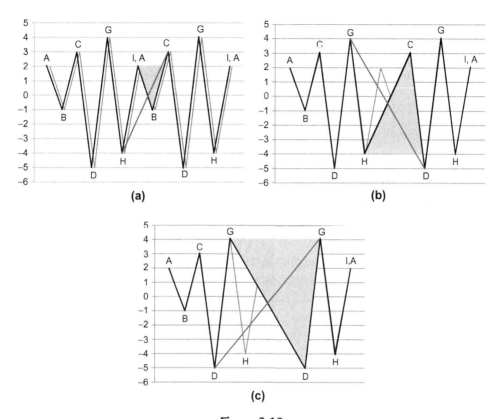

Figure 3.10
The four-point cycle counting technique applied to a sequence of [residue + residue]
from the previous counting residue in Figure 3.9: (a) residue duplication—cycle
A–B extracted; (b) cycle H–C extracted; (c) cycle G–D extracted—residue
A, B, C, D, G, H, and I.

- Consideration of H, A, B, and C (Figure 3.10(a)). X = |C–B|, Y = |B–A|, Z = |A–H|. X ≥ Y and Z ≥ Y, therefore count a cycle from A–B and connect H and C. Designate G as a new starting point.

- Consideration of G, H, C, and D (Figure 3.10(b)). X = |D–C|, Y = |C–H|, Z = |H–G|. X ≥ Y and Z ≥ Y, therefore count a cycle from H–C and connect G and D. Designate D as a new starting point.

- Consideration of D, G, D, and G (Figure 3.10(c)). X = |G–D|, Y = |D–G|, Z = |G–D|. X ≥ Y and Z ≥ Y, therefore count a cycle from G–D and connect D and G. Designate D as a new starting point.

- Consideration of D, G, H, and A (Figure 3.10(c)). X = |A–H|, Y = |H–G|, Z = |G–D|. X < Y and Z ≥ Y, therefore no cycle is counted.

- Consideration of the remaining points will not identify any more cycles. This completes the closure of the residue.

- Add the newly counted cycles (A–B, H–C, G–D) to the original cycle count (E–F).

The summary of the equivalent three-point cycle counting result obtained by the four-point technique with "closure" of the residue is given in Table 3.3. Note that Table 3.6 yields an identical range-mean output to Table 3.3, except that the largest cycle count G–D has an opposite from-to count (D–G) in Table 3.3.

Multiaxial Rainflow Reversal Counting Techniques

There are two commonly accepted approaches to assess fatigue damage of a structure subjected to variable amplitude multiaxial loading. One is the critical plane approach where the fatigue damage parameter of each potential failure plane has been identified and the total accumulated damage is calculated based on the cycle counting results by using the uniaxial cycle counting technique as described earlier on the signed damage parameter time history.

The other is the equivalent stress or strain approach where the fatigue damage parameter is defined as an equivalent stress or strain value. For proportional loading, the uniaxial cycle counting technique can be used to extract cycles from a signed equivalent stress or strain parameter time history.

For nonproportional loading, a multiaxial rainflow reversal counting method, an extension of the Matsuishi–Endo rainflow cycle counting technique along with the equivalent strain amplitude concept, was first proposed by Wang and Brown (1996) to extract the reversals from a complicated variable amplitude multiaxial load time history.

Table 3.6: Equivalent Three-Point Cycle Counting Result Based on the Four-Point Counting Technique

No. of Cycles	From	To	From	To	Range	Mean
1	E	F	1	−3	4	−1
1	A	B	2	−1	3	0.5
1	H	C	−4	3	7	−0.5
1	G	D	4	−5	9	−0.5

Two variants of the Wang–Brown cycle counting technique were later developed, namely, the Lee–Tjhung–Jordan equivalent stress cycle counting (Lee et al., 2007) and the path dependent maximum range (PDMR) cycle counting method (Dong et al., 2010). It is the objective of this section to introduce fundamentals of the multiaxial counting technique.

Following the rainflow reversal counting rule, the multiaxial rainflow reversal counting technique defines a reversal or half-cycle based on the maximum relative equivalent stress or strain range. All the points in the stress or strain space with a monotonic increasing relative stress range with respect to a turning point are considered part of the reversal identified. The counting rule can be illustrated by a plane stress condition; the normal stress (σ_x) and the shear stress (τ_{xy}) time histories are shown in Figure 3.11.

If the equivalent stress is defined by the von Mises stress, the normal stress (σ_x) and the shear stress $\left(\sqrt{3}\tau_{xy}\right)$ can be cross plotted as in Figure 3.12(a), and the maximum equivalent stress range with respect to the first turning point A and the maximum equivalent stress is shown in Figure 3.12(b). The first multiaxial reversal A–B–B*–D can be identified either by the monotonically increasing

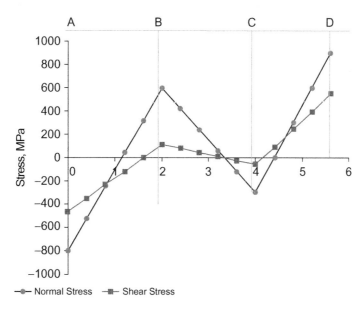

Figure 3.11

Normal stress versus time and shear stress versus time plots.

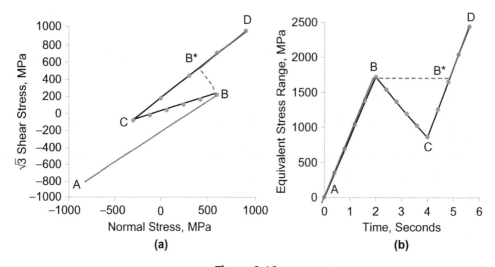

Figure 3.12
Definition of a multiaxial reversal A–B–B*–D in a $\sqrt{3}\tau_{xy} - \sigma_x$ space in (a)
and an equivalent stress range plot in (b).

distance in the $\sqrt{3}\tau_{xy} - \sigma_x$ space as in Figure 3.12(a) or by the increasing equivalent stress range in Figure 3.12(b). The intermediate point B* is defined as the point where the relative distance or equivalent stress range is identical to Point B.

The concept of using the maximum von Mises stress or strain range as the counting criterion has been proposed (Wang & Brown, 1996; Lee et al., 2007), and the concept of using the maximum distance with a crack growth related effective stress range was developed (Dong et al., 2010). Figures 3.12(a) and (b) are equivalent, with a different way to define the reversal.

The following describes the multiaxial reversal counting algorithm for a plane stress condition:

1. From the time histories of σ_x (t), σ_y (t), and $\tau_{xy}(t)$, calculate the equivalent stress time history of $\sigma_{eq}(t)$. Based on the von Mises criterion, the equivalent stress can be defined as

$$\sigma_{eq}(t) = \sqrt{\sigma_x^2(t) + \sigma_y^2(t) - \sigma_x(t)\sigma_y(t) + 3\tau_{xy}^2(t)} \qquad (3.1)$$

2. Reorder the data to begin with the point, t_o, that has the maximum equivalent stress.

3. Calculate the relative equivalent stress time history of $\Delta\sigma_{eq}(t)$.

$$\Delta\sigma_{eq}(t) = \sqrt{\Delta\sigma_x^2(t) + \Delta\sigma_y^2(t) - \Delta\sigma_x(t)\Delta\sigma_y(t) + 3\Delta\tau_{xy}^2(t)} \qquad (3.2)$$

where

$$\Delta\sigma_x = \sigma_x(t) - \sigma_x(t_o) \qquad (3.3)$$

$$\Delta\sigma_y = \sigma_y(t) - \sigma_y(t_o) \qquad (3.4)$$

$$\Delta\tau_{xy} = \tau_{xy}(t) - \tau_{xy}(t_o) \qquad (3.5)$$

4. Collect all points that cause $\Delta\sigma_{eq}(t)$ to increase. This block of points constitutes the first major reversal.

5. Store the remaining points consisting of one or more blocks that start and end with the same value of $\Delta\sigma_{eq}(t)$ and a trailing block.

6. For each of these "uncounted" blocks, treat the first point as the reference point with which to calculate the relative equivalent stress $\Delta\sigma_{eq}(t)$. Proceed to collect points that cause the new $\Delta\sigma_{eq}(t)$ to increase. This process will yield additional reversals (and possibly more uncounted blocks).

7. Repeat step 6 until all the data are counted.

Example 3.6

Assume the two service stress time histories (σ_x and τ_{xy}) in Figure 3.13(a) are periodic and the equivalent stress is defined by the von Mises stress criterion. The points from A through I are the peaks and valleys identified when at least one of the slopes is changing. Reverse count the two service stress time histories.

Solution

- Mapping the $\sigma_x(t)$ and $\tau_{xy}(t)$ time histories onto the $\sigma_x - \sqrt{3}\tau_{xy}$ coordinate from Point A to Point I, shown in Figure 3.13(b).

- Extraction of a reversal A–B–B*–H based on the maximum distance or equivalent stress range with respect to Point A and identification of the two

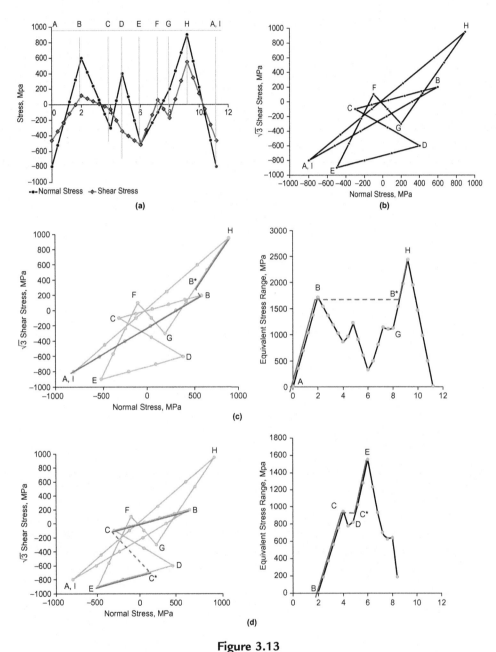

Figure 3.13

Illustration of multiaxial reversal counting method: (a) normal and shear stress time histories, (b) $\sqrt{3}$ shear and normal stress plot, (c) extracted reversal A–B–B*–H, (d) extracted reversal B–C–C*–E, (e) extracted reversal C–D–C*, (f) extracted reversal E–F–FI–B*, (g) extracted reversal F–G–F*, and (h) extracted reversal H–A.

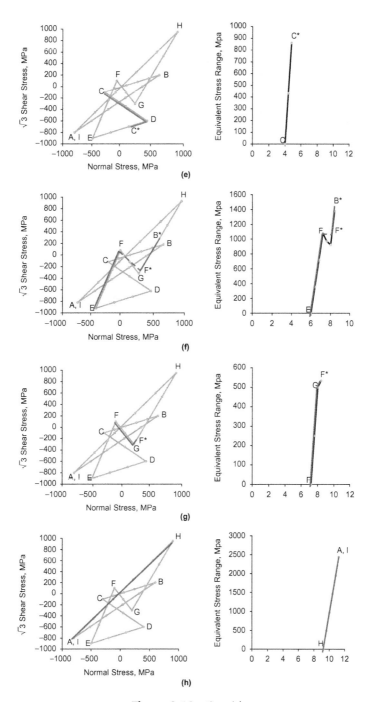

Figure 3.13 Cont'd

Table 3.7: Multiaxial Reversal Counting Results

No. of Cycles	Data Points
0.5	A–B–B*–H
0.5	B–C–C*–E
0.5	C–D–C*
0.5	E–F–F*–B*
0.5	F–G–F*
0.5	H–A

groups of uncounted points from B to B* and from H to I, as shown in Figure 3.13(c). Point B* is the intermediate point where the relative distance or equivalent stress range starts to increase.

• Extraction of a reversal B–C–C*–E from the first group of uncounted points from B to B*, based on the maximum distance or equivalent stress range with respect to Point B and identification of the two groups of uncounted points from C–C* and E–B*, as shown in Figure 3.13(d). Point C* is the intermediate point where the relative distance or equivalent stress range starts to increase.

• Extraction of a reversal C–C* from the group of uncounted points from C to C*, based on the maximum distance or equivalent stress range with respect to Point C, as shown in Figure 3.13(e).

• Extraction of a reversal E–F–F*–B* from the group of uncounted points from E–B*, based on the maximum distance or equivalent stress range with respect to Point E and identification of one group of uncounted points from F to F*, as shown in Figure 3.13(f).

• Extraction of a reversal F–G–F* from the group of uncounted points from F to F*, based on the maximum distance or equivalent stress range with respect to Point F, as shown in Figure 3.13(g).

• Extraction of a reversal H–A(or I) from the group of uncounted points from H to A (or I), based on the maximum distance or equivalent stress range with respect to Point H, as shown in Figure 3.13(h). This finally completes the counting process for all the data points in this service stress time history.

• The summary of the multiaxial reversal counting result is given in Table 3.7.

Summary

The techniques available for the rate independent process of extracting cycles or reversals from a complicated loading history have been introduced. The original rainflow cycle counting technique, the three-point cycle counting method, and the four-point cycle counting algorithm can be used to extract cycles from complicated uniaxial loading.

For complicated multiaxial loading, the choice of a cycle counting technique depends on the multiaxial fatigue damage assessment method. If a critical plane approach is adopted, any of the three uniaxial cycle counting techniques can be used to calculate the fatigue damages for each potential failure plane. If an equivalent stress or strain approach is chosen, a multiaxial cycle counting technique would be used, developed based on the assumption that the fatigue damage or life can be evaluated from the cycles identified from the complicated equivalent loading history.

The Wang–Brown method, the Lee–Tjhung–Jordan approach, and the path-dependent maximum range technique are the commonly used multiaxial reversal counting methods. They have been reviewed in detail in this chapter.

References

Amzallag, C., Gerey, J. P., Robert, J. L., & Bahuaud, J. (1994). Standardization of the rainflow counting method for fatigue analysis. *International Journal of Fatigue, 16*, 287–293.

ASTM Standard E 1049-85. (2005). *Standard practices for cycle counting in fatigue analysis*. Philadelphia: ASTM.

Conle, A. L., Grenier, G., Johnson, H., Kemp, S., Kopp, G., & Morton, M. (1997). Service history determination. In *SAE Fatigue Design Handbook*, AE-22, 3rd edition, 115–158.

Dreßler, K., Kottgen, V., & Speckert, M. (1995). Counting methods in fatigue analysis. LMS-TECMATH technical report. Kaiserlautern: Tecmath GmbH.

Dong, P., Wei, Z., & Hong, J. K. (2010). A path-dependent cycle counting method for variable-amplitude multi-axial loading. *International Journal of Fatigue, 32*, 720–734.

Downing, S. D., & Socie, D. F. (1982). Simplified rainflow counting algorithms. *International Journal of Fatigue, 4*(1), 31–40.

Lee, Y. L., & Taylor, D. (2005). Cycle counting techniques. In Lee, et al, *Fatigue testing and analysis: Theory and practice* (pp. 77–102). Boston: Elsevier/Butterworth-Heinemann.

Lee, Y. L., Tjhung, T., & Jordan, A. (2007). A life prediction model for welded joints under multiaxial variable amplitude loading. *International Journal of Fatigue, 29*, 1162–1173.

Matsuishi, M., & Endo, T. (1968). *Fatigue of metals subjected to varying stress*, presented to the Japan Society of Mechanical Engineers, Fukuoka, Japan. See also Endo, T., Mitsunaga, K., Takahashi, K., Kobayashi, K., & Matsuishi, M (1974). Damage evaluation of metals for random or varying loading, *Proceedings of the 1974 Symposium on Mechanical Behavior of Materials,* Society of Materials Science, *1,* 371–380.

Miner, M. A. (1945). Cumulative damage in fatigue. *Journal of Applied Mechanics, 67,* A159–A164.

Palmgren, A. (1924). Die lebensdauer von kugellagern. *Zeitschrift des Vereinesdeutscher Ingenierure, 68*(14), 339–341.

Richards, F. D., LaPointe, N. R., & Wetzel, R. M. (1974). *A cycle counting algorithm for fatigue damage analysis.* SAE Technical Paper No. 740278. Warrendale, PA: SAE International.

Wang, C. H., & Brown, M. W. (1996). Life prediction techniques for variable amplitude multiaxial fatigue—Part 1: Theories. *Journal of Engineering Materials and Technology, 118,* 367–1370.

CHAPTER 4

Stress-Based Uniaxial Fatigue Analysis

Yung-Li Lee
Chrysler Group LLC

Mark E. Barkey
The University of Alabama

Chapter Outline

Introduction

The uniaxial fatigue analysis is used to estimate the fatigue life of a component under cycling loading when the crack is initiated due to a uniaxial state of stress. The fatigue life of a component refers to the fatigue initiation life defined as the number of cycles (N) or reversals (2N) to a specific crack initiation length of the component under cyclic stress controlled tests. Note that one cycle consists of two reversals.

The stress in a cycle can be described either by stress amplitude (S_a) and mean stress (S_m) or by maximum stress (S_{max}) and minimum stress (S_{min}), as shown in Figure 4.1. Since S_a is the primary factor affecting N, it is often chosen as the controlled or independent parameter in fatigue testing, and consequently, N is the dependent variable on S_a.

The choice of the dependent and independent variables places an important role in performing a linear regression analysis to define the stress and life relation. As the stress amplitude becomes larger, the fatigue life is expected to be shorter. The stress and life relation (namely, the constant amplitude S-N curve) can be generated by fatigue testing material specimens or real components at various load/stress levels. For this type of fatigue testing, the mean stress is usually held as a constant, and commonly is equal to zero.

The constant amplitude S-N curve is often plotted by a straight line on log-log coordinates, representing fatigue data in the high cycle fatigue (HCF) regime where fatigue damage is due to little plastic deformation. In German, the constant

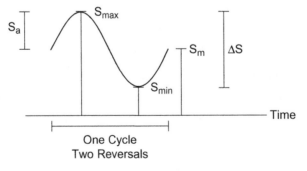

Figure 4.1
Symbols used with cyclic stresses and cycles.

amplitude S-N curve is often named the Wöhler curve, to honor Mr. Wöhler for his contribution to the first fatigue study in the world.

The term of an S-N curve is used as the abbreviation of a constant amplitude S-N curve. Depending on the test objects, there can be the material S-N curve or the component S-N curve. Also, depending on the definition of stress, the real component S-N curve can be categorized as the nominal S-N curve or the pseudo σ^e-N curve.

Generally, an S-N curve can be constructed as a piecewise-continuous curve consisting of two distinct linear regimes when plotted on log-log coordinates. For the typical S-N curve of a component made of steels, as schematically illustrated in Figure 4.2, there is one inclined linear segment for the HCF regime and one horizontal asymptote for the fatigue limit.

The parameters used to define the inclined linear segment of an S-N curve are the fatigue properties. The slope of an S-N curve in the HCF regime can be denoted as b (the height-to-base ratio) or as k (the negative base-to-height ratio).

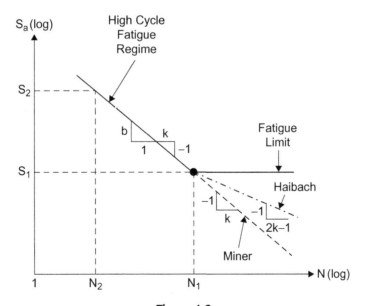

Figure 4.2
Schematic constant amplitude S-N curve of a component made of steels.

The parameter k is the slope factor. The two slopes are related in the following expression:

$$k = -\frac{1}{b}.$$ (4.1)

Any two S-N data points (S_1, N_1) and (S_2, N_2) in the HCF regime can be related by the slope b or the slope factor k in the following equation:

$$\frac{N_2}{N_1} = \left(\frac{S_1}{S_2}\right)^k = \left(\frac{S_1}{S_2}\right)^{-1/b}.$$ (4.2)

Equation (4.2) also means any data point (S_2, N_2) can be obtained by a reference point (S_1, N_1) and a given b or k. The S-N curve is commonly expressed as

$$NS_a^k = A$$ (4.3)

or

$$S_a = S_f'(2N)^b$$ (4.4)

where

   A = the fatigue parameter
   S_f' = the fatigue strength coefficient defined as the fatigue strength at one reversal

The fatigue limit of a component made of steels and cast irons can be defined as the fully reversed stress amplitude at which the fatigue initiation life becomes infinite or when fatigue initiation failure does not occur. The fatigue limit can be interpreted from the physical perspective of the fatigue damage phenomenon under constant amplitude loading. Due to cyclic operating stresses, a microcrack will nucleate within a grain of material and grow to the size of about the order of a grain width until the grain boundary barrier impedes its growth.

If the grain barrier is not strong enough, the microcrack will eventually propagate to a macrocrack and may lead to final failure. However, if the grain barrier is very strong, the microcrack will be arrested and become a nonpropagating crack. The minimum stress amplitude to overcome the crack growth barrier for further crack propagation is referred to as the fatigue limit.

The fatigue limit might be negatively influenced by other factors such as periodic overloads, elevated temperatures, or corrosion. When Miner's rule (Miner, 1945) is applied in variable amplitude loading, the stress cycles with amplitudes below the fatigue limit could become damaging if some of the subsequent stress amplitudes exceed the original fatigue limit. It is believed that the increase in crack driving force due to periodic overloads will overcome the original grain barrier strength and help the crack to propagate until failure.

Therefore, two methods such as the Miner rule and the Miner–Haibach model (Haibach, 1970), as shown in Figure 4.2, were proposed to include the effect of periodic overloads on the stress cycle behavior below the original fatigue limit. The Miner rule extends the S-N curve with the same slope factor k to approach zero stress amplitude, while the Miner–Haibach model extends the original S-N curve below the fatigue limit to the zero stress amplitude with a flatter slope factor 2k–1. Stanzl et al. (1986) concluded that a good agreement is found for measured and calculated results according to the Miner–Haibach model.

For a component made of aluminum alloys and austenitic steel, the fatigue limit does not exist and fatigue testing must be terminated at a specified large number of cycles. This nonfailure stress amplitude is often referred to as the endurance limit, which need not be the fatigue limit. However, in this chapter the endurance limit is defined as the fully reversed stress amplitude at the endurance cycle limit ($N_E = 10^6$ cycles) for all materials. The particular model for fatigue data in the HCF regime and beyond the endurance cycle limit will be described later.

The detailed procedures to generate the synthetic nominal S-N and the pseudo σ^e-N curves for fatigue designs are the focus of this chapter, and will be addressed in the following sections. The techniques used to conduct S-N testing and perform data analysis for fatigue properties are beyond the scope of our discussion, and can be found elsewhere (Lee et al., 2005; ASTM E 739-91, 2006).

Through many years of experience and testing, empirical relationships that relate the data among ultimate tensile strengths and endurance limits at 10^6 cycles have been developed. These relationships are not scientifically based but are simple and useful engineering tools for generating the synthetic component stress-life curves for various materials. It is also worth mentioning that the materials presented here have been drawn heavily from FKM-Guideline (Haibach, 2003).

Ultimate Tensile Strength of a Component

The mean ultimate tensile strength of a standard material specimen can be determined by averaging the static test results of several smooth, polished round test specimens of 7.5 mm diameter. If the test data are not available, the ultimate tensile strength of a standard test specimen can be estimated by a hardness value. There has been a strong correlation between hardness and mean ultimate tensile strength of standard material test specimens. Several models have been proposed to estimate mean ultimate tensile strength from hardness.

Lee and Song (2006) reviewed most of them and concluded that Mitchell's equation (Mitchell, 1979) provides the best results for both steels and aluminum alloys:

$$S_{t,u,std}(\text{MPa}) = 3.45\,\text{HB} \tag{4.5}$$

where $S_{t,u,std}$ is the mean ultimate tensile strength of a standard material test specimen and HB is the Brinell hardness.

It has been found that the surface treatment/roughness and the local notch geometry have little effect on the ultimate tensile strength of a notched component and that the size of the real component has some degree of influence on the strength of the component. Therefore, based on the estimated or measured mean ultimate tensile strength of a "standard" material specimen, the ultimate tensile strength of a real component ($S_{t,u}$) with a survival rate (reliability) of $R_r\%$ is estimated as follows:

where
$$S_{t,u} = C_D C_R S_{t,u,std} \tag{4.6}$$

C_R = the reliability correction factor
C_D = the size correction factor

If a component exposed to an elevated temperature condition is subjected to various loading modes, the ultimate strength values of a notched, rod-shaped component in axial ($S_{S,ax,u}$), bending ($S_{S,b,u}$), shear ($S_{S,s,u}$), and torsion ($S_{S,t,u}$) can be estimated as follows:

$$S_{S,ax,u} = C_\sigma C_{u,T} S_{t,u} \tag{4.7}$$

$$S_{S,b,u} = C_{b,L} C_\sigma C_{u,T} S_{t,u} \tag{4.8}$$

$$S_{S,s,u} = C_\tau C_{u,T} S_{t,u} \tag{4.9}$$

$$S_{S,t,u} = C_{t,L} C_\tau C_{u,T} S_{t,u}. \tag{4.10}$$

Similarly, the ultimate strength values of a notched, shell-shaped component for normal stresses in x and y directions ($S_{S,x,u}$ and $S_{S,y,u}$) and for shear stress ($S_{S,\tau,u}$) can be determined as follows:

$$S_{S,x,u} = C_\sigma C_{u,T} S_{t,u} \tag{4.11}$$

$$S_{S,y,u} = C_\sigma C_{u,T} S_{t,u} \tag{4.12}$$

$$S_{S,\tau,u} = C_\tau C_{u,T} S_{t,u} \tag{4.13}$$

where

$C_{u,T}$ = the temperature correction factor C_σ
C_τ = the stress correction factors in normal and shear stresses
$C_{b,L}$ and $C_{t,L}$ = the load correction factors in bending and torsion

Figure 4.3 shows the schematic effects of these corrections factors on the component ultimate tensile strength and its endurance limit. These correction factors will be discussed in the following sections.

Reliability Correction Factor

If test data are not available, a statistical analysis cannot be performed to account for variability of the ultimate tensile strength. In the absence of the statistical analysis, the suggested reliability values for various reliability levels

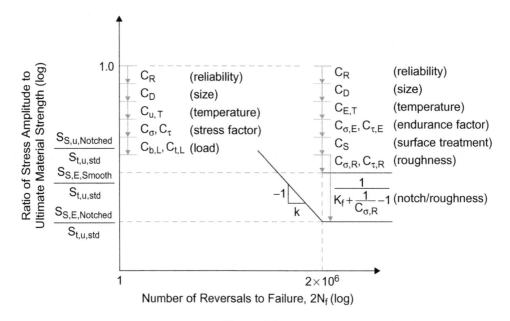

Figure 4.3
Correction factors for the component ultimate strength and the endurance limit.

Table 4.1: Reliability Correction Factors, C_R

Reliability	C_R
0.5	1.000
0.90	0.897
0.95	0.868
0.975	0.843
0.99	0.814
0.999	0.753
0.9999	0.702
0.99999	0.659

are given in Table 4.1, derived on the assumptions of a normally distributed ultimate tensile strength and the coefficient of variations (COV_S) of 8%. The derivation of these C_R values can be obtained by using the following equation:

$$C_R = 1 - |\Phi^{-1}(1 - R_r)|COV_S \qquad (4.14)$$

where

$\Phi(-)$ = the standard normal density function
$R_r\%$ = a reliability value

Note that FKM-Guideline (Haibach, 2003) specifies that the ultimate strength of a component for a design should be based on the probability of a 97.5% survival rate, meaning a corresponding C_R value of 0.843.

Size Correction Factor

The size correction factor (C_D) is used to account for the fact that the strength of a component reduces as the size increases with respect to that of the standard material test specimen (a diameter of 7.5 mm) due to the possibility of a weak link increasing with a larger material volume. Based on FKM-Guideline (Haibach, 2003), the size correction factor, dependent on the cross-sectional size and the type of material, can be obtained as follows:

- For wrought aluminum alloys:

$$C_D = 1.0 \tag{4.15}$$

- For cast aluminum alloys:

$$C_D = 1.0 \quad \text{for} \quad d_{eff} \le 12\,\text{mm} \tag{4.16}$$

$$C_D = 1.1(d_{eff}/7.5\,\text{mm})^{0.2} \quad \text{for} \quad 12\,\text{mm} < d_{eff} \le 150\,\text{mm} \tag{4.17}$$

$$C_D = 0.6 \quad \text{for} \quad d_{eff} \ge 150\,\text{mm} \tag{4.18}$$

- For grey cast irons:

$$C_D = 1.207 \quad \text{for} \quad d_{eff} \le 7.5\,\text{mm} \tag{4.19}$$

$$C_D = 1.207(d_{eff}/7.5\,\text{mm})^{-0.1922} \quad \text{for} \quad d_{eff} > 7.5\,\text{mm} \tag{4.20}$$

- For all steels, steel castings, ductile irons, and malleable cast iron:

$$C_D = 1.0 \quad \text{for} \quad d_{eff} \le d_{eff,min} \tag{4.21}$$

Table 4.2: Constants Used to Estimate the Size Correction Factors

Material Type	$d_{eff,min}$ (mm)	a_d	Case of d_{eff}
Plain carbon steel	40	0.15	Case 2
Fine-grained steel	70	0.2	Case 2
Steel, quenched and tempered	16	0.3	Case 2
Steel, normalized	16	0.1	Case 2
Steel, case hardened	16	0.5	Case 1
Nitriding steel, quenched and tempered	40	0.25	Case 1
Forging steel, quenched and tempered	250	0.2	Case 1
Forging steel, normalized	250	0	Case 1
Steel casting	100	0.15	Case 2
Steel casting, quenched and tempered	200	0.15	Case 1
Ductile irons	60	0.15	Case 1
Malleable cast iron	15	0.15	Case 1

Source: Adapted from FKM-Guideline, published by Forschungskuratorium Maschinenebau, 2003.

$$C_D = \frac{1 - 0.7686 \cdot a_d \cdot \log(d_{eff}/7.5\,\text{mm})}{1 - 0.7686 \cdot a_d \cdot \log(d_{eff,min}/7.5\,\text{mm})} \quad \text{for} \quad d_{eff} > d_{eff,min} \quad (4.22)$$

where

d_{eff} = the effective diameter of a cross section

$d_{eff,min}$ and a_d = the constants tabulated in Table 4.2

Depending on the type of material as listed in Table 4.2, two cases are required to be distinguished to determine d_{eff}. In Case 1, d_{eff} is defined by

$$d_{eff} = \frac{4V}{O} \quad (4.23)$$

where V and O are the volume and surface area of the section of the component of interest. In Case 2, d_{eff} is equal to the diameter or wall thickness of the component, and applies to all components made of aluminum alloys. Examples of d_{eff} calculation are illustrated in Table 4.3.

Temperature Correction Factor for Ultimate and Yield Strengths

The temperature factor ($C_{u,T}$) is used to take into account the ultimate and the yield strength reductions in the field of elevated temperatures. FKM-Guideline

Table 4.3: Calculation of the Effective Diameter d_{eff}

No.	Cross Section	d_{eff} Case 1	d_{eff} Case 2
1		d	d
2		2s	s
3		2s	s
4		$\dfrac{2bs}{b+s}$	s
5		b	b

Source: Adapted from FKM-Guideline, published by Forschungskuratorium Maschinenebau, 2003.

(Haibach, 2003) specifies the following temperature effects for various materials:

- For age-hardening (or heat treatable) aluminum alloys where $T > 50°C$:

$$C_{u,T} = 1 - 4.5 \cdot 10^{-3}(T - 50) \geq 0.1 \tag{4.24}$$

- For nonage-hardening aluminum alloys where $T > 100°C$:

$$C_{u,T} = 1 - 4.5 \cdot 10^{-3}(T - 100) \geq 0.1 \tag{4.25}$$

- For fine-grained steel where $T > 60°C$:

$$C_{u,T} = 1 - 1.2 \cdot 10^{-3}T \tag{4.26}$$

- For all steels except fine-grained steel where $T > 100°C$:

$$C_{u,T} = 1 - 1.7 \cdot 10^{-3}(T - 100) \qquad (4.27)$$

- For steel castings where $T > 100°C$:

$$C_{u,T} = 1 - 1.5 \cdot 10^{-3}(T - 100) \qquad (4.28)$$

- For ductile irons where $T > 100°C$:

$$C_{u,T} = 1 - 2.4 \cdot 10^{-3}T \qquad (4.29)$$

Note that the temperature must be given in degrees Celsius.

Stress Correction Factor

The stress correction factor is used to correlate the different material strengths in compression or shear with respect to that in tension, and can be found in Table 4.4. Note that $C_\sigma = 1.0$ for tension.

Load Correction Factor

The stress gradient of a component in bending or torsion can be taken into account by the load correction factor, also called the "section factor" or the

Table 4.4: Stress Correction Factors C_σ and C_τ in Compression and in Shear

Materials	C_σ	C_τ
Case hardening steel	1	$1/\sqrt{3} = 0.577$*
Stainless steel	1	0.577
Forging steel	1	0.577
Steel other than above types	1	0.577
Steel castings	1	0.577
Ductile irons	1.3	0.65
Malleable cast iron	1.5	0.75
Grey cast iron	2.5	0.85
Aluminum alloys	1	0.577
Cast aluminum alloys	1.5	0.75

*Note that $1/\sqrt{3} = 0.577$ is based on the von Mises yield criterion.
Source: Adapted from FKM-Guideline, published by Forschungskuratorium Maschinenebau, 2003.

Table 4.5: Load Correction Factors $C_{b,L}$ and $C_{t,L}$

Cross Section	$C_{b,L}$	$C_{t,L}$
Rectangle	1.5	—
Circle	1.7	1.33
Tubular	1.27	1

Source: Adapted from FKM-Guideline, published by Forschungskuratorium Maschinenebau, 2003.

"plastic notch factor" in FKM. The load correction factor is defined as the ratio of the nominal stress at global yielding to the nominal stress at the initial notch yielding.

Alternatively, the load correction factors in Table 4.5 are derived from the ratio of fully plastic yielding force, moment, or torque to the initial yielding force, moment, or torque. For example, a component has the tensile yield strength $(S_{t,y})$ and a rectangular section with a width of b_W and a height of $2h_T$. Its initial yielding moment is calculated as $M_i = 2/3b_W h_T^2 S_{t,y}$, and the fully plastic yielding moment is $M_o = b_W h_T^2 S_{t,y}$. Thus, the corresponding load modifying factor for bending is found to be $M_o/M_i = 1.5$.

The load correction factor also depends on the type of materials according to FKM-Guideline (Haibach, 2003). For surface hardened components, the load factors are not applicable and $C_{b,L} = C_{t,L} = 1.0$. For high ductility of austenitic steels in a solution annealed condition, $C_{b,L}$ and $C_{t,L}$ follow the values in Table 4.5.

Also, for other steels, steel castings, ductile irons, and aluminum alloys,

$$C_{b,L} = \text{minimum of} \left(\sqrt{S_{t,y,max}/S_{t,y}}; C_{b,L} \right) \tag{4.30}$$

$$C_{t,L} = \text{minimum of} \left(\sqrt{S_{t,y,max}/S_{t,y}}; C_{t,L} \right) \tag{4.31}$$

where

$S_{t,y}$ = the tensile yield strength with $R_r 97.5$ in MPa
$S_{t,y,max}$ = the maximum tensile yield strength in MPa

This is shown in Table 4.6.

Table 4.6: Maximum Tensile Yield Strength $S_{t,y,max}$ for Various Materials

Type of Material	Steels, Steel Castings	Ductile Irons	Aluminum Alloys
$S_{t,y,max}$ (MPa)	1050	320	250

Source: Adapted from FKM-Guideline, published by Forschungskuratorium Maschinenebau, 2003.

Component Endurance Limit under Fully Reversed Loading

The endurance limit is defined as the stress amplitude for a fully reversed loading at an endurance cycle limit ($N_E = 10^6$ cycles). Since R is defined as the ratio of minimum stress to maximum stress, a fully reversed loading is also called an R = −1 loading. Even though the endurance limit is occasionally expressed in terms of range in some references, it is worth noting that the endurance limit in this chapter is clearly defined in amplitude.

With the probability of an $R_r\%$ survival rate, the endurance limit for a smooth, polished component at an elevated temperature condition and under fully reversed tension or shear stress can be estimated from $S_{t,u}$, which has already taken into account the factors for size and reliability:

$$S_{S,\sigma,E} = C_{\sigma,E}C_{E,T}S_{t,u} \qquad (4.32)$$

$$S_{S,\tau,E} = C_\tau S_{S,\sigma,E} \qquad (4.33)$$

where

  $C_{E,T}$ = the temperature correction factor for the endurance limit
  $C_{\sigma,E}$ = the endurance limit factor for normal stress
  C_τ = the shear stress correction factor

It has been found that the endurance limit of a notched component is affected by the residual stress/surface hardened layer due to surface treatment and by the high stress concentration/stress gradient due to surface roughness and the local geometrical change. These effects have been empirically quantified by FKM-Guideline (Haibach, 2003).

For example, the endurance limit values of a notched, rod-shaped component under fully reversed loading in axial ($S_{S,ax,E}$), bending ($S_{S,b,E}$), shear ($S_{S,s,E}$), and torsion ($S_{S,t,E}$) can be obtained as follows:

$$S_{S,ax,E} = \frac{C_S S_{S,\sigma,E}}{K_{ax,f} + \dfrac{1}{C_{\sigma,R}} - 1} \tag{4.34}$$

$$S_{S,b,E} = \frac{C_S S_{S,\sigma,E}}{K_{b,f} + \dfrac{1}{C_{\sigma,R}} - 1} \tag{4.35}$$

$$S_{S,s,E} = \frac{C_S S_{S,\tau,E}}{K_{s,f} + \dfrac{1}{C_{\tau,R}} - 1} \tag{4.36}$$

$$S_{S,t,E} = \frac{C_S S_{S,\tau,E}}{K_{t,f} + \dfrac{1}{C_{\tau,R}} - 1}. \tag{4.37}$$

Similarly, the endurance limit values of a notched, shell-shaped component under fully reversed normal stresses in x and y directions and under shear stress can be obtained as follows:

$$S_{S,x,E} = \frac{C_S S_{S,\sigma,E}}{K_{x,f} + \dfrac{1}{C_{\sigma,R}} - 1} \tag{4.38}$$

$$S_{S,y,E} = \frac{C_S S_{S,\sigma,E}}{K_{y,f} + \dfrac{1}{C_{\sigma,R}} - 1} \tag{4.39}$$

$$S_{S,\tau,E} = \frac{C_S S_{S,\tau,E}}{K_{s,f} + \dfrac{1}{C_{\tau,R}} - 1} \tag{4.40}$$

where

$$C_S = \text{the surface treatment factor}$$

$$C_{\sigma,R} \text{ and } C_{\tau,R} = \text{the roughness correction factors for normal and shear stresses}$$

$$K_{ax,f}, \ K_{b,f}, \ K_{s,f}, \ K_{t,f}, \ K_{x,f}, K_{y,f}, \text{ and } K_{\tau,f} = \text{the fatigue notch factors for various loading modes}$$

Note that the endurance limit of a smooth component can be calculated by using the preceding equations with $K_f = 1$.

Refer to Figure 4.3, which shows the schematic effects of these corrections factors on the endurance limits of smooth and notched components. The correction factors for temperature, endurance limit for tension, surface treatment, roughness, and the fatigue notch factor are discussed in the following sections.

Temperature Correction Factor

It has been observed that at an elevated temperature, the component fatigue strength is reduced with increasing temperature. The temperature reduction factor for the endurance limit is different from the factor applied to the ultimate tensile strength ($C_{u,T}$).

Depending on the type of materials, FKM-Guideline (Haibach, 2003) specifies these temperature correction factors as follows:

- For aluminum alloys where $T > 50°C$:

$$C_{E,T} = 1 - 1.2 \cdot 10^{-3}(T - 50)^2 \tag{4.41}$$

- For fine-grained steel where $T > 60°C$:

$$C_{E,T} = 1 - 10^{-3}T \tag{4.42}$$

- For all steels except fine-grained steel where $T > 100°C$:

$$C_{E,T} = 1 - 1.4 \cdot 10^{-3}(T - 100) \tag{4.43}$$

- For steel castings where $T > 100°C$:

$$C_{E,T} = 1 - 1.2 \cdot 10^{-3}(T - 100) \tag{4.44}$$

- For ductile irons where $T > 100°C$:

$$C_{E,T} = 1 - 1.6 \cdot (10^{-3} \cdot T)^2 \tag{4.45}$$

- For malleable cast iron where $T > 100°C$:

$$C_{E,T} = 1 - 1.3 \cdot (10^{-3} \cdot T)^2 \tag{4.46}$$

- For grey cast iron where $T > 100°C$:

$$C_{E,T} = 1 - 1.0 \cdot (10^{-3} \cdot T)^2 \tag{4.47}$$

Note that the temperature must be given in degrees Celsius.

Endurance Limit Factor

The endurance limit factor ($C_{\sigma,E}$) for normal stress (found in Table 4.7) is an empirical factor to estimate the endurance limit based on the ultimate tensile strength of a component with the chance of a survival rate of $R_r\%$.

Table 4.7: Endurance Limit Factors for Various Materials

Material Type	$C_{\sigma,E}$
Case-hardening steel	0.40
Stainless steel	0.40
Forging steel	0.40
Steel other than above types	0.45
Steel casting	0.34
Ductile iron	0.34
Malleable cast iron	0.30
Grey cast iron	0.30
Wrought aluminum alloys	0.30
Cast aluminum alloys	0.30

Source: Adapted from FKM-Guideline, published by Forschungskuratorium Maschinenebau, 2003.

Surface Treatment Factor

The surface treatment factor, C_S, which takes into account the effect of a treated surface layer on the fatigue strength of a component, is defined as the ratio of the endurance limit of a surface layer to that of the core material. C_S depends on whether the crack origin is expected to be located at the surface or in the core.

According to FKM-Guideline (Haibach, 2003), the upper and lower limits of the surface treatment factors for steel and cast iron materials are tabulated in Table 4.8. The values in the table are applicable to components of 30 to 40 mm diameter, while the values in the parenthesis are for 8 to 15 mm diameter. C_S can also account for the effect of a surface coating such as electrolytically formed anodic coatings on the endurance limit of a component made of aluminum alloys, and is specified as follows:

$$C_S = 1 - 0.271 \cdot \log(t_c) \tag{4.48}$$

where

t_c = the coating layer thickness in μm

Roughness Correction Factor

Surface roughness or irregularity acts as a stress concentration and results in crack initiation on the surface as well as fatigue strength reduction. The roughness correction factors $C_{\sigma,R}$ and $C_{\tau,R}$ account for the effect of surface roughness on the component endurance limit in tension and shear.

According to FKM-Guideline (Haibach, 2003), the two roughness correction factors under normal and shear stresses are defined as follows:

$$C_{\sigma,R} = 1 - a_R \log(R_Z) \log(2S_{t,u}/S_{t,u,min}) \tag{4.49}$$

and

$$C_{\tau,R} = 1 - C_\tau a_R \log(R_Z) \log(2S_{t,u}/S_{t,u,min}) \tag{4.50}$$

where

a_R = a roughness constant listed in Table 4.9
R_Z = the average roughness value of the surface in μm
$S_{t,u,min}$ = the minimum ultimate tensile strength in MPa in Table 4.9

An average roughness value ($R_Z = 200\,\mu m$) applies for a rolling skin, a forging skin, and the skin of cast irons. For steels, the roughness value of a ground

Table 4.8: Surface Treatment Factors for Various Materials

Surface Treatment	Unnotched Components	Notched Components
Steel		
Chemo-Thermal Treatment		
Nitriding Depth of case 0.1–0.4 mm Surface hardness 700–1000 HV10	1.10–1.15 (1.15–1.25)	1.30–2.00 (1.90–3.00)
Case hardening Depth of case 0.2–0.8 mm Surface hardness 670–750 HV10	1.10–1.50 (1.20–2.00)	1.20–2.00 (1.50–2.50)
Carbo-nitriding Depth of case 0.2–0.8 mm Surface hardness 670–750 HV10	(1.80)	
Mechanical Treatment		
Cold rolling	1.10–1.25 (1.20–1.40)	1.30–1.80 (1.50–2.20)
Shot peening	1.10–1.20 (1.10–1.30)	1.10–1.50 (1.40–2.50)
Thermal Treatment		
Inductive hardening Flame-hardening Depth of case 0.9–1.5 mm Surface hardness 51–64 HRC	1.20–1.50 (1.30–1.60)	1.50–2.50 (1.60–2.8)
Cast Iron Materials		
Nitriding	1.10 (1.15)	1.3 (1.9)
Case hardening	1.1 (1.2)	1.2 (1.5)
Cold rolling	1.1 (1.2)	1.3 (1.5)
Shot peening	1.1 (1.1)	1.1 (1.4)
Inductive hardening, flame-hardening	1.2 (1.3)	1.5 (1.6)

Source: Adapted from FKM-Guideline, published by Forschungskuratorium Maschinenebau, 2003.

Table 4.9: a_R and $S_{t,u,min}$ for Various Materials

Materials	a_R	$S_{t,u,min}$ MPa
Steel	0.22	400
Steel castings	0.20	400
Ductile iron	0.16	400
Malleable cast iron	0.12	350
Grey cast iron	0.06	100
Wrought aluminum alloys	0.22	133
Cast aluminum alloys	0.20	133

Source: Adapted from FKM-Guideline, published by Forschungskuratorium Maschinenebau, 2003.

surface varies from 1 μm to 12 μm, and the value of a finished surface ranges from 6.3 μm to 100 μm.

Fatigue Notch Factor

It was once believed that at the same crack initiation life near the endurance cycle limit of 10^6 cycles, the pseudo surface stress (σ_E^e) at the stress concentration location of a notched component would be identical to the surface stress of a smooth component ($S_{S,E,Smooth}$). Since this belief provides $\sigma_E^e = S_{S,E,Smooth}$ and $\sigma_E^e = K_t \cdot S_{S,E,Notched}$ where K_t and $S_{S,E,Notched}$ are the elastic stress concentration factor and the nominal stress of a notched component, respectively, we can conclude that $S_{S,E,Notched}$ is smaller than $S_{S,E,Smooth}$ by a factor of K_t.

Tryon and Dey (2003), however, presented a study revealing the effect of fatigue strength reduction for Ti-6Al-4V in the HCF regime shown in Figure 4.4. The test has indicated at the same endurance cycle, the presence of a notch on a component under cyclic stressing reduces the nominal stress of a smooth component by a factor K_f instead of K_t. The K_f is termed the fatigue notch factor or fatigue strength reduction factor defined as follows:

$$K_f = \frac{S_{S,E,Smooth}}{S_{S,E,Notched}} \leq K_t.$$ (4.51)

Equation (4.51) can be interpreted as that when $K_f S_{S,E,Notched} = S_{S,E,Smooth}$, both the notched and smooth components would have the same endurance cycle limit, as shown in Figure 4.5.

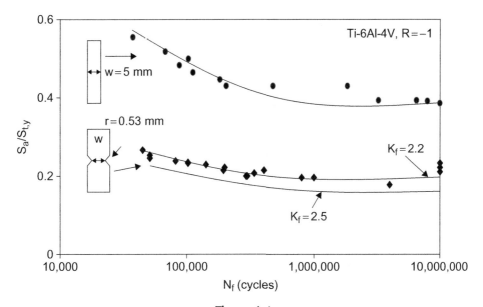

Figure 4.4
Effect of a notch on S-N behavior for Ti-6Al-4V in the HCF regime.

The smaller K_f than K_t can be explained either by the local cyclic yielding behavior or by the stress field intensity theory (Yao, 1993; Qylafku et al., 1999; Adib & Pluvinage, 2003). The local yielding theory suggests the cyclic material yielding at a notch root reduces the peak pseudo surface stress, while the stress field intensity theory postulates that the fatigue strength of a notched component depends on the average stress in a local damage zone, instead of the peak pseudo surface stress at a notch root.

The stress field intensity theory is valid in the endurance cycle limit regime where the peak pseudo surface stress is approximately equal to the true surface stress. According to the stress field intensity theory, the average stress is responsible for the crack initiation life, and associated with the stress distribution and the local damage volume at the notch. The average stress is defined as $K_f S_C$ as opposed to the peak pseudo surface stress, $K_t S_C$, where S_C is the nominal stress of a notched component.

Figure 4.6 schematically shows two notched components with the same peak pseudo surface stress and steel material. Note that the subscripts numbers 1 and 2

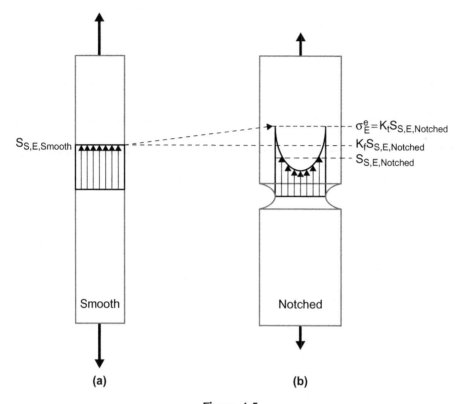

Figure 4.5
Identical crack initiation life for smooth and notched components.

denote the notched components 1 and 2. For illustration, the damage zone of steel material is assumed to be of the order of two grain sizes.

As the notch radius decreases, the stress gradient becomes steeper, resulting in a lower average stress level. So the notch component with a smaller notch radius in Figure 4.6(b) would have a lower K_f value, a longer fatigue initiation life, and be less damaging than the component with a larger notch radius in Figure 4.6(a).

Figure 4.7 schematically illustrates another example of the same notched components made of mild strength and high strength steels. Note that the subscripts 1 and 2 denote the notched components 1 and 2. Again, the damage zone of steel material is assumed to be of the order of two grain sizes. Since the high strength steel has

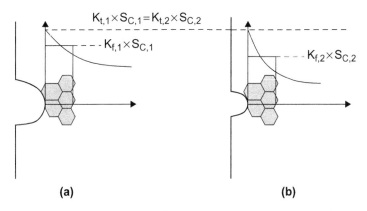

Figure 4.6
Effect of notch size and stress gradient on K_f: (a) large notch radius and mild stress gradient and (b) small notch radius and steep stress gradient.

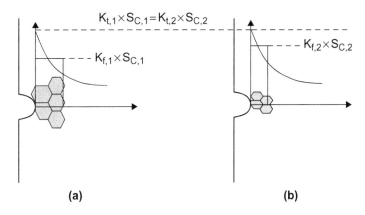

Figure 4.7
Effect of strength of materials on K_f: (a) mild-strength steel and (b) high-strength steel on.

smaller grain size than the mild strength steel, it suggests that the damage zone for high strength steel is smaller than that for mild strength steel.

Under the same peak pseudo surface stress and distribution, the component made of mild strength steel in Figure 4.7(a) would have a lower average stress in a larger damage zone, a lower K_f value, a longer fatigue initiation life, and be less damaging than the component made of high strength steel in Figure 4.7(b).

Based on the stress field intensity theory, the K_f factor is closely related to a notch root radius (or a stress gradient) and the strength of materials (or the grain size). Therefore, several empirical methods have been developed to determine the $K_t - K_f$ relationship based on any combination of the earlier two parameters. For example, a notch sensitivity factor (q) was introduced by Peterson (1959) as follows:

$$q = \frac{K_f - 1}{K_t - 1} \tag{4.52}$$

where q is a function of a notch root radius and the ultimate tensile strength of a material. Also the K_t/K_f ratio or the supporting factor (n_K) was developed:

$$n_K = \frac{K_t}{K_f} \tag{4.53}$$

where n_K depends either on a relative stress gradient and tensile yield strength (Siebel & Stieler, 1955) or on a notch root radius and ultimate tensile strength (Haibach, 2003).

The three approaches will be discussed in the following sections; we recommend the one based on FKM-Guideline.

Notch sensitivity factor

Based on Equation (4.52), the formula for K_f can be written as follows:

$$K_f = 1 + (K_t - 1)q. \tag{4.54}$$

When $q = 1$ or $K_t = K_f$, the material is considered to be fully notch sensitive. On the other hand, when $q = 0$ and $K_f = 1.0$, the material is considered not to be notch sensitive (the so-called "notch blunting" effect).

Peterson (1959) assumed fatigue damage occurs when the stress at a critical distance (a_P) away from the notch root is equal to the fatigue strength of a smooth component. Based on the assumption that the stress near a notch reduces linearly, Peterson obtained the following empirical equation for q:

$$q = \frac{1}{1 + \frac{a_P}{r}} \tag{4.55}$$

where r is the notch root radius and a_P is Peterson's material constant related to the grain size (or $S_{t,u}$) and the loading mode. A plot by Peterson is provided in Figure 4.8 to determine the notch sensitivity factor for high and mild strength steels.

Furthermore, Neuber (1946) postulated that fatigue failure occurs if the average stress over a length from the notch root equals the fatigue strength of a smooth component, and proposed the following empirical equation for q:

$$q = \frac{1}{1 + \sqrt{\dfrac{a_N}{r}}} \tag{4.56}$$

where

a_N = Neuber's material constant related to the grain size or the ultimate tensile strength

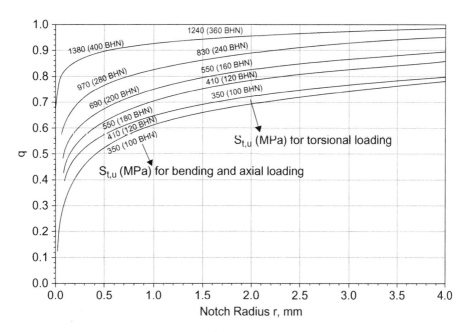

Figure 4.8
Peterson's notch sensitivity curves for steels.

Relative Stress Gradient

Siebel and Stieler (1955) introduced a new parameter $(\overline{G})$ in a unit of 1/mm, termed the relative stress gradient, which is defined as follows:

$$\overline{G} = \frac{(G)_{x=0}}{\sigma^e_{max}} = \frac{1}{\sigma^e_{max}}\left(\frac{d\sigma^e(x)}{dx}\right)_{x=0} \tag{4.57}$$

where

$x =$ the distance from the notch root
$G =$ the stress gradient along x
$\sigma^e(x) =$ the calculated pseudo stress distribution along x
$\sigma^e_{max} =$ the maximum pseudo stress at $x = 0$, as illustrated in Figure 4.9

By testing many smooth and notched components for the endurance cycle limit at 2×10^7 cycles, they generated a series of empirical curves relating the K_t/K_f ratios to $\overline{G}$ values for various materials in terms of tensile yield

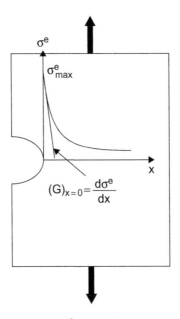

Figure 4.9
Pseudo stress distribution and the stress gradient at a notch root.

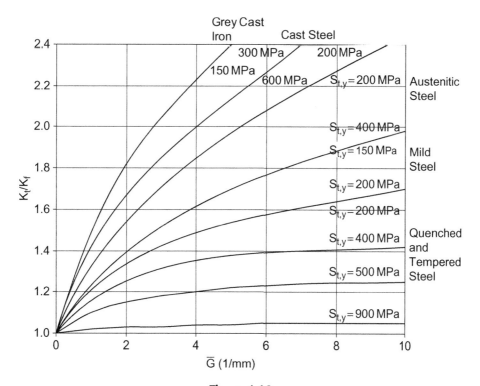

Figure 4.10

Relative stress gradient effect on $K_t - K_f$ ratios for various materials in terms of tensile yield strength.

strength ($S_{t,y}$ in MPa). These empirical curves, as illustrated in Figure 4.10, can be expressed by the following generic formula:

$$n_K = \frac{K_t}{K_f} = 1 + \sqrt{a_{ss} \cdot \overline{G}} \qquad (4.58)$$

where

a_{ss} = the Siebel and Stieler material parameter

FKM-Guideline

The fatigue notch factors for a notched shaft under axial, bending, shear, and torsional stress ($K_{ax,f}$, $K_{b,f}$, $K_{s,f}$, $K_{t,f}$) can be calculated from the corresponding elastic stress concentration factors ($K_{ax,t}$, $K_{b,t}$, $K_{s,t}$, $K_{t,t}$) and the K_t/K_f ratios ($n_{K,\sigma}(r)$, $n_{K,\sigma}(d)$, $n_{K,\tau}(r)$, $n_{K,\tau}(d)$) as follows:

$$K_{ax,f} = \frac{K_{ax,t}}{n_{K,\sigma}(r)} \tag{4.59}$$

$$K_{b,f} = \frac{K_{b,t}}{n_{K,\sigma}(r) \cdot n_{K,\sigma}(d)} \tag{4.60}$$

$$K_{s,f} = \frac{K_{s,t}}{n_{K,\tau}(r)} \tag{4.61}$$

$$K_{t,f} = \frac{K_{t,t}}{n_{K,\tau}(r) \cdot n_{K,\tau}(d)} \tag{4.62}$$

where

r = the notch radius and d is the net diameter or net width of a section

Similarly, the fatigue notch factors for a notched shell-shaped component under normal stresses in x and y directions and shear stress ($K_{x,f}$, $K_{y,f}$, $K_{s,f}$) can be calculated from the corresponding elastic stress concentration factors ($K_{x,t}$, $K_{y,t}$, $K_{s,t}$) and the K_t/K_f ratios ($n_{K,\sigma,x}(r)$, $n_{K,\sigma,y}(r)$, $n_{K,\tau}(r)$) as follows:

$$K_{x,f} = \frac{K_{x,t}}{n_{K,\sigma,x}(r)} \tag{4.63}$$

$$K_{y,f} = \frac{K_{y,t}}{n_{K,\sigma,y}(r)} \tag{4.64}$$

$$K_{s,f} = \frac{K_{s,t}}{n_{K,\tau}(r)}. \tag{4.65}$$

The K_t/K_f ratios ($n_{K,\sigma}(r)$ and $n_{K,\sigma}(d)$) for normal stress are calculated from the relative normal stress gradients $\overline{G}_\sigma(r)$ and $\overline{G}_\sigma(d)$:

$$n_{K,\sigma} = 1 + \overline{G}_\sigma \cdot 10^{-(a_G - 0.5 + S_{t,u}/b_G)} \quad \text{for} \quad \overline{G}_\sigma \leq 0.1 \text{ mm}^{-1} \tag{4.66}$$

$$n_{K,\sigma} = 1 + \sqrt{\overline{G}_\sigma} \cdot 10^{-(a_G + S_{t,u}/b_G)} \quad \text{for} \quad 0.1 \text{ mm}^{-1} < \overline{G}_\sigma \leq 1 \text{ mm}^{-1} \tag{4.67}$$

$$n_{K,\sigma} = 1 + \sqrt[4]{\overline{G}_\sigma} \cdot 10^{-(a_G + S_{t,u}/b_G)} \quad \text{for} \quad 1\,\text{mm}^{-1} < \overline{G}_\sigma \le 100\,\text{mm}^{-1}. \tag{4.68}$$

Likewise, the K_t/K_f ratios for shear stress are calculated from the relative shear stress gradients $\overline{G}_\tau(r)$ and $\overline{G}_\tau(d)$:

$$n_{K,\tau} = 1 + \overline{G}_\tau \cdot 10^{-(a_G - 0.5 + C_\tau S_{t,u}/b_G)} \quad \text{for} \quad \overline{G}_\tau \le 0.1\,\text{mm}^{-1} \tag{4.69}$$

$$n_{K,\tau} = 1 + \sqrt{\overline{G}_\tau} \cdot 10^{-(a_G + C_\tau S_{t,u}/b_G)} \quad \text{for} \quad 0.1\,\text{mm}^{-1} < \overline{G}_\tau \le 1\,\text{mm}^{-1} \tag{4.70}$$

$$n_{K,\tau} = 1 + \sqrt[4]{\overline{G}_\tau} \cdot 10^{-(a_G + C_\tau S_{t,u}/b_G)} \quad \text{for} \quad 1\,\text{mm}^{-1} < \overline{G}_\tau \le 100\,\text{mm}^{-1} \tag{4.71}$$

where

$S_{t,u}$ = the ultimate strengths with R97.5 in tension in the unit of MPa
C_τ = the shear stress correction factor
a_G and b_G = the material constants listed in Table 4.10

The relative stress gradient of a notched component under a specific loading mode really depends on the diameter (or net width) of the component and its notch radius. The relative stress gradients for bending and torsion as a function of

Table 4.10: a_G and b_G for Various Materials

Materials	a_G	b_G MPa
Stainless steel	0.40	2400
Steels except for stainless steel	0.50	2700
Steel castings	0.25	2000
Ductile irons	0.05	3200
Malleable cast iron	−0.05	3200
Grey cast iron	−0.05	3200
Wrought aluminum alloys	0.05	850
Cast aluminum alloys	−0.05	3200

Source: Adapted from FKM-Guideline, published by Forschungskuratorium Maschinenebau, 2003.

Table 4.11: Relative Stress Gradients $\overline{G}_{K,\sigma}(r)$ and $\overline{G}_{K,\tau}(r)$ for Various
Notched Geometries

Notched Components	$\overline{G}_{K,\sigma}(r)$ mm^{-1}	$\overline{G}_{K,\tau}(r)$ mm^{-1}
A groove shaft in Figure 4.11(a)	$\frac{2}{r}(1+\varphi)$	$\frac{1}{r}$
A shoulder shaft in Figure 4.11(b)	$\frac{2.3}{r}(1+\varphi)$	$\frac{1.15}{r}$
A groove plate in Figure 4.11(c)	$\frac{2}{r}(1+\varphi)$	—
A shoulder plate in Figure 4.11(d)	$\frac{2.3}{r}(1+\varphi)$	—
A central hole plate in Figure 4.11(e)	$\frac{2.3}{r}$	—

Notes: (1) $\varphi = 1/(4\sqrt{t/r}+2)$ for $t/d \leq 0.25$ or $t/b_{nw} \leq 0.25$; (2) $\varphi = 0$ for $t/d > 0.25$ or $t/b_{nw} > 0.25$.
Source: Adapted from FKM-Guideline, published by Forschungskuratorium Maschinenebau, 2003.

the net diameter (d) of a notched shaft or net width (b_{nw}) of a notched plate can be obtained:

$$\overline{G}_{K,\sigma}(d) = \overline{G}_{K,\tau}(d) = \frac{2}{d} \qquad (4.72)$$

$$\overline{G}_{K,\sigma}(b_{nw}) = \overline{G}_{K,\tau}(b_{nw}) = \frac{2}{b_{nw}}. \qquad (4.73)$$

Also, the relative stress gradients ($\overline{G}_{K,\sigma}(r)$ and $\overline{G}_{K,\tau}(r)$) can be found in Table 4.11 for various notched geometries.

FKM-Guideline also specifies that the resulting fatigue notch factors for superimposed notches (e.g., $K_{1,f}$ and $K_{2,f}$) can be estimated as

$$K_f = 1 + (K_{1,f} - 1) + (K_{2,f} - 1). \qquad (4.74)$$

Superposition does not need to be considered if the distance of notches is equal to $2r_{max}$ or greater, where r_{max} is the larger one of both notch radii.

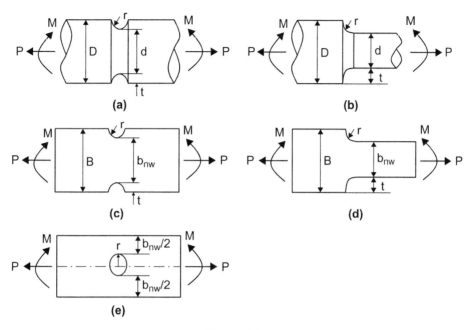

Figure 4.11
Definition of notched components in FKM-Guideline.

Constant Amplitude Stress-Life Curve for a Notched Component under Fully Reversed Loading

Based on the definition of stress (namely nominal stress or pseudo stress), the procedure to generate a constant amplitude stress-life curve for a notched component under fully reversed loading is discussed. (See Figure 4.11.)

Constant Amplitude Nominal Stress-Life Curve

This section presents the FKM method to construct the synthetic S-N curve for a notched component, based on a reference point and a specified slope factor (k). The endurance limit (S_E) at an endurance cycle limit ($N_E = 10^6$ cycles) is the reference point. Thus, the S-N equation can be obtained as follows:

$$NS^k = N_E S_E^k = \text{constant.} \tag{4.75}$$

The specified slope factor (k) depends on the type of material and stress. The following are the specifications from FKM-Guideline:

- For surface nonhardened components made of steels and cast irons, except austenitic steel, the component constant amplitude S-N curves based on normal stress and shear stress are illustrated in Figures 4.12(a) and (b), respectively. The endurance limit value ($S_{\sigma,E}$ or $S_{\tau,E}$) at 10^6 cycles is the fatigue limit at which the fatigue initiation life becomes infinite or when fatigue initiation failure does not occur. The specific slope factors for normal stress and shear stress ($k_{1,\sigma}$ and $k_{1,\tau}$) are defined as 5 and 8, respectively.

- For surface hardened components made of steels and cast irons, the component constant amplitude S-N curves for both normal and shear stresses have larger slope factors than the nonhardened components by a factor close to 3.0. As shown in Figure 4.13, the specific slope factors for normal stress and shear stress ($k_{1,\sigma}$ and $k_{1,\tau}$) are defined as 15 and 25, individually.

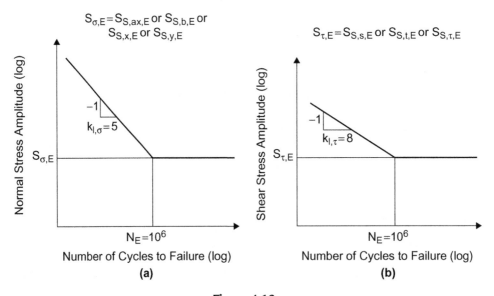

Figure 4.12

Synthetic component constant amplitude S-N curve for surface nonhardened *components* made of steels and cast irons.

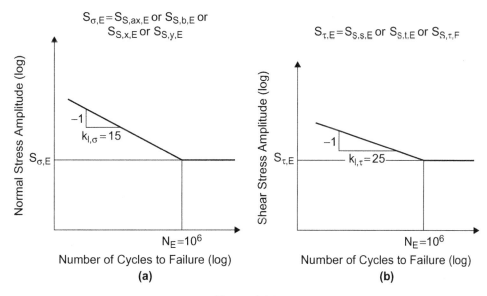

Figure 4.13
Synthetic component constant amplitude S-N curve for *surface hardened components* made of steels and cast irons.

- For components made of aluminum alloys and austenitic steel, the component constant amplitude S-N curves based on normal stress and shear stress are illustrated in Figures 4.14(a) and (b), respectively. The stress amplitude at 10^8 cycles is defined as the fatigue limit ($S_{\sigma,FL}$ or $S_{\tau,FL}$). The S-N curve between 10^6 and 10^8 cycles are defined by the same reference point, but with a different slope factor such as 15 for normal stress or 25 for shear stress. The fatigue limit can be calculated by using Equation (4.75).

Constant Amplitude Pseudo Stress-Life Curve

The synthetic constant amplitude pseudo stress-life curve for a notched component is preferable if the local stress is determined by a linear elastic finite element analysis. This section presents the method to convert the nominal stress-life curve mentioned earlier to the local pseudo stress-life curve.

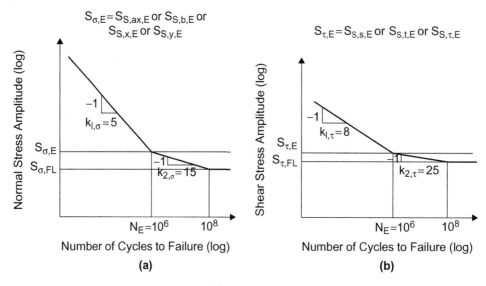

Figure 4.14
Synthetic component constant amplitude S-N curve for *components* made
of aluminum alloys and austenitic steel.

The following equations are valid:

$$\sigma_E^e = K_t S_{S,E,Notched} \tag{4.76}$$

$$K_f = \frac{S_{S,E,Smooth}}{S_{S,E,Notched}} \tag{4.77}$$

$$n_K = \frac{K_t}{K_f} \tag{4.78}$$

where

σ_E^e = the pseudo endurance limit

n_K = the K_t/K_f factor or the supporting factor

$S_{S,E,Notched}$ = the endurance limit of a notched component, calculated by the equations in Section 4.3

$S_{S,E,Smooth}$ = the endurance limit of a smooth component, calculated by the equations in Section 4.3 with the exception of $K_f = 1$

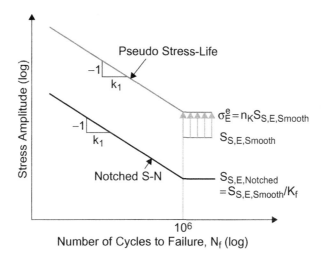

Figure 4.15
Synthetic constant amplitude S-N and σ^e-N curves for a notched
component made of steel.

These equations lead to a new reference stress point at 10^6 cycles as

$$\sigma_E^e = n_K S_{S,E,Smooth}.\qquad(4.79)$$

With the assumption of the identical slope factor specified by FKM-Guideline,
the synthetic constant amplitude pseudo stress-life curve can then be determined.
Figure 4.15 shows the concept of defining the pseudo endurance limit with respect
to the endurance limit of a smooth component as well as the comparison between
the constant amplitude S-N and σ_E^e-N curves for a notched component made of
steels.

Stress-Life Curve for a Component under Variable Amplitude Loading

For a component subjected to variable amplitude loading over time, a rainflow
cycle counting technique as addressed in Chapter 3 is typically used to convert
a complicated time-varying stress history to a series of discrete simple constant
amplitude stress events that consist of a mean stress level and a number of stress
cycles (n_i).

The fatigue life ($N_{f,i}$) corresponding to the number of cycles to failure at the specific stress event can be estimated from the component constant amplitude S-N curve. In this case, the fatigue damage is defined as the cycle ratio ($=n_i/N_{f,i}$).

The Palmgren–Miner (Palmgren, 1924; Miner, 1945) linear damage rule is adopted to calculate the accumulated damage, which assumes fatigue damage occurs when the sum of the cycle ratios at each constant amplitude stress event reaches a critical damage value (D_{PM}). In mathematics, fatigue failure is predicted when

$$\sum \frac{n_i}{N_{f,i}} \geq D_{PM}. \qquad (4.80)$$

Palmgren and Miner found the critical damage value of 1.0 in their studies. But since their work was conducted, it has been shown (Wirshing et al., 1995; Lee et al., 2005; Lalanne, 2002) that the critical damage value is a random variable varying from 0.15 to 1.06. For mechanical designs, FKM-Guideline (Haibach, 2003) recommends $D_{PM} = 0.3$ for steels, steel castings, and aluminum alloys, and $D_{PM} = 1.0$ for ductile irons, grey cast irons, and malleable cast irons. For electronic equipment design, Steinberg (1973) suggests $D_{PM} = 0.7$.

The component constant amplitude S-N curve is supposed to be used for estimating the fatigue life of a component at a given constant amplitude stress event. But when the Palmgen–Miner linear damage rule is applied to a component in variable amplitude loading, the stress cycles with amplitudes below the fatigue limit could become damaging if some of the subsequent stress amplitudes exceed the original fatigue limit. It is believed that the increase in crack-driving force due to the periodic overloads will overcome the original grain barrier strength and help the crack to propagate until failure.

Therefore, there is a need to modify the fatigue limit for a component subjected to variable amplitude loading history because the fatigue limit obtained from constant amplitude loading might be negatively influenced by periodic overloads. Two methods such as the Miner rule and the Miner–Haibach model (Haibach, 1970), as shown in Figure 4.16, were proposed to include the effect of periodic overloads on the stress cycle behavior below the original fatigue limit.

The Miner rule extends the S-N curve with the same slope factor k to approach zero stress amplitude, while the Miner–Haibach model extends the original S-N curve below the fatigue limit to the zero stress amplitude with a flatter slope

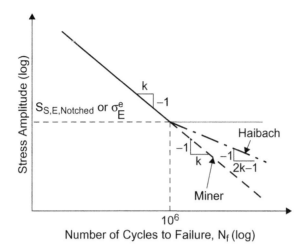

Figure 4.16
Constant amplitude S-N curve for a component made of steels and subjected to variable amplitude loading.

factor 2k–1. Stanzl et al. (1986) concluded that a good agreement is found for measured and calculated results according to the Miner–Haibach model.

Mean Stress Effect

From the perspective of applied cyclic stresses, the fatigue damage of a component strongly correlates with the applied stress amplitude or applied stress range, and is secondarily influenced by the mean stress. The mean stress effect should be seriously considered in fatigue analyses. In the HCF regime, normal mean stresses have a significant effect on fatigue behavior of components.

Mean normal stresses are responsible for the opening and closing state of microcracks. Since the opening of microcracks accelerates the rate of crack propagation and the closing of microcracks retards the growth of cracks, tensile mean normal stresses are detrimental and compressive mean normal stresses are beneficial in terms of fatigue strength. There is very little or no effect of mean stress on fatigue strength in the low cycle fatigue (LCF) regime where the large amount of plastic deformation significantly reduces any beneficial or detrimental effect of the mean stress.

The mean normal stress effect can be represented by the mean stress ($S_{\sigma,m}$) or the stress ratio (R). Both are defined as follows:

$$S_{\sigma,m} = \frac{(S_{\sigma,max} + S_{\sigma,min})}{2} \tag{4.81}$$

$$R = \frac{S_{\sigma,min}}{S_{\sigma,max}} \tag{4.82}$$

where

$S_{\sigma,max}$ and $S_{\sigma,min}$ = the maximum and minimum normal stresses in a stress cycle

For an example of a fully reversed stress condition, it can be found that $S_{\sigma,m} = 0$ and $R = -1$.

The early models to account for the mean stress effect, such as Gerber (1874), Goodman (1899), Haigh (1917), Soderberg (1930), and Morrow (1968), were usually plotted against empirical data in constant life plots of stress amplitude ($S_{\sigma,a}$) versus mean stress ($S_{\sigma,m}$). In Germany, these constant life plots are called Haigh's diagram; in North America they are commonly referred as Goodman's diagram.

As schematically illustrated in Figure 4.17, Haigh's diagram can be determined from a family of constant amplitude $S_{\sigma,a} - N$ curves (Wohler curves) with various mean stress values (0, $S_{\sigma,m1}$, $S_{\sigma,m2}$, and $S_{\sigma,m3}$). The equivalent fully reversed stress amplitude ($S_{\sigma,ar}$) is the generic interception point of the $S_{\sigma,a}$ axis, which is used to determine the fatigue life (N_i) from a corresponding component S-N curve.

According to Goodman's and Morrow's models as illustrated in Figure 4.17(c), the ultimate tensile strength ($S_{t,u}$) and the fatigue strength coefficient ($S'_{\sigma,f}$) are the physical limits to $S_{\sigma,m}$ and the interception of the $S_{\sigma,m}$ axis, respectively. Alternatively, Haibach in FKM-Guideline introduces the mean stress sensitivity factor (M_σ) to define the Haigh diagram, which is the absolute value of the slope of the constant life plot. The M_σ factor depends on the type of materials and loading condition.

Even though numerous models have been developed to account for the mean stress effect on fatigue strength and lives, the four commonly used formulas are

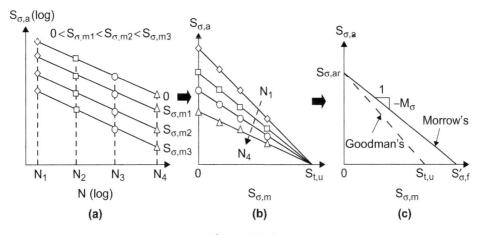

Figure 4.17
Construction of constant life plots on normal stress in $S_{\sigma,a}$ and $S_{\sigma,m}$ coordinates.

chosen for discussion: Goodman's (1899), Morrow's (Morrow, 1968), Smith–Watson–Topper's (SWT; Smith et al., 1970), and Walker's (Walker, 1970).

The differences among the four models can be observed from the following expressions of the fully reserved stress amplitude in the case of moderate mean stress values:

- Goodman's:

$$S_{\sigma,ar} = \frac{S_{\sigma,a}}{1 - \dfrac{S_{\sigma,m}}{S_{t,u}}} \tag{4.83}$$

- Morrow's:

$$S_{\sigma,ar} = \frac{S_{\sigma,a}}{1 - \dfrac{S_{\sigma,m}}{S'_{\sigma,f}}} \tag{4.84}$$

- SWT's:

$$S_{\sigma,ar} = \sqrt{S_{\sigma,max} S_{\sigma,a}} = \sqrt{(S_{\sigma,a} + S_{\sigma,m}) S_{\sigma,a}} \tag{4.85}$$

- Walker's:

$$S_{\sigma,ar} = S_{\sigma,max}^{1-\gamma w} S_{\sigma,a}^{\gamma w} = (S_{\sigma,a} + S_{\sigma,m})^{1-\gamma w} S_{\sigma,a}^{\gamma w} \tag{4.86}$$

where

γ_W = a mean stress fitting parameter

The SWT and the Walker equations predict that fatigue crack will not initiate if the maximum normal stress in a cycle is less than or equal to zero, meaning $S_{\sigma,max} \leq 0$.

The following conclusions are extracted from the extensive studies by Dowling et al. (2008) on experimental fatigue data for steels, aluminum alloys, and titanium alloy where the R ratio ranges from –2 to 0.45. Goodman's model for life predictions is highly inaccurate and should not be used.

Walker's model gives the superior results if an additional mean stress fitting parameter (γ_W) is provided. Otherwise, both Morrow's and SWT's models yield reasonable life estimates for steels. The SWT model is the one recommended for aluminum alloys. In summary, the SWT method provides good results in most cases and is a good choice for general use.

If there are no experimental data available for materials or the R ratio is beyond the range of the previous studies, FKM-Guideline (Haibach, 2003) is recommended for use. According to FKM-Guideline, Haigh's diagram based on a normal stress can be classified as four regimes, as shown in Figure 4.18 and summarized next:

* Regime I is applied for the stress ratio $R > 1$ where the maximum and minimum stresses are under compression.

* Regime II is applicable to the case of $-\infty \leq R \leq 0$ where $R = -\infty$ is the zero compression stress; $R = -1$, the fully reversed stress; $R = 0$, the alternating tension stress with a zero minimum stress.

* Regime III is for $0 < R < 0.5$ where the maximum and minimum stresses are under tension.

* Regime IV is for $R \geq 0.5$, the regime of high alternating tension stress.

FKM-Guideline specifies various mean stress sensitivity factors due to normal stress for all four regimes and any material. The fully reversed fatigue strength $S_{\sigma,ar}$ in the classified regimes can be written as follows:

* Regimes I and IV:

$$S_{\sigma,ar} = S_{\sigma,a} \qquad (4.87)$$

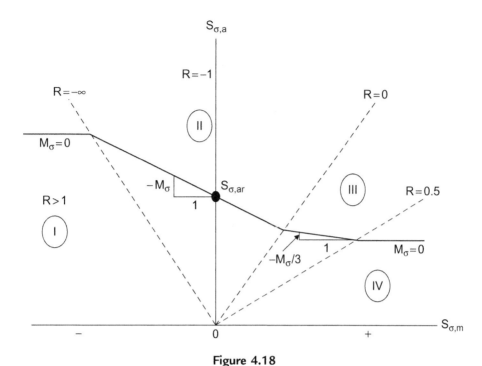

Figure 4.18
Haigh's diagram for mean stress effect on normal stress.

- Regime II:

$$S_{\sigma,ar} = S_{\sigma,a} + M_\sigma S_{\sigma,m} \qquad (4.88)$$

- Regime III:

$$S_{\sigma,ar} = (1 + M_\sigma)\frac{S_{\sigma,a} + (M_\sigma/3)S_{\sigma,m}}{1 + M_\sigma/3}. \qquad (4.89)$$

Haigh's diagram based on shear stress can be classified as three regimes as shown in Figure 4.19 because the negative mean stress in shear is always regarded as positive and treated the same as the positive mean stress. For the case of ambient or elevated temperatures, the mean stress sensitivity factors for normal and shear stresses can be obtained as follows:

$$M_\sigma = a_M S_{t,u} + b_M \qquad (4.90)$$

$$M_\tau = C_\tau M_\sigma \qquad (4.91)$$

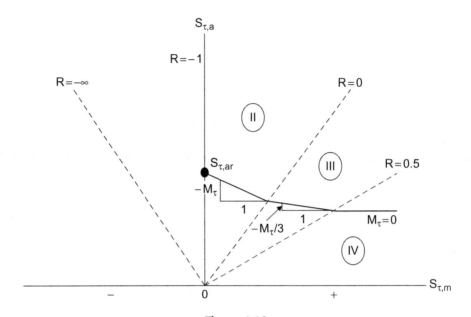

Figure 4.19
Haigh's diagram for mean stress effect on shear stress.

Table 4.12: a_M and b_M for Various Materials

Materials	a_M	b_M
Steel	0.00035	−0.1
Steel casting	0.00035	0.05
Ductile irons	0.00035	0.08
Malleable cast iron	0.00035	0.13
Grey cast iron	0	0.5
Wrought aluminum alloys	0.001	−0.04
Cast aluminum alloys	0.001	0.2

*Source: Adapted from FKM-Guideline, published by Forschungskuratorium
Maschinenebau, 2003.*

where

a_M and b_M = the material parameters listed in Table 4.12
$S_{t,u}$ and C_τ = the ultimate tensile strength in MPa and the shear stress
correction factor

For mechanical designs, Wilson and Haigh (1923) introduced the line of constant
tensile yield strength as an additional design constraint for materials, termed as the

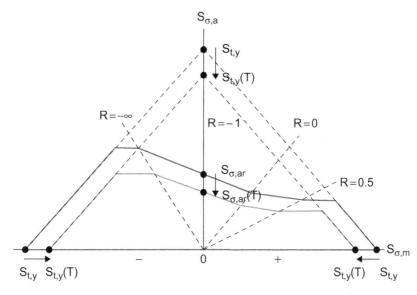

Figure 4.20
Safe design regions for fatigue strength and ultimate tensile strength at both ambient and elevated temperatures.

safe design region for the consideration of both the fatigue strength and the tensile yield strength. The line of constant tensile yield strength was constructed by the line connecting the two tensile yield strength values in $S_{\sigma,a}$ and $S_{\sigma,m}$ axes. Any combination of $S_{\sigma,a}$ and $S_{\sigma,m}$ that falls into the inside of the enclosed area is considered a safe design that will meet both of the yield strength and fatigue strength criteria.

Figure 4.20 shows the isothermal temperature effect on the safe design regions for the fatigue strength and tensile yield strength at both ambient and elevated temperatures. The temperature correction factors for tensile yield strength and fatigue strength follow the guidelines as previously described.

Summary

According to FKM-Guideline, the empirical procedures to generate the synthetic nominal S-N and the pseudo σ^e-N curves have been introduced in this chapter. These procedures are not scientifically based but are simple and useful engineering tools for generating the synthetic component stress-life curves for various materials.

If the test data are not available, the ultimate strength of a real component at a specific temperature with a survival rate can be estimated based on an estimated strength value of standard smooth, polished round test specimens of 7.5 mm diameter with the correction factors for temperature, reliability, size, load, and stress.

The endurance limit of a real component at a specific temperature with a survival rate can be estimated based on an estimated strength value of standard smooth, polished round test specimens of 7.5 mm diameter with the correction factors for temperature, reliability, size, endurance limit in normal or shear stress, surface treatment, roughness, and fatigue notch factor.

The stress field intensity theory has been used to explain why the fatigue notch factor or fatigue strength reduction factor is less than the elastic stress concentration factor. There are three popular approaches (notch sensitivity factor by Peterson and Neuber, relative stress gradient by Siebel and Stieler, and FKM-Guideline by Haibach) to estimate the fatigue notch factor, among which we recommend FKM-Guideline.

The constant amplitude nominal stress-life approach and the local pseudo stress-life approach for a notched component under fully reversed loading were introduced. Either one can be derived on a reference point, the endurance limit at an endurance cycle limit, and a suggested slope factor by FKM-Guideline.

The application of the Palmgren–Miner linear damage rule to a component subjected to variable amplitude loading over time was discussed. It has been shown that the critical damage value is a random variable varying from 0.15 to 1.06. So for mechanical designs, FKM-Guideline recommends $D_{PM} = 0.3$ for steels, steel castings, and aluminum alloys, and $D_{PM} = 1.0$ is recommended for ductile irons, grey cast irons, and malleable cast irons. Also for electronic equipment designs, Steinberg suggests $D_{PM} = 0.7$.

The mean stress effect on the fatigue strength and lives of a component was addressed. Even though there are numerous models developed to account for the mean stress effect, five commonly used formulas such as Goodman's, Morrow's, Smith–Watson–Topper's, Walker's, and the one by FKM-Guideline were discussed.

According to the extensive studies (Dowling et al., 2008) on experimental fatigue data for steels, aluminum alloys, and titanium alloy where the R ratio ranges from

–2 to 0.45, Goodman's model for life predictions is highly inaccurate and should not be used. Walker's model gives the superior results if an additional mean stress fitting parameter (γ_W) is provided. Both Morrow's and SWT's models yield reasonable life estimates for steels and the Smith–Watson–Topper model is the one recommended for aluminum alloys. If there are no experimental data available for materials or the R ratio is beyond the range of the previous studies, FKM-Guideline is recommended.

References

ASTM E 739-91. (2006). *Standard practice for statistical analysis of linear or linearized stress-life (S-N) and strain-life (e-N) fatigue data.* Philadelphia: ASTM.

Adib, H., & Pluvinage, G. (2003). Theoretical and numerical aspects of the volumetric approach for fatigue life prediction in notched components. *International Journal of Fatigue, 25,* 67–76.

Dowling, N. E., Calhoun, C. A., & Arcari, A. (2008). Mean stress effects in stress-life fatigue and the Walker equation. *Fatigue and Fracture of Engineering Materials and Structures, 32,* 163–179.

Gerber, W. Z. (1874). Calculation of the allowable stresses in iron structures. *Z. Bayer ArchitIngVer, 6*(6), 101–110.

Goodman, J. (1899). *Mechanics applied to engineering* (1st ed.). London: Longmans, Green and Company.

Haibach, E. (2003). *FKM-guideline: Analytical strength assessment of components in mechanical engineering* (5th Rev ed.), English version. Germany, Frankfurt/Main: Forschungskuratorium Maschinenebau.

Haibach, E. (1970). *Modifizierte Lineare Schadensakkumulations-hypothese zur Berücksichtigung des Dauerfestigkeitsabfalls mit Fortschreitender Schädigung.* LBF TM No 50/70 (Lab fur Betriebsfestigkeit Darmstadt, FRG).

Haigh, B. P. (1917). Experiments on the fatigue of brasses. *Journal of the Institute of Metals, 18,* 55–86.

Lalanne, C. (2002). *Mechanical vibration & shock: Fatigue damage,* Volume IV. Paris: Hermes Penton Ltd.

Lee, Y. L., Pan, J., Hathaway, R., & Barkey, M. (2005). *Fatigue testing and analysis: Theory and practice.* Boston: Elsevier/Butterworth-Heinemann.

Lee, P. S., & Song, J. H. (2006). Estimation methods for strain-life fatigue properties from hardness. *International Journal of Fatigue, 28,* 386–400.

Miner, M. A. (1945). Cumulative damage in fatigue. *Journal of Applied Mechanics, 67,* A159–A164.

Mitchell, M. R. (1979). *Fundamentals of modern fatigue analysis for design: Fatigue and microstructure* (pp. 385–437). Metals Park, OH: American Society for Metals.

Morrow, J. (1968). Fatigue properties of metals. In J. A. Graham, J. F. Millan, & F. J. Appl (Eds.), *Fatigue design handbook, advances in engineering* (pp. 21–29), AE-4, Warrendale, PA: SAE International.

Neuber, H. (1946). *Theory of notch stress*. Ann Arbor, MI: J. W. Edwards.

Palmgren, A. (1924). Die Lebensdauer von Kugellagern. *Zeitschrift des Vereinesdeutscher Ingenierure, 68*(14), 339–341.

Peterson, R. E. (1959). *Analytical approach to stress concentration effects in aircraft materials*. Technical Report 59-507, U. S. Air Force—WADC Symp. Fatigue Metals, Dayton, Ohio.

Qylafku, G., Azari, Z., Kadi, N., Gjonaj, M., & Pluvinage, G. (1999). Application of a new model proposal for fatigue life prediction on notches and key-seats. *International Journal of Fatigue, 21*, 753–760.

Siebel, E., & Stieler, M. (1955). Significance of dissimilar stress distributions for cycling loading. *VDI-Zeitschrift, 97*(5), 121–126 (in German).

Smith, K. N., Watson, P., & Topper, T. H. (1970). A stress-strain function for the fatigue of metals. *Journal of Materials, 5*(4), 767–778.

Soderberg, C. R. (1930). Factor of safety and working stress. *Transactions of the American Society of Mechanical Engineers. 52* (Part APM-52-2), 13–28.

Stanzl, S. E., Tschegg, E. K., & Mayer, H. (1986). Lifetime measurements for random loading in the very high cycle fatigue range. *International Journal of Fatigue, 8*(4), 195–200.

Steinberg, D. S. (1973). *Vibration analysis for electronic equipment*. New York: John Wiley & Sons, Inc.

Tryon, R. G., & Dey, A. (2003). *Reliability-based model for fatigue notch effect*. SAE Paper No. 2003-01-0462. Warrendale, PA: SAE International.

Walker, K. (1970). The effect of stress ratio during crack propagation and fatigue for 2024-T3 and 7075-T6 aluminum. In M. S. Rosefeld (Ed.), *Effects of environment and complex load history on fatigue life*, ASTM STP 462 (pp. 1–14). Philadelphia, PA: American Society of Testing and Materials.

Wilson, J. S., & Haigh, B. P. (1923). Stresses in bridges. *Engineering* (London), 116:411–413, 446–448.

Wirshing, P. H., Paez, T. L., & Ortiz, H. (1995). *Random vibration: Theory and practice*. New York: John Wiley & Sons, Inc.

Yao, W. (1993). Stress field intensity approach for predicting fatigue life. *International Journal of Fatigue, 15*(3), 243–245.

CHAPTER 5

Stress-Based Multiaxial Fatigue Analysis

Yung-Li Lee
Chrysler Group LLC

Mark E. Barkey
The University of Alabama

Chapter Outline

Introduction

It is common to perform a finite element analysis (FEA) to calculate stresses at the stress concentration sites of a complex structure. There are three popular FEA strategies, depending on the need and application. First of all, if material yielding is expected as in the case of the thermal mechanical fatigue assessment, a non-linear, elastic-visco-plastic FEA is preferable because it renders accurate stress results and has been favored for use in a structure with complicated boundary, material, and loading conditions. However, it is not practical for use in a structure under long duration, multiaxial variable amplitude loading histories, due to expensive computational CPU time.

Second, a linear elastic FEA in conjunction with a multiaxial notch stress–strain analysis is employed to estimate the true notch stresses and strains. Traditionally, this approach has been used with the strain-based multiaxial fatigue analysis. Lastly, based on the assumption that there is very little plastic deformation in the high cycle fatigue (HCF) regime, the pseudo or fictitious stress components calculated from a linear elastic FEA are employed to estimate the fatigue damage parameter and to predict life from the synthesized pseudo stress-life curve as described in Chapter 4.

The previous two approaches have been commonly coded in some commercial fatigue analysis tools. The last approach, excluding the additional step for the multiaxial notch stress–strain analysis, is the most efficient fatigue analysis for use in the HCF regime. Therefore, it is the focus of this chapter.

Please note that all the stress components described herein are referred to as the pseudo stresses calculated from a linear elastic FEA such as the stress influence superposition approach, the inertia relief method, or the modal transient analysis. The fundamentals of the three pseudo stress analysis methods can be found in Chapter 2.

Multiaxial cyclic stresses at high stress concentration areas can be commonly found in structures under multiaxial cyclic loads. Due to geometric constraints at notches, multiaxial loads can result in either a uniaxial stress state as applied to a notched thin metal sheet or in a multiaxial stress state as applied to a notched shaft or rod. Overall, multiaxial cyclic loading can be classified into two categories: nonproportional and proportional loading.

Nonproportional loading is the multiaxial loading paths that cause the principal stress axis or maximum shear stress axis of a local element to rotate with time and with respect to a local coordinate system. On the other hand, proportional loading will result in a stationary principal stress axis or maximum shear stress axis.

If out-of-phase loading causes local sinusoidal normal and shear stress paths with a phase angle, or phase shift, this type of out-of-phase loading is nonproportional loading. For example, as shown in Figure 5.1, a smooth round shaft is subjected to a 90° out-of-phase normal-shear stress time history, where σ_x and τ_{xy} are the normal and shear surface stresses in a local x–y coordinate system. In this example, the normal stress and shear stress time histories are expressed in terms of a cosine function.

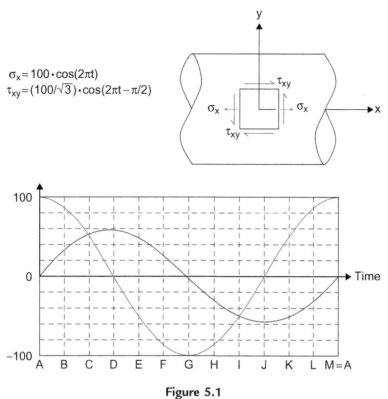

$$\sigma_x = 100 \cdot \cos(2\pi t)$$
$$\tau_{xy} = (100/\sqrt{3}) \cdot \cos(2\pi t - \pi/2)$$

Figure 5.1
A surface element of a round shaft subjected to 90° out-of-phase normal and shear stress time histories.

The shear stress time history has a lagging phase angle (Φ) of $\pi/2$ in radians, or of 90° with respect to the normal stress history. Thus, the 90° out-of-phase stressing is termed due to the phase angle of 90°. As a time instant progresses from A to M, the maximum principal stress (σ_1) axis rotates with respect to the local x axis in an angle of θ, as illustrated in Figures 5.2 through 5.4.

Nonproportional loading causes equal (Archer, 1987), or present more damage (Siljander et al., 1992; Sonsino, 1995) than, proportional loading, based on the same von Mises stress range. The cause of this phenomenon has been explained by an additional nonproportional strain hardening due to slip behavior of the material (Itoh et al., 1995; Socie & Marquis, 2000).

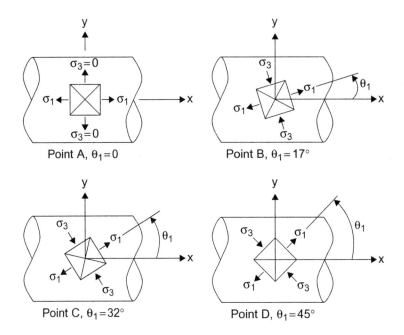

Figure 5.2
Maximum and minimum principal stress axes with respect to the local
x axis in time steps A, B, C, and D.

As shown in Figure 5.5, due to nonproportional normal and shear stressing, the continuous change of the principal stress plane, or maximum shear stress plane, increases the interaction between slip systems resulting in plastic deformation along different slip systems. The cross slip interaction due to plastic deformation can induce an additional strain hardening as compared to that observed in proportional loading.

This strain-hardening phenomenon is illustrated in Figure 5.6, which shows the effective stress–strain curves for the same material under in-phase and 90° out-of-phase loadings. The additional strain hardening due to 90° out-of-phase loading can be described by the nonproportional hardening coefficient, α_{NP}, defined by

$$\alpha_{NP} = \frac{\sigma_{VM,a}(\Phi = 90°)}{\sigma_{VM,a}(\Phi = 0°)} - 1 \tag{5.1}$$

where

$$\sigma_{VM,a}(\Phi = 90°) = \text{the } 90° \text{ out-of-phase von Mises stress amplitude}$$
$$\sigma_{VM,a}(\Phi = 0°) = \text{the in-phase von Mises stress amplitude at the same strain amplitude } (\varepsilon_{VM,a})$$

The severity of nonproportional hardening is dependent on the ease in which slip systems interact and the type of loading path. Materials such as aluminum alloys have weak interactions and show wavy slips because dislocations can

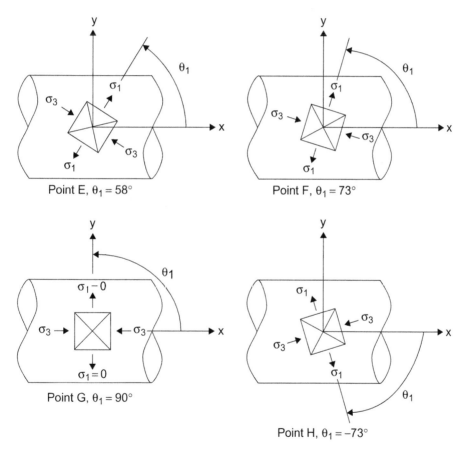

Figure 5.3
Maximum and minimum principal stress axes with respect to the local x axis in time steps E, F, G, and H.

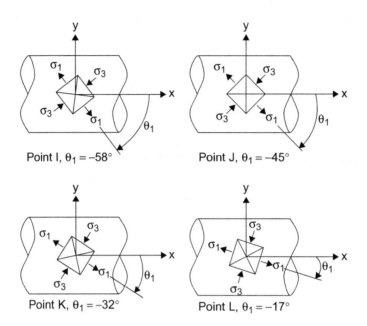

Figure 5.4
Maximum and minimum principal stress axes with respect to the local
x axis in time steps I, J, K, and L.

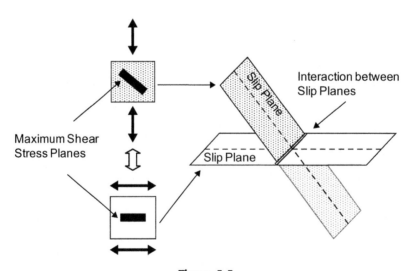

Figure 5.5
Interaction between slip planes due to nonproportional loading.

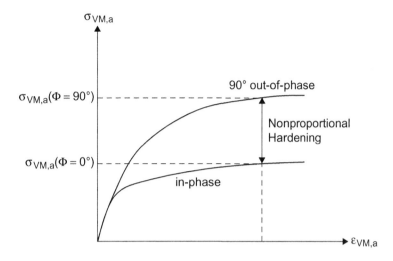

Figure 5.6

Comparison of equivalent stress–strain curves for the same material under in-phase and 90° out-of-phase loadings.

easily change their slip planes as the maximum shear stress plane rotates, resulting in no or small additional strain hardening. However, a typical material, such as Type 304 stainless steel (Doong et al., 1990; Itoh et al., 1995), shows higher hardening and damaging effects than aluminum alloys (Krempl & Lu, 1983; Doong et al., 1990; Itoh et al., 1997) based on the identical nonproportional loading tests.

Also the study (Itoh et al., 1995) showed that different nonproportional loading paths produce different degrees of nonproportional hardening, among which the 90° out-of-phase loading path has the largest degree of nonproportional hardening. Therefore, the fatigue life reduction of a material under nonproportional loading is strongly connected to additional nonproportional hardening due to both loading history and material.

A reliable fatigue model (theory or criterion) should be capable of assessing different fatigue damage values for proportional and nonproportional loadings. Existing stress-based multiaxial fatigue models will be reviewed in the following sections.

Fatigue Damage Models

The existing stress-based multiaxial fatigue damage models can be categorized into four different classifications: empirical formula approach, equivalent stress approach, critical plane approach, and Dang Van multiscale approach.

The empirical formula approach is referred to as the best-fit equation to the experimental fatigue limit data tested under combined normal and shear stresses, and it is only applicable to the biaxial stress state in a fully reversed loading condition.

The equivalent stress approach, based on either the von Mises or Tresca yield theory, was originally developed to define fatigue limit criteria of various materials under proportional loading. Some of the formulations for nonproportional loading have been developed by modifying the von Mises equivalent stress amplitude to account for the nonproportional hardening effect.

The critical plane approach provides a physical interpretation of the damage initiation process. The crack orientation can be identified by searching for the most damaging plane among numerous potential crack initiation planes, where the damage parameter is usually defined as a function of the shear and normal stresses on such a plane.

The Dang Van multiscale approach is a popular approach for assessing mesoscopic fatigue damage at the fatigue limit. It is assumed that an elastic shakedown in a macroscopic state occurs before the fatigue limit and that both mesoscopic and macroscopic plastic strains and residual stresses are stabilized.

The four multiaxial fatigue approaches are presented in the following sections. Although these approaches are expressed to assess the damage at fatigue limit, they can be extended for fatigue life predictions in the HCF regime by simply replacing the fatigue limit with an S-N material equation.

Empirical Formula Approach

The empirical formula approach is only applicable to the biaxial, fully reversed stress state.

Gough and Pollard

The concept of using a failure surface for bending and torsional stresses was presented by Gough and Pollard (1935; Gough, 1950) who found that the fatigue limits of ductile and brittle materials under in-phase bending-torsional loading have the following separate empirical expressions:

$$\left(\frac{\sigma_a}{\sigma_{E,R=-1}}\right)^2 + \left(\frac{\tau_a}{\tau_E}\right)^2 = 1 \quad \text{for ductile materials} \tag{5.2}$$

$$\left(\frac{\sigma_a}{\sigma_{E,R=-1}}\right) + \left(\frac{\tau_a}{\tau_E}\right)^2 = 1 \quad \text{for brittle materials} \tag{5.3}$$

where

σ_a and τ_a = the applied in-phase normal and shear stress amplitudes, respectively

$\sigma_{E,R=-1}$ = the fully reversed fatigue limit for normal stress

τ_E = the fully reversed fatigue limit for shear stress

In addition to correlating well with in-phase fatigue limit data, these equations can be easily used for design purposes. Studies have shown that the Gough–Pollard equations have general applicability when expressed in terms of principal stresses (Hashin, 1981; Rotvel, 1970), and they have a physical interpretation on fatigue damage mechanism if expressed in terms of shear stress and normal stress on the maximum shear stress plane (McDiarmid, 1974).

Dietmann, and Socie and Marquis

Dietmann (1973a,b) and Socie and Marquis (2000) developed the following general empirical formulas for fatigue limits under fully reversed in-phase loading, respectively:

- Dietmann:

$$\left(\frac{\sigma_a}{\sigma_{E,R=-1}}\right)^k + \left(\frac{\tau_a}{\tau_E}\right)^2 = 1 \tag{5.4}$$

- Socie and Marquis:

$$(k-1)\left(\frac{\sigma_a}{\sigma_{E,R=-1}}\right)^2 + (2-k)\left(\frac{\sigma_a}{\sigma_{E,R=-1}}\right) + \left(\frac{\tau_a}{\tau_E}\right)^2 = 1 \qquad (5.5)$$

where

$$k = \frac{\sigma_{E,R=-1}}{\tau_E}$$

These two formulas are very generic such that the Gough–Pollard formulas become special cases of them. For example, for a ductile material where the k value is 2.0, Equations (5.4) and (5.5) are equivalent and can be reduced to Equation (5.2). And for a brittle material where the k value is 1.0, Equations (5.4) and (5.5) are identical to Equation (5.3).

Lee and Lee and Chiang

Taking the Gough–Pollard ellipse quadrant formula as a frame of work, Lee (1985) as well as Lee and Chiang (1991) developed the following criteria for fatigue limits under fully reversed out-of-phase loading:

- Lee:

$$\left(\frac{\sigma_a}{\sigma_{E,R=-1}}\right)^{\eta\cdot(1+\alpha_{NP}\cdot\sin\Phi)} + \left(\frac{\tau_a}{\tau_E}\right)^{\eta\cdot(1+\alpha_{NP}\cdot\sin\Phi)} = 1 \qquad (5.6)$$

- Lee and Chiang:

$$\left(\frac{\sigma_a}{\sigma_{E,R=-1}}\right)^{k\cdot(1+\alpha_{NP}\cdot\sin\Phi)} + \left(\frac{\tau_a}{\tau_E}\right)^{2\cdot(1+\alpha_{NP}\cdot\sin\Phi)} = 1 \qquad (5.7)$$

where

$\eta = 2$ for ductile materials
$\eta = 1.5$ for brittle materials
α_{NP} = the parameter to account for nonproportional hardening due to the phase shaft Φ between normal and shear stresses

Equivalent Stress Approach

The existing equivalent stress models are reviewed in this section, among which the effective equivalent stress amplitude approach (Sonsino, 1995) and the equivalent nonproportional stress amplitude method (Lee et al., 2007) are the promising ones to account for the nonproportional hardening effect.

The latter approach, as proposed by Lee et al. (2007), is the only one that can be employed to calculate the fatigue damage of a structure under variable amplitude, nonproportional loading with both the Miner–Palmgren linear damage rule (Miner, 1945; Palmgren, 1924), and the Wang–Brown rainflow cycle counting technique (Wang & Brown, 1996).

Maximum Principal Stress Theory

Fatigue initiation occurs when

$$\sigma_{PS,a} = \sigma_{1,a} \geq \sigma_{E,R=-1} \tag{5.8}$$

where $\sigma_{1,a}$ is the maximum principal stress amplitude and the subscripts PS and a represent the principal stress and amplitude, respectively.

Maximum Shear Stress Theory (Tresca Theory)

Fatigue initiation occurs if

$$\tau_{MS,a} = \sigma_{1,a} - \sigma_{3,a} \geq \tau_E \tag{5.9}$$

where

$\sigma_{3,a} =$ the minimum principal stress amplitude and the subscript MS
represents the maximum shear

von Mises

Fatigue initiation happens if

$$\sigma_{VM,a} + \alpha_{VM}\sigma_{VM,m} \geq \sigma_{E,R=-1} \tag{5.10}$$

where

$\alpha_{VM} =$ the mean stress sensitivity factor
$\sigma_{VM,m}$ and $\sigma_{VM,a} =$ the mean stress and von Mises stress amplitude

The last two are defined as follows:

$$\sigma_{VM,m} = \sigma_{1,m} + \sigma_{2,m} + \sigma_{3,m} = \sigma_{x,m} + \sigma_{y,m} + \sigma_{z,m} \tag{5.11}$$

$$\sigma_{VM,a} = \frac{1}{\sqrt{2}} \sqrt{\left(\sigma_{1,a} - \sigma_{2,a}\right)^2 + \left(\sigma_{2,a} - \sigma_{3,a}\right)^2 + \left(\sigma_{1,a} - \sigma_{3,a}\right)^2} \qquad (5.12)$$

or

$$\sigma_{VM,a} = \frac{1}{\sqrt{2}} \sqrt{\left(\sigma_{x,a} - \sigma_{y,a}\right)^2 + \left(\sigma_{y,a} - \sigma_{z,a}\right)^2 + \left(\sigma_{x,a} - \sigma_{z,a}\right)^2 + 6\left(\tau_{xy,a}^2 + \tau_{yz,a}^2 + \tau_{xz,a}^2\right)}$$

$$(5.13)$$

where

x, y, and z (subscripts) = the local coordinate axes
1, 2, and 3 = the principal stress axes

Sines

Sines (1959) defined the damage at the fatigue limit in terms of the octahedral shear stress amplitude ($\tau_{oct,a}$) and the hydrostatic stress (σ_h) for mean stresses. The Sines theory, also named the octahedral shear stress theory, shows satisfactory correlation with experimental investigations. It states that the fatigue failure occurs when

$$\tau_{Sines,a} = \tau_{oct,a} + \alpha_{oct}\left(3\sigma_h\right) \geq \tau_E \qquad (5.14)$$

where

α_{oct} = the hydrostatic stress sensitivity factor
σ_h and $\tau_{oct,a}$ = the following:

$$\sigma_h = \frac{1}{3}\left(\sigma_{1,m} + \sigma_{2,m} + \sigma_{3,m}\right) = \frac{1}{3}\left(\sigma_{x,m} + \sigma_{y,m} + \sigma_{z,m}\right) \qquad (5.15)$$

$$\tau_{oct,a} = \frac{1}{3} \sqrt{\left(\sigma_{1,a} - \sigma_{2,a}\right)^2 + \left(\sigma_{2,a} - \sigma_{3,a}\right)^2 + \left(\sigma_{1,a} - \sigma_{3,a}\right)^2} \qquad (5.16)$$

or

$$\tau_{oct,a} = \frac{1}{3} \sqrt{\left(\sigma_{x,a} - \sigma_{y,a}\right)^2 + \left(\sigma_{y,a} - \sigma_{z,a}\right)^2 + \left(\sigma_{x,a} - \sigma_{z,a}\right)^2 + 6\left(\tau_{xy,a}^2 + \tau_{yz,a}^2 + \tau_{xz,a}^2\right)}.$$

$$(5.17)$$

Example 5.1

The staircase method for fatigue limit testing was conducted on thin-walled tubular specimens subjected to axial sinusoidal loading with R = −1 and R = 0. The two fatigue limits are found as $\sigma_{E,R=-1}$ = 700 MPa and $\sigma_{E,R=0}$ = 560 MPa. Determine the mean stress sensitivity factor α_{VM} used in the von Mises stress theory and the hydrostatic stress sensitivity factor α_{oct} in the Sines theory.

Solution

1. The von Mises stress theory:

   (a) For R=−1 loading (fully reversed loading), it is given $\sigma_{E,R=-1}$ = 700 MPa

   (b) For R = 0 loading ($\sigma_{1,a} = \sigma_{1,m} = \sigma_{E,R=0} = 560$)

$$\sigma_{VM,a} = \frac{1}{\sqrt{2}}\sqrt{(\sigma_{1,a}-0)^2+(0-0)^2+(\sigma_{1,a}-0)^2} = \sigma_{1,a} = \sigma_{E,R=0} = 560$$

$$\sigma_{VM,m} = \sigma_{1,m}+0+0 = \sigma_{E,R=0} = 560\,\text{MPa}$$

$$\sigma_{VM,a}+\alpha_{VM}\sigma_{VM,m} = \sigma_{E,R=-1}$$

Thus, $\alpha_{VM} = \dfrac{(\sigma_{E,R=-1}-\sigma_{E,R=0})}{\sigma_{E,R=0}} = \dfrac{(700-560)}{560} = 0.25$

2. The Sines theory:

   (a) For R=−1 loading (fully reversed loading),

$$\tau_{oct,a} = \frac{1}{3}\sqrt{(\sigma_{1,a}-0)^2+(0-0)^2+(0-\sigma_{1,a})^2} = \frac{\sqrt{2}}{3}\sigma_{1,a} = \frac{\sqrt{2}}{3}\sigma_{E,R=-1}$$

$$\sigma_{h,R=-1} = \frac{1}{3}(\sigma_{1,m}+\sigma_{2,m}+\sigma_{3,m}) = 0$$

$$\tau_{oct,a}+\alpha_{oct}(3\sigma_h) = \frac{\sqrt{2}}{3}\sigma_{E,R=-1}+\alpha_{oct}(0) = \tau_E$$

$$\tau_E = \frac{\sqrt{2}}{3}\sigma_{E,R=-1} = \frac{\sqrt{2}}{3}700 = 330\,\text{MPa}$$

   (b) For R = 0 loading

$$\tau_{oct,a} = \frac{1}{3}\sqrt{(\sigma_{1,a}-0)^2+(0-0)^2+(0-\sigma_{1,a})^2} = \frac{\sqrt{2}}{3}\sigma_{1,a} = \frac{\sqrt{2}}{3}\sigma_{E,R=0}$$

$$3\sigma_{h,R=0} = \sigma_{E,R=0}$$

$$\tau_{oct,a} + \alpha_{oct}(3\sigma_h) = \frac{\sqrt{2}}{3}\sigma_{E,R=0} + \alpha_{oct}(\sigma_{E,R=0}) = \frac{\sqrt{2}}{3}\sigma_{E,R=-1}$$

$$\text{Thus, } \alpha_{oct} = \frac{\frac{\sqrt{2}}{3}(\sigma_{E,R=-1} - \sigma_{E,R=0})}{\sigma_{E,R=0}} = \frac{\frac{\sqrt{2}}{3}(700 - 560)}{560} = 0.12$$

Sonsino

Sonsino (1995) developed a theory, so-called "the effective equivalent stress amplitude method," to account for nonproportional hardening as a result of out-of-phase
loading paths. It is assumed that the Case A crack growing along a free surface—as
illustrated in Figure 5.7(a)—is the typical fatigue failure of a ductile material under
multiaxial loading, and is induced by the shear stress ($\tau_n(\theta)$) on an interference
plane with an inclination angle (θ) to a local x axis, as shown in Figure 5.7(b).

The interaction of shear stresses in various interference planes, representing the
severity of nonproportional loading, is taken into account by the following effective
shear stress:

$$\tau_{arith} = \frac{1}{\pi}\int_0^\pi \tau_n(\theta)d\theta. \tag{5.18}$$

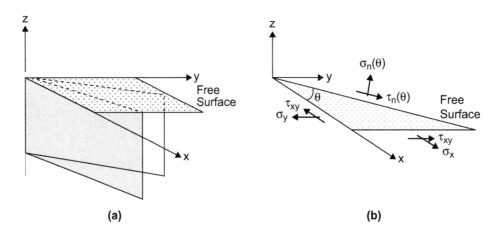

Figure 5.7
(a) Potential Case A surface cracks and (b) normal and shear stresses
on an interference plane.

For an out-of-phase sinusoidal loading in a plane stress condition with a phase shift of Φ, the effective shear stress derived on the von Mises stress criterion is used to determine the effective equivalent stress amplitude ($\sigma_{VM,a}(\Phi)$) as follows:

$$\sigma_{VM,a}(\Phi) = \sigma_{VM,a} \frac{\tau_{arith}(\Phi)}{\tau_{arith}(\Phi = 0°)} \sqrt{G \cdot \exp\left[1 - \left(\frac{\Phi - 90°}{90°}\right)^2\right]} \qquad (5.19)$$

where

$$\sigma_{VM,a} = \sqrt{\sigma_{x,a}^2 + \sigma_{y,a}^2 - \sigma_{x,a}\sigma_{y,a} + 3\alpha_S^2\tau_{xy,a}^2} \qquad (5.20)$$

$$\alpha_S = \frac{\sqrt{\sigma_{x,a}^2 + \sigma_{y,a}^2 - \sigma_{x,a} \cdot \sigma_{y,a}}}{\sqrt{3}\tau_{xy,a}} \qquad (5.21)$$

$$G = \frac{1 + K_{b,t}}{1 + K_{t,t}} \qquad (5.22)$$

where

$K_{b,t}$ and $K_{t,t}$ = the elastic stress concentration factors due to bending
and torsion, respectively

The ratio G takes into account the stress gradient effect. α_S is called the size effect factor, representing the sensitivity of the shear stress amplitude on the normal stress amplitude. This factor can be determined by comparing the S-N curve for bending-only stress with that for pure torsional stress such that the data should lie on top of each other.

Given the fact that the effective equivalent stress amplitude is increased by out-of-phase loading, the square root in Equation (5.19) considers the nonproportional hardening effect due to an out-of-phase sinusoidal loading with a phase shift of Φ. It was developed based on empirical observations on ductile materials. And the ratio $\tau_{arith}(\Phi)/\tau_{arith}(\Phi = 0°)$ represents the nonproportional hardening effect due to the type of materials.

Finally, the effective equivalent stress amplitude indicates that with the identical von Mises equivalent stress amplitude, 90° out-of-phase loading is more damaging than in-phase loading due to the nonproportional hardening effect. However, the Sonsino stress theory is only applicable to in-phase or out-of-phase loading.

Lee, Tjhung, and Jordan

Lee et al. (2007) proposed the concept of equivalent nonproportional stress amplitude in 2007, analogous to the equivalent nonproportional strain amplitude by Itoh et al. (1995). The equivalent nonproportional stress amplitude ($\sigma_{VM,a,NP}$) is defined as follows:

$$\sigma_{VM,a,NP} = \sigma_{VM,a}(1 + \alpha_{NP}f_{NP}) \tag{5.23}$$

where

$\sigma_{VM,a}$ and $\sigma_{VM,a,NP}$ = the equivalent proportional and nonproportional stress amplitudes

$(1 + \alpha_{NP}f_{NP})$ = the term accounts for the additional strain hardening observed during nonproportional cyclic loading

α_{NP} = the nonproportional hardening coefficient for the material dependence

f_{NP} = the nonproportional loading path factor for the severity of loading paths

In the previous equation, $\sigma_{VM,a}$ is based on the von Mises hypothesis, but employs the maximum stress amplitude between two arbitrary stress points among all multiple points in a cycle. For the example of a plane stress condition, $\sigma_{VM,a}$ is maximized with respect to time, and defined as follows to account for the mean stress effect:

$$\sigma_{VM,a} = \max\left\{\sqrt{\sigma_{x,a}^2 + \sigma_{y,a}^2 - \sigma_{x,a}\sigma_{y,a} + 3\alpha_S^2\tau_{xy,a}^2} \times \left(\frac{\sigma_f'}{\sigma_f' - \sigma_{eq,m}}\right)\right\} \tag{5.24}$$

where

α_S = the sensitivity shear-to-normal stress parameter

σ_f' = the fatigue strength coefficient determined from the best fit of the proportional loading data with a stress ratio R = −1

$\sigma_{eq,m}$ = the equivalent mean stress, ignoring the effect of torsional mean stress on fatigue lives

$\sigma_{eq,m}$ is calculated in the following equation:

$$\sigma_{eq,m} = \sigma_{x,m} + \sigma_{y,m}. \tag{5.25}$$

The nonproportional material coefficient α_{NP} as defined in Equation (5.1) is related to the additional hardening of the materials under 90° out-of-phase loading. Alternatively, this coefficient can be obtained from the von Mises stress amplitude versus life curves of the same material under in-phase and 90° out-of-phase fatigue testing.

As shown in Figure 5.8, at a same $\sigma_{VM,a}(\Phi = 0°)$ value, the life ($N_{90°}$) for 90° out-of-phase loading is shorter or more damaging than that ($N_{0°}$) for in-phase loading. Since the in-phase $\sigma_{VM,a} - N$ curve is the baseline S-N curve for life predictions, the higher stress amplitude $\sigma_{VM,a}(\Phi = 90°)$ than $\sigma_{VM,a}(\Phi = 0°)$ is found to produce an equivalent damage or $N_{90°}$ life to the 90° out-of-phase loading, which is assumed to be attributable to the nonproportional strain-hardening phenomenon. Thus, ($\alpha_{NP} + 1$) can be determined by the ratio of $\sigma_{VM,a}(\Phi = 90°)$ to $\sigma_{VM,a}(\Phi = 0°)$.

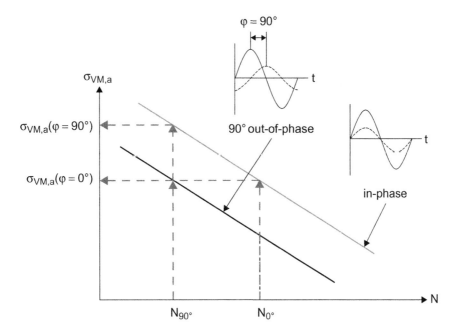

Figure 5.8
Graphical representation of nonproportional hardening due to 90° out-of-phase loading in S-N curves.

If there are no experimental data available, α_{NP} could be estimated by various empirical formulas. According to the experimental study (Doong et al., 1990), there is almost no additional hardening ($\alpha_{NP} \approx 0$) for aluminum alloys; $\alpha_{NP} \approx 0.3$ for coppers, and $\alpha_{NP} \approx 1.0$ for stainless steels.

This study also stated that the nonproportional hardening is related to the material stacking fault energy (SFE). The materials with a low SFE level do exhibit additional cyclic hardening under nonproportional loading, whereas those with a high SFE are susceptible to little hardening. It is difficult to quantify the nonproportional hardening coefficient with the material SFE parameter.

Recent efforts (Borodii & Strizhalo, 2004; Borodii & Shukaev, 2007) to develop an analytical description from the macro material perspective were done. An empirical relation was proposed as follows:

$$\log|\alpha_{NP}| = 0.705 \left(\frac{\sigma_{t,u}}{\sigma_{t,y}} \right) - 1.22 \qquad (5.26)$$

where

$\sigma_{t,u}$ and $\sigma_{t,y}$ = ultimate tensile strength and tensile yield strength, respectively

Moreover, Shamsaei, and Fatemi (2010) observed that under nonproportional loading, cyclic hardening materials exhibit nonproportional hardening and cyclic softening materials have little nonproportional hardening. Since both cyclic hardening and nonproportional hardening are associated with the SFE, they developed the following empirical strain-based equation for α_{NP}:

$$\alpha_{NP} = 1.6 \left(\frac{K}{K'} \right)^2 \left(\frac{\Delta\varepsilon}{2} \right)^{2(n-n')} - 3.8 \left(\frac{K}{K'} \right) \left(\frac{\Delta\varepsilon}{2} \right)^{(n-n')} + 2.2 \qquad (5.27)$$

where

$\Delta\varepsilon$ = the strain range that can be approximated by $\Delta\sigma/E$ in the HCF regime

K, K', n, and n' = the monotonic strength coefficient, cyclic strength coefficient, monotonic strain-hardening exponent, and cyclic strain-hardening exponent, respectively

The nonproportional loading factor (f_{NP}), varying from zero to one, represents the effect of a loading path on nonproportional hardening. In-phase loading generates the value of f_{NP} equal to zero, whereas 90° out-of-phase loading produces the value of f_{NP} equal to one, indicating the most damaging loading condition.

As reported by Itoh et al. (1995), this factor is calculated by integrating the contributions of all maximum principal stresses ($\overrightarrow{\sigma}_{1,max}(t)$) on the plane being perpendicular to the plane of the largest absolute principal stress ($\overrightarrow{\sigma}_{1,max}^{ref}$). This factor is mathematically represented as follows:

$$f_{NP} = \frac{C}{T|\overrightarrow{\sigma}_{1,max}^{ref}|} \int_0^T \left(\left| \sin \xi(t) \times |\overrightarrow{\sigma}_{1,max}(t)| \right| \right) dt \tag{5.28}$$

where

$|\overrightarrow{\sigma}_{1,max}(t)| =$ the maximum absolute value of the principal stress at time t, depending on the maximum's larger magnitude and the minimum principal stresses at time t (maximum of $|\overrightarrow{\sigma}_1(t)|$ and $|\overrightarrow{\sigma}_3(t)|$)

$\xi(t) =$ the angle between $\overrightarrow{\sigma}_{1,max}^{ref}$ and $\overrightarrow{\sigma}_{1,max}(t)$, as shown in Figure 5.9

$\overrightarrow{\sigma}_{1,max}(t) =$ the orientation at an angle of θ with respect to the x axis

$f_{NP} =$ normalized to $|\overrightarrow{\sigma}_{1,max}^{ref}|$ and T (the time for a cycle)

The constant C is chosen to make f_{NP} unity under 90° out-of-phase loading.

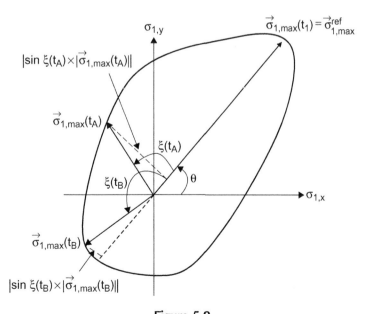

Figure 5.9
Schematic plot of principal stress vectors at various time steps.

Example 5.2

A cantilever, tubular structural is seam welded to a fixed plate at one end and is subjected to out-of-phase nominal bending (S_σ) and torsional (S_τ) stresses at the other end. The tubular structural is made of a thin-walled, rectangular section, as shown in Figure 5.10. Testing indicates that most of the fatigue cracks are initiated at the weld toe near a corner of the tube.

Therefore, the stress state in the surface element (in the x–y coordinate) at this crack initiation location is assumed to be a plane stress condition, and is chosen for our fatigue assessment. With the given elastic stress concentration factors for bending and torsion ($K_{b,t}$ and $K_{t,t}$), the local pseudo stresses (σ_x^e and τ_{xy}^e) of interest can be related to the nominal bending and torsional stresses (S_σ and S_τ) as

$$\sigma_x^e = K_{b,t}S_\sigma \qquad\qquad (5.29)$$

$$\tau_{xy}^e = K_{t,t}S_\tau. \qquad\qquad (5.30)$$

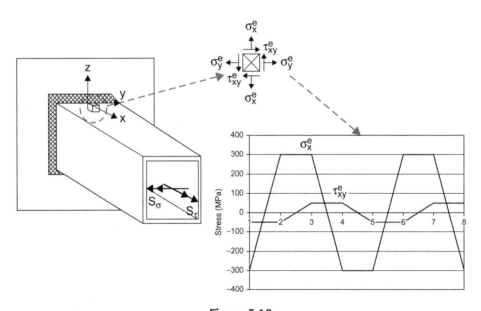

Figure 5.10
A crack initiation site with a local coordinate system at a structural component subjected to external loads.

Also, the assumption of the constraint $\varepsilon_y = 0$ leads to the following relation

$$\sigma_y^e = \nu\sigma_x^e \tag{5.31}$$

where ν = the Poisson ratio ($\nu = 0.3$).

The local pseudo stresses and their induced time history plots are illustrated in Figure 5.10.

1. Determine the constant C in f_{NP}

2. The fatigue limit for normal stress ($\sigma_{E,R=-1} = 700$ MPa) and the nonproportional coefficient for the material ($\alpha_{NP} = 0.3$) are given. Determine whether the welded structure under the nonproportional loading as shown in Figure 5.10 will have an infinite life.

Solution

1. Since the constant C is obtained to make f_{NP} unity under 90° out-of-phase loading, there is a need to generate a 90° out-of-phase relation between σ_x^e and τ_{xy}^e having the same magnitude of the von Mises stress where $\sigma_y^e = \nu\sigma_x^e$. Therefore, a fictitious 90° out-of-phase stressing as illustrated in Figure 5.11 is generated by

$$0.79\left(\sigma_x^e(t)\right)^2 + 3\left(\tau_{xy}^e(t)\right)^2 = 1. \tag{5.32}$$

The elliptic stress path is then divided into 72 points with a 5° angular increment. For each data point, the maximum and minimum principal stresses can be calculated as follows:

$$\sigma_1^e(t) = \frac{\sigma_x^e(t) + \sigma_y^e(t)}{2} + \sqrt{\left(\frac{\sigma_x^e(t) - \sigma_y^e(t)}{2}\right)^2 + \left(\tau_{xy}^e(t)\right)^2} \tag{5.33}$$

$$\sigma_3^e(t) = \frac{\sigma_x^e(t) + \sigma_y^e(t)}{2} - \sqrt{\left(\frac{\sigma_x^e(t) - \sigma_y^e(t)}{2}\right)^2 + \left(\tau_{xy}^e(t)\right)^2}. \tag{5.34}$$

And $\sigma_{1,max}^e(t) = |\sigma_1^e(t)|$ if $|\sigma_1^e(t)| \geq |\sigma_3^e(t)|$ or $\sigma_{1,max}^e(t) = |\sigma_3^e(t)|$ if $|\sigma_3^e(t)| > |\sigma_1^e(t)|$. The orientation of the absolute maximum principal stress with respect to the x axis can be calculated as

$$\theta(t) = \frac{1}{2} \cdot \arctan\left(\frac{2\tau_{xy}^e(t)}{\sigma_x^e(t) - \sigma_y^e(t)}\right). \tag{5.35}$$

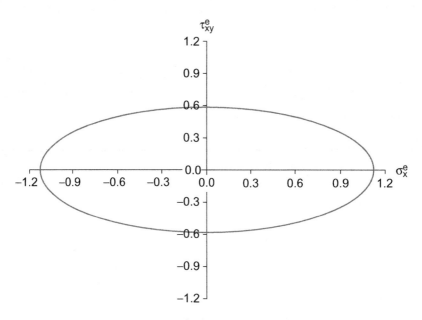

Figure 5.11
A fictitious 90° out-of-phase relation between σ_x^e and τ_{xy}^e having the unit magnitude of the von Mises stress.

Thus, the maximum principal stress plane at any time instant $(\vec{\sigma}_{1,max}^{e}(t))$ can be expressed in terms of its x and y components $(\sigma_{1,x,max}^e(t), \sigma_{1,y,max}^e(t))$ obtained as follows:

$$\sigma_{1,x,max}^e(t) = -\sigma_{1,max}^e(t)\sin\left(\theta(t)\right) \qquad (5.36)$$

$$\sigma_{1,y,max}^e(t) = \sigma_{1,max}^e(t)\cos\left(\theta(t)\right). \qquad (5.37)$$

Figure 5.12 shows plots of principal stresses versus θ angles. The largest magnitude of $\sigma_{1,max}^e(t)$ is found to be 1.125 at three θ angles (0°, 180°, 360°). The perpendicular plane to $\vec{\sigma}_{1,max}^{ref}$ at $\theta = 0°$ is used to determine the contributions of all absolute maximum principal stresses on various interference planes, representing the severity of nonproportional loading. The cosine and sine angle between $\vec{\sigma}_{1,max}^{e}(t)$ and $\vec{\sigma}_{1,max}^{ref}$ can be determined as

$$\cos\xi(t) = \frac{\vec{\sigma}_{1,max}^{ref}\,\vec{\sigma}_{1,max}^{e}(t)}{|\vec{\sigma}_{1,max}^{ref}||\vec{\sigma}_{1,max}^{e}(t)|} = \frac{\sigma_{1,max,x}^{ref}\sigma_{1,max}^{e}(t) + \sigma_{1,max,y}^{ref}\sigma_{1,y}^{e}(t)}{|\vec{\sigma}_{1,max}^{ref}||\vec{\sigma}_{1,max}^{e}(t)|} \qquad (5.38)$$

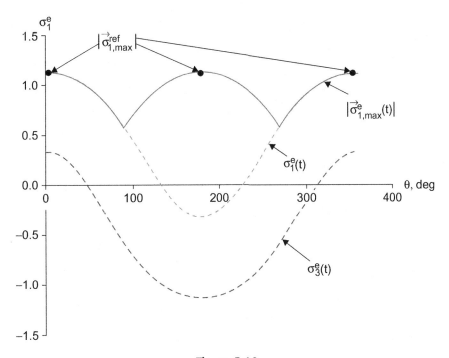

Figure 5.12
Plots of maximum elastic principal stresses and angles due to the fictitious 90° out-of-phase stressing.

$$\sin \xi(t) = \sqrt{1 - \left(\cos \xi(t)\right)^2}. \tag{5.39}$$

Figure 5.13 shows the plots of $|\vec{\sigma}^{\,e}_{1,max}(t)|$, $|\sin \xi(t)|$, and $|\sin \xi(t) \times |\vec{\sigma}_{1,max}(t)||$ due to the fictitious 90° out-of-phase stressing. The area under the curve of $|\sin \xi(t) \times |\vec{\sigma}_{1,max}(t)||$ equals to $\sum_1^T |\sin \xi(t_i) \times |\vec{\sigma}^{\,e}_{1,max}(t_i)|| = 27.26$. By setting $f_{NP} = 1$ in Equation (5.28), C can be determined with the following equation:

$$C = \frac{f_{NP} \cdot T \cdot \sigma^{ref}_{1,max}}{\sum_1^T |\sin \xi(t_i) \times |\vec{\sigma}^{\,e}_{1,max}(t_i)||} = \frac{1 \times 72 \times 1.125}{27.26} = 2.97.$$

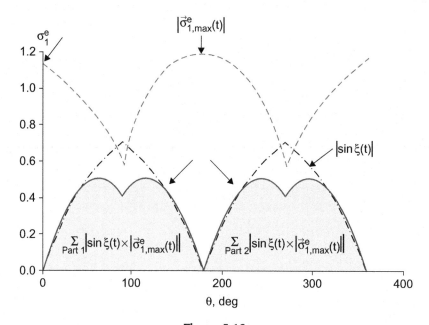

Figure 5.13
Plots of $|\vec{\sigma}_{1,max}^{e}(t)|$, $|\sin\xi(t)|$, and $|\sin\xi(t)\times|\vec{\sigma}_{1,max}^{e}(t)||$ due to a fictitious
90° out-of-phase stressing.

2. The local pseudo stress time histories in Figure 5.10 can be cross plotted
   as shown in Figure 5.14. Thus, the one cycle of the out-of-phase stressing
   history can be described by four points, Points 1 through 4, as tabulated
   in Table 5.1. Choose an extreme stress point as a reference point at which
   the stress components are algebraically either a maximum or minimum.
   If an extreme point is not easily identified, it will require trying different
   alternatives for the largest stress amplitude. Because of no mean stress
   found in this history and the assumption of $\alpha_S = 1.0$, Equation (5.28) is
   reduced to

$$\sigma_{VM,a}^{e} = \max\left\{\sqrt{(\sigma_{x,a}^{e})^{2} + (\sigma_{y,a}^{e})^{2} - \sigma_{x,a}^{e}\sigma_{y,a}^{e} + 3(\tau_{xy,a}^{e})^{2}}\right\}. \qquad (5.40)$$

This table shows the calculation procedures for the value of $\sigma_{VM,a}^{e}$, based
on Point 1 as a reference point. So as a result, $\sigma_{VM,a}^{e} = 280.3$ MPa.

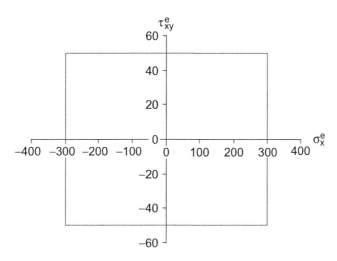

Figure 5.14
The applied out-of-phase relation between σ_x^e and τ_{xy}^e.

Table 5.1: Tabulated Data to Calculate $\sigma_{VM,a}^e$

Point		σ_x^e	σ_y^e	τ_{xy}^e	$\sigma_{x,a}^e$	$\sigma_{y,a}^e$	$\tau_{xy,a}^e$	$\sigma_{VM,a}^e$
1	Ref.	−300	−90	−50	−	−	−	−
2	−	300	90	−50	300	90	0	266.6
3	−	300	90	50	300	90	50	280.3
4	−	−300	−90	50	0	0	50	86.6

The next step is to calculate the f_{NP} factor due to the applied out-of-phase stressing path, which is divided into 40 data points. According to the calculated maximum principal stress components in x and y axes, $|\overrightarrow{\sigma}_{1,max}^e(t)|$ and $|\sin\xi(t)\times|\overrightarrow{\sigma}_{1,max}^e||$ can be determined and plotted against θ in Figure 5.15. The area under the curve of $|\sin\xi(t)\times|\overrightarrow{\sigma}_{1,max}^e||$ is equal to $\sum_1^T|\sin\xi(t_i)\times|\overrightarrow{\sigma}_{1,max}^e(t_i)||=2468.5$. Given C = 2.97 and T = 40, f_{NP} can be found to be 0.59 by using Equation (5.28).

The equivalent stress amplitude is then calculated as follows:

$$\sigma_{VM,a,NP}=\sigma_{VM,a}(1+\alpha_{NP}f_{NP})=280.3(1+0.3\times0.59)=330\text{ MPa}.$$

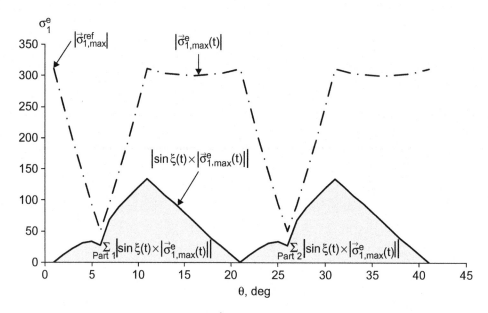

Figure 5.15
Plots of $|\vec{\sigma}_{1,max}^{e}(t)|$ and $|\sin\xi(t)\times|\vec{\sigma}_{1,max}^{e}||$ versus θ angles due to the applied out-of-phase loading.

This is found to be less than the fatigue limit for normal stress ($\sigma_{E,R=-1} = 700\,MPa$) by a safety factor of 2.1. Therefore, we can conclude that the welded structural component under the nonproportional stressing will survive for infinite life.

Critical Plane Approach

The critical plane approach has been commonly used for fatigue analyses of components under variable amplitude, multiaxial loading. It is known that nonproportional loading results in rotating the maximum principal and the shear stress planes at a crack initiation location where the potential crack orientation is changing as well.

A final crack will eventually initiate on a certain orientation where the fatigue damage parameter representing the crack nucleation and growth is maximized. Therefore, this approach essentially involves the critical plane searching technique to identify the orientation of the highest damage plane.

The fatigue damage parameter can be a function of shear stress and normal stress on the plane to evaluate the degree of a damage process. The dominant stress used in a fatigue damage parameter depends on the fatigue damage mechanism. For example, fatigue initiation life of ductile materials is typically dominated by crack initiation and growth on maximum shear stress planes; therefore, shear stress and normal stress are the primary and the secondary damage parameters to use, respectively. On the other hand, the fatigue initiation life of brittle materials is typically controlled by crack growth along maximum tensile stress plane; normal stress and shear stress are the primary and secondary parameters to use, respectively.

Based on nucleation and growth of fatigue cracks, two cracks termed Case A and Case B cracks by Brown and Miller (1973) are considered. Case A crack is the surface crack, which tends to be very shallow and has a small aspect ratio. Case B crack is the in-depth crack, propagating into the surface. As illustrated in Figures 5.16 and 5.17, Case A cracks grow on the planes perpendicular to the z axis, whereas Case B cracks grow on the planes perpendicular to the y axis. The critical plane searching techniques based on Case A and Case B cracks are addressed next.

First, the critical plane search for Case A cracks is described. Figure 5.16(b) shows two possible critical planes that are perpendicular to the z axis, each of which locates at an interference angle θ* with respect to the y–z surface. And the typical normal and shear stresses acting on an interference plane with the angle of θ* are depicted in Figure 5.18. For the plane of maximum damage, it is necessary

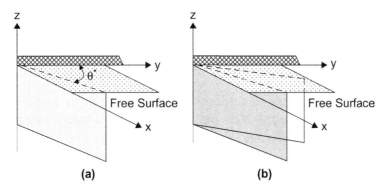

Figure 5.16
Case A cracks: (a) critical plane orientation and (b) potential critical planes at a crack initiation site.

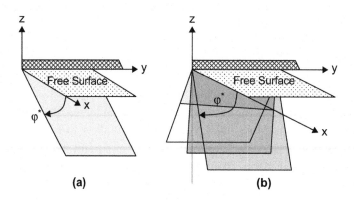

Figure 5.17
Case B cracks: (a) critical plane orientation and (b) potential critical planes at a crack initiation site.

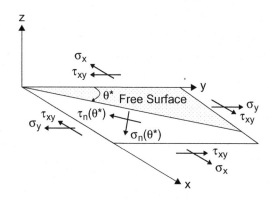

Figure 5.18
Normal and shear stresses on an interference plane θ^* from the y–z plane for Case A cracks.

to assess the fatigue damage parameter on each critical plane by varying θ^* from $0°$ to $180°$ at every $5°$ interval. Note that a $5°$ interval is common practice, but it could lead to discretization errors in some cases.

Second, for Case B cracks, each critical plane locates at an angle φ^* with respect to the x–y plane and has its normal vector defined as the local x′ axis, as illustrated in Figure 5.19. For the plane of maximum damage, it is necessary to assess a damage parameter on each critical plane by varying φ^* from $0°$ to $180°$ at every $5°$ interval.

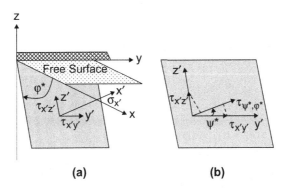

Figure 5.19
(a) Normal and shear stresses on an interference plane ψ^* from the x–y
surface for Case B cracks (b) coordinate transformation of shear
stresses on a potential critical plane.

Using an appropriate coordinate transformation matrix, the normal stress and
shear stresses acting on the plane orientated at φ^* can be determined. The two
shear stress components ($\tau_{x'y'}$ and $\tau_{x'z'}$) along this critical plane produce a resultant shear stress, $\tau_{\psi^*,\phi}$, as shown in Figure 5.19. Both magnitude and direction
of the resultant shear stress vary with time.

Thus, an angle of ψ^* with respect to the local z' axis is introduced for the direction
of the resultant shear stress. At each time instant and at a specific φ^*, the magnitude
of the resultant shear stress at ψ^* varying from $0°$ to $180°$ at every $5°$ interval is calculated. During one load cycle, the most damaging plane at the angles of φ^* and ψ^*
can be identified by searching for the largest of the fatigue damage parameter.

In summary, a critical plane searching approach would require a stress transformation technique and the definition of fatigue damage parameter. Both are discussed
next.

Stress Transformation

The state of stresses at a point relative to a global xyz coordinate system is given
by the stress matrix

$$[\sigma]_{xyz} = \begin{bmatrix} \sigma_{xx} & \tau_{xy} & \tau_{zx} \\ \tau_{xy} & \sigma_{yy} & \tau_{yz} \\ \tau_{zx} & \tau_{yz} & \sigma_{zz} \end{bmatrix} \tag{5.41}$$

and the state of stresses at the same point relative to a local x'y'z' coordinate system is noted as

$$[\sigma]_{x'y'z'} = \begin{bmatrix} \sigma'_{xx} & \tau'_{xy} & \tau'_{zx} \\ \tau'_{xy} & \sigma'_{yy} & \tau'_{yz} \\ \tau'_{zx} & \tau'_{yz} & \sigma'_{zz} \end{bmatrix}. \tag{5.42}$$

It is shown that the stress transformation is

$$[\sigma]_{x'y'z'} = [T] \cdot [\sigma]_{xyz} \cdot [T]^T \tag{5.43}$$

where

$[T]$ = the coordinate transformation matrix that transforms a vector in the xyz system to a vector in the x'y'z' system

$[T]^T$ = the transpose of $[T]$

Consider a tetrahedron composed of four triangular faces, as shown in Figure 5.20(a), three of which meet at a vertex (0) and are perpendicular to each other. The common vertex and three sides of the tetrahedron define a global xyz coordinate system, whereas a local x'y'z' coordinate system is introduced along the fourth face, where x' is normal to the surface. Since the x'y'z' system can be related to the xyz system by two inclination angles (θ and φ), it can be expressed as

$$\begin{Bmatrix} x' \\ y' \\ z' \end{Bmatrix} = [T] \cdot \begin{Bmatrix} x \\ y \\ z \end{Bmatrix}. \tag{5.44}$$

The [T] matrix is obtained by two subsequent coordinate rotations. The first counterclockwise rotation of the x–y plane about the z axis by an inclination angle of θ, as shown in Figure 5.20(b), results in the following coordinate relationship:

$$\begin{Bmatrix} x_1 \\ y_1 \\ z_1 \end{Bmatrix} = \begin{bmatrix} \cos\theta & \sin\theta & 0 \\ -\sin\theta & \cos\theta & 0 \\ 0 & 0 & 1 \end{bmatrix} \begin{Bmatrix} x \\ y \\ z \end{Bmatrix}. \tag{5.45}$$

Then the x'y'z' coordinates can be established by rotating the x_1–z_1 plane clockwise about y_1 with an inclination angle of φ, as shown in Figure 5.20(c), and

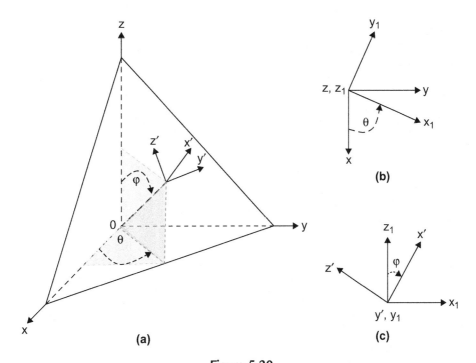

Figure 5.20
Transformation of global and local coordinates: (a) global and local coordinates;
(b) first rotation; (c) second rotation.

they are related to the $x_1 y_1 z_1$ coordinates by

$$\begin{Bmatrix} x' \\ y' \\ z' \end{Bmatrix} = \begin{bmatrix} \sin\varphi & 0 & \cos\varphi \\ 0 & 1 & 0 \\ -\cos\varphi & 0 & \sin\varphi \end{bmatrix} \begin{Bmatrix} x_1 \\ y_1 \\ z_1 \end{Bmatrix}. \tag{5.46}$$

Combining the two coordinate transformation matrices leads to

$$\begin{Bmatrix} x' \\ y' \\ z' \end{Bmatrix} = \begin{bmatrix} \sin\varphi & 0 & \cos\varphi \\ 0 & 1 & 0 \\ -\cos\varphi & 0 & \sin\varphi \end{bmatrix} \begin{bmatrix} \cos\theta & \sin\theta & 0 \\ -\sin\theta & \cos\theta & 0 \\ 0 & 0 & 1 \end{bmatrix} \begin{Bmatrix} x \\ y \\ z \end{Bmatrix}$$

$$= \begin{bmatrix} \cos\theta \cdot \sin\varphi & \sin\theta \cdot \sin\varphi & \cos\varphi \\ -\sin\theta & \cos\theta & 0 \\ -\cos\theta \cdot \cos\varphi & -\sin\theta \cdot \cos\varphi & \sin\varphi \end{bmatrix} \begin{Bmatrix} x \\ y \\ z \end{Bmatrix}. \tag{5.47}$$

Thus,

$$[T] = \begin{bmatrix} \cos\theta \cdot \sin\varphi & \sin\theta \cdot \sin\varphi & \cos\varphi \\ -\sin\theta & \cos\theta & 0 \\ -\cos\theta \cdot \cos\varphi & -\sin\theta \cdot \cos\varphi & \sin\varphi \end{bmatrix}. \tag{5.48}$$

Fatigue Damage Parameters

Three popular stress-based damage parameters are reviewed in this section, among which the Findley (1959) and the Gaier–Dannbauer (2008) theories are recommended for fatigue damage assessments of structures under variable amplitude, multiaxial loading conditions.

Findley

Findley (1959) first proposed the concept of the critical plane approach and assumed a fatigue crack will form on the plane where the damage parameter (the shear stress amplitude τ_a with some contributions from the maximum normal stress $\sigma_{n,max}$ on that plane) is maximized and exceeds the fatigue limit in shear τ_E. It can be expressed as follows:

$$\left(\tau_{Findley,a}\right)_{max} = \tau_E \tag{5.49}$$

$$\tau_{Findley,a} = \tau_a + k_F\sigma_{n,max} \tag{5.50}$$

where

k_F = the normal stress sensitivity factor, representing the influence of the maximum normal stress on the maximum shear stress amplitude

k_F can be determined by fatigue limits from two loading conditions. If $\sigma_{n,max}$ exceeds the tensile yield strength of the material, then $\sigma_{n,max} = \sigma_{t,y}$.

McDiarmid

McDiarmid (1987) developed a generalized failure criterion including the crack initiation modes as follows:

$$\tau_{McDiarmid,a} = \tau_{A,B,E} \tag{5.51}$$

$$\tau_{McDiarmid,a} = \tau_a + \left(\frac{\tau_{A,B,E}}{2\sigma_{t,u}}\right)\sigma_{n,max} \tag{5.52}$$

where

$$\tau_{A,B,E} = \text{the shear fatigue limit for Case A or B crack}$$
$$\sigma_{t,u} = \text{the ultimate tensile strength}$$
$$\tau_a \text{ and } \sigma_{n,max} = \text{the shear stress amplitude and maximum normal stress on}$$
$$\text{the critical plane with the maximum shear stress amplitude}$$

Gaier and Dannbauer

After reviewing Gough's experimental data for combined bending and torsional loads, Gaier and Dannbauer (2008) proposed that a fatigue crack will likely initiate on the critical plane where the scaled normal stress on the plane (f_{σ_n}) is maximized and exceeds a failure criterion. The failure criterion is defined as

$$(f_{GD}\sigma_n)_{max} = \sigma_{E,R=-1} \tag{5.53}$$

$$f_{GD} = 1 + (1-k)V \tag{5.54}$$

where

f_{GD} = the scaled normal stress factor
$k = \sigma_{E,R=-1}/\tau_E$
V = the ratio of the minimum to maximum principal stresses, defined as
$V = \sigma_3/\sigma_1$ for $|\sigma_1| > |\sigma_3|$ and as $V = \sigma_1/\sigma_3$ for $|\sigma_3| > |\sigma_1|$

It has been found that the k values are 2.0 and 1.0 for ductile and brittle materials, respectively. The V value ranges between -1 and $+1$. $V = -1$ for shear loading where $\sigma_3 = -\sigma_1$. $V = 0$ for tension/bending loading where $\sigma_3 = 0$. $V = +1$ for hydrostatic loading where $\sigma_1 = \sigma_2 = \sigma_3$.

It has been proven that Equation (5.53) is identical to Equation (5.5) proposed by Socie and Marquis, which agrees very well with experimental data for fully reversed combined in-phase normal and shear stresses. The most damaging effect of torsion load on ductile materials is considered by scaling up the stress tensor with a factor $f_{GD} > 1$. The magnitude of the hydrostatic stress is scaled down, which is in good agreement with distortion energy criterion. Furthermore, brittle materials result in $k = 1$ and $f_{GD} = 1$, which matches up with the normal stress hypothesis.

Example 5.3

The staircase method for fatigue limit testing was conducted on thin-walled tubular specimens subjected to axial sinusoidal loading with R = −1 and R = 0. Two fatigue limits are found to be $\sigma_{E,R=-1}$ = 700 MPa and $\sigma_{E,R=0}$ = 560 MPa.

1. Determine the material parameters used in the Findley theory.

2. Determine whether the component under the nonproportional loads as shown in Figure 5.10 will survive for infinite life, using the critical plane approach with the Findley damage parameter.

3. Determine whether the component under the nonproportional loads as shown in Figure 5.10 will survive for an infinite life, using the critical plane approach with the Gaier–Dannbauer damage parameter.

Solution

1. Based on the applied axial stress (σ_x), the normal and shear stresses (σ_θ and τ_θ) on a plane oriented at an angle θ from a local x axis can be computed as follows:

$$\sigma_\theta = \frac{\sigma_x}{2} + \frac{\sigma_x}{2}\sin(2\theta) \tag{5.55}$$

$$\tau_\theta = \frac{\sigma_x}{2}\cos(2\theta) \tag{5.56}$$

$$(\tau_a + k_F\sigma_n)_{max} = \frac{1}{2}\left(\sigma_a\cos2\theta + \sigma_x k_F(1+\sin2\theta)\right)_{max}$$

$$= \frac{1}{2}\left(\sqrt{\sigma_a^2 + k_F^2\sigma_{max}^2} + k_F\sigma_{max}\right) \tag{5.57}$$

(a) For R = −1 loading, $\sigma_a = \sigma_{max} = \sigma_{E,R=-1}$ = 700 MPa

$$\left(\sqrt{\sigma_{E,R=-1}^2 + k_F^2\sigma_{E,R=-1}^2} + k_F^2\sigma_{E,R=-1}\right) = 2\tau_E \tag{5.58}$$

(b) For R = 0 loading, $\sigma_a = \sigma_{E,R=0}$ = 560 MPa and $\sigma_{max} = 2\sigma_{E,R=0}$ = 1120 MPa

$$\left(\sqrt{\sigma_{E,R=0}^2 + (2k_F^2\sigma_{E,R=0})^2} + 2k_F^2\sigma_{E,R=0}\right) = 2\tau_E \tag{5.59}$$

Dividing Equation (5.58) by Equation (5.59) leads to

$$\frac{k_F + \sqrt{1 + k_F^2}}{2k_F + \sqrt{1 + (2k_F)^2}} = \frac{560}{700} = 0.8$$

from the equation k_F = 0.23 and τ_E = 440 MPa.

2. As shown in Figures 5.16 and 5.17, the inclination angles of interference planes for Case A and Case B cracks, θ^* and φ^*, are defined, respectively. For the case of welded joints, the majority of cracks typically initiate along the weld toes line. Therefore, Case B cracks at $\theta^* = 0°$ along the weld toe line are the cracks of interest, and critical planes into the surface will be searched. For the damage assessment of Case B cracks at $\theta^* = 0°$, the normal stress and shear stresses acting on an interference plane with an inclination angle of φ^* can be determined by an appropriate stress transformation. In this case, the coordinate transformation matrix $[T]$ is modified by setting $\theta = 0°$ and $\varphi = \varphi^* + 90°$ as

$$[T] = \begin{bmatrix} \cos\theta \cdot \sin\varphi & \sin\theta \cdot \sin\varphi & \cos\varphi \\ -\sin\theta & \cos\theta & 0 \\ -\cos\theta \cdot \cos\varphi & -\sin\theta \cdot \cos\varphi & \sin\varphi \end{bmatrix} = \begin{bmatrix} \cos\varphi^* & 0 & -\sin\varphi^* \\ 0 & 1 & 0 \\ \sin\varphi^* & 0 & \cos\varphi^* \end{bmatrix}. \quad (5.60)$$

Therefore, the stress state corresponding to a new coordinate system is

$$[\sigma]_{x'y'z'} = [T][\sigma]_{xyz}[T]^T$$

$$= \begin{bmatrix} \sigma^e_{xx} \cdot \cos^2\varphi^* & \tau^e_{xy} \cdot \cos\varphi^* & \sigma^e_{xx} \cdot \sin\varphi^* \cdot \cos\varphi^* \\ \tau^e_{xy} \cdot \cos\varphi^* & \sigma^e_{yy} & \tau^e_{xy} \cdot \sin\varphi^* \\ \sigma^e_{xx} \cdot \sin\varphi^* \cdot \cos\varphi^* & \tau^e_{xy} \cdot \sin\varphi^* & \sigma^e_{xx} \cdot \sin^2\varphi^* \end{bmatrix}. \quad (5.61)$$

As a result, the stress components on the interference plane are

$$\sigma^e_{x'x'} = \sigma^e_{xx} \cdot \cos^2\varphi^* \quad (5.62)$$

$$\tau^e_{x'y'} = \tau^e_{xy} \cdot \cos\varphi^* \quad (5.63)$$

$$\tau^e_{x'z'} = \sigma^e_{xx} \cdot \sin\varphi^* \cdot \cos\varphi^*. \quad (5.64)$$

The two shear stresses on the specific plane can be combined to produce a resultant shear stress on the plane. Since the direction and magnitude of the resultant shear stress change with time, there is a need to determine the critical shear stress plane at an angle ψ^* with respect to the y' axis. As shown in Figure 5.19(b), the magnitude of the resultant shear stress corresponding to ψ^* from $0°$ to $180°$ is then computed as follows:

$$\begin{aligned} \tau^e_{\psi^*,\varphi^*} &= \tau^e_{x'y'}\cos\psi^* + \tau^e_{x'z'}\sin\psi^* \\ &= \tau^e_{xy}\cos\varphi^*\cos\psi^* + \sigma^e_{xx}\sin\varphi^*\cos\varphi^*\sin\psi^*. \end{aligned} \quad (5.65)$$

Using Equations (5.62) and (5.65), the normal and shear stresses on interference planes in every $20°$ increment are computed throughout the loading history as described by 4 points, as illustrated in Table 5.2. Actually every $5°$ increment should be used for actual practice, but the $20°$ increment is adopted in this example for illustration. Maximum shear stress

Table 5.2: Resultant Shear Stresses ($\tau^e_{\psi*,\varphi*}$) and Normal Stresses ($\sigma^e_{x'x'}$) on Interference Planes ($0° \leq \varphi*$ and $\psi* \leq 180°$)

Point 1: $\sigma^e_{xx} = -300\,\text{MPa}$; $\tau^e_{xy} = -50\,\text{MPa}$; $\sigma^e_{yy} = -90\,\text{MPa}$
$\sigma^e_1 = 0\,\text{MPa}$; $\sigma^e_2 = -78.70\,\text{MPa}$; $\sigma^e_3 = -311.30\,\text{MPa}$; V = 0

$\tau^e_{\psi*,\varphi*}$

$\psi*\backslash\varphi*$	0	20	40	60	80	100	120	140	160
0	−50	−47	−38	−25	−9	9	25	38	47
20	−47	−77	−87	−68	−26	26	68	87	77
40	−38	−98	−124	−103	−40	40	103	124	98
60	−25	−107	−147	−125	−49	49	125	147	107
80	−9	−103	−152	−132	−52	52	132	152	103
100	9	−87	−139	−124	−49	49	124	139	87
120	25	−60	−109	−100	−40	40	100	109	60
140	38	−26	−66	−64	−26	26	64	66	26
160	47	11	−15	−21	−9	9	21	15	−11

$\sigma^e_{x'x'}$

	−300	−265	−176	−75	−9	−9	−75	−176	−265

Point 2: $\sigma^e_{xx} = 300\,\text{MPa}$; $\tau^e_{xy} = -50\,\text{MPa}$; $\sigma^e_{yy} = 90\,\text{MPa}$
$\sigma^e_1 = 311.30\,\text{MPa}$; $\sigma^e_2 = 78.70\,\text{MPa}$; $\sigma^e_3 = 0.0\,\text{MPa}$; V = 0

$\tau^e_{\psi*,\varphi*}$

$\psi*\backslash\varphi*$	0	20	40	60	80	100	120	140	160
0	−50	−47	−38	−25	−9	9	25	38	47
20	−47	−11	15	21	9	−9	−21	−15	11
40	−38	26	66	64	26	−26	−64	−66	−26
60	−25	60	109	100	40	−40	−100	−109	−60
80	−9	87	139	124	49	−49	−124	−139	−87
100	9	103	152	132	52	−52	−132	−152	−103
120	25	107	147	125	49	−49	−125	−147	−107
140	38	98	124	103	40	−40	−103	−124	−98
160	47	77	87	68	26	−26	−68	−87	−77

$\sigma^e_{x'x'}$

	300	265	176	75	9	9	75	176	265

Point 3: $\sigma^e_{xx} = 300\,\text{MPa}$; $\tau^e_{xy} = 50\,\text{MPa}$; $\sigma^e_{yy} = 90\,\text{MPa}$
$\sigma^e_1 = 311.30\,\text{MPa}$; $\sigma^e_2 = 78.70\,\text{MPa}$; $\sigma^e_3 = 0.0\,\text{MPa}$; V = 0

$\tau^e_{\psi*,\varphi*}$

$\psi*\backslash\varphi*$	0	20	40	60	80	100	120	140	160
0	50	47	38	25	9	−9	−25	−38	−47
20	47	77	87	68	26	−26	−68	−87	−77

Table 5.2: —Cont'd

Point 3: $\sigma_{xx}^e = 300$ MPa; $\tau_{xy}^e = 50$ MPa; $\sigma_{yy}^e = 90$ MPa
$\sigma_1^e = 311.30$ MPa; $\sigma_2^e = 78.70$ MPa; $\sigma_3^e = 0.0$ MPa; $V = 0$

$\tau_{\psi*,\varphi*}^e$

40	38	98	124	103	40	−40	−103	−124	−98
60	25	107	147	125	49	−49	−125	−147	−107
80	9	103	152	132	52	−52	−132	−152	−103
100	−9	87	139	124	49	−49	−124	−139	−87
120	−25	60	109	100	40	−40	−100	−109	−60
140	−38	26	66	64	26	−26	−64	−66	−26
160	−47	−11	15	21	9	−9	−21	−15	11

$\sigma_{x'x'}^e$

300	265	176	75	9	9	75	176	265

Point 4: $\sigma_{xx}^e = -300$ MPa; $\tau_{xy}^e = 50$ MPa; $\sigma_{yy}^e = -90$ MPa
$\sigma_1^e = 0$ MPa; $\sigma_2^e = -78.70$ MPa; $\sigma_3^e = -311.30$ MPa; $V = 0$

$\tau_{\psi*,\varphi*}^e$

$\psi*\backslash\varphi*$	0	20	40	60	80	100	120	140	160
0	−50	47	38	25	9	−9	−25	−38	−47
20	47	11	−15	−21	−9	9	21	15	−11
40	38	−26	−66	−64	−26	26	64	66	26
60	25	−60	−109	−100	−40	40	100	109	60
80	9	−87	−139	−124	−49	49	124	139	87
100	−9	−103	−152	−132	−52	52	132	152	103
120	−25	−107	−147	−125	−49	49	125	147	107
140	−38	−98	−124	−103	−40	40	103	124	98
160	−47	−77	−87	−68	−26	26	68	87	77

$\sigma_{x'x'}^e$

−300	−265	−176	−75	−9	−9	−75	−176	−265

amplitudes, maximum normal stresses, and Findley's shear stress amplitudes with $k_F = 0.23$ on various interference planes are summarized in Table 5.3. It is found that the maximum Findley stress amplitude $(\tau_{Findley,a}^e)_{max}$ is located on two planes ($\varphi* = 20°$ and $140°$) with the magnitude of 193 MPa. The factor of safety is then computed as 2.3, based on the ratio of τ_E to $(\tau_{Findley,a}^e)_{max}$.

3. According to the Gaier–Dannbauer damage parameter in the critical plane approach, it is necessary to determine the scaled normal stress on the interference plane. By definition, $k = \sigma_{E,R=-1}/\tau_E = 1.59$ is obtained from the two given fatigue limits and $V = 0$ is found for each stress point in Table 5.2. The normal stresses at

Table 5.3: Maximum Shear Stress Amplitudes ($\tau^e_{a,max}$), Maximum Normal Stresses ($\sigma^e_{x'x',max}$), and Findley's Equivalent Shear Stress Amplitudes ($\tau^e_{Findley,a}$) on Interference Planes ($0° \leq \phi^*$ and $\psi^* \leq 180°$)

	$\tau^e_{a,max}$								
$\psi^*\backslash\varphi^*$	0	20	40	60	80	100	120	140	160
0	50	47	38	25	9	9	25	38	47
20	47	77	87	68	26	26	68	87	77
40	38	98	124	103	40	40	103	124	98
60	25	107	147	125	49	49	125	147	107
80	9	103	152	132	52	52	132	152	103
100	9	103	152	132	52	52	132	152	103
120	25	107	147	125	49	49	125	147	107
140	38	98	124	103	40	40	103	124	98
160	47	77	87	68	26	26	68	87	77
$\tau^e_{a,max}$	50	107	152	132	52	52	132	152	107
$\sigma^e_{x'x',max}$	300	265	176	75	9	9	75	176	265
$\tau^e_{Findley,a}$	119	168	**193**	150	54	54	150	**193**	168

Table 5.4: Normal Stresses $\sigma^e_{x'x'}$ and Maximum Normal Stresses $\sigma^e_{x'x',max}$ on Interference Planes ($0° \leq \phi^* \leq 180°$)

	$\sigma^e_{x'x'}$								
Point$\backslash\varphi^*$	0	20	40	60	80	100	120	140	160
1	−300	−265	−176	−75	−9	−9	−75	−176	−265
2	300	265	176	75	9	9	75	176	265
3	300	265	176	75	9	9	75	176	265
4	−300	−265	−176	−75	−9	−9	−75	−176	−265
	$\sigma^e_{x'x',max}$								
	300	265	176	75	9	9	75	176	265

each stress point on every interference plane are tabulated in Table 5.4, where the maximum normal stress of 300 MPa is found on the plane of $\varphi^* = 0°$. So the maximum scaled normal stress is determined

$$\left[\left(1 + (1-k)V\right) \cdot \sigma^e_{x'x'}\right]_{max} = 300 \text{ MPa} < \sigma_{E,R=-1} = 700 \text{ MPa}$$

and the factor of safety is obtained as 2.3, based on the ratio of 700 MPa to 300 MPa.

Dang Van's Multiscale Approach

This is a multiscale fatigue initiation approach based on the use of mesoscopic stresses instead of engineering macroscopic quantities. The fatigue initiation process can be described by three different scales: the microscopic scale of dislocations, the mesoscopic scale of plastic slip bands localized in some crystalline grains, and the macroscopic scale of short crack development. The macroscopic scale is characterized by an element of finite element mesh.

As shown in Figure 5.21, Dang Van (Dang Van et al., 1982; Dang Van, 1993) postulated that for an infinite lifetime (near the fatigue limit), crack nucleation in slip bands may occur in the most unfavorably oriented grains, which are subjected to plastic deformation even if the macroscopic stress is elastic. Residual stresses in these plastically deformed grains will be induced due to the restraining effect of the adjacent grains.

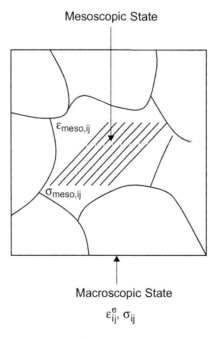

Figure 5.21
Stresses and strains in the macroscopic and mesoscopic states.

Dang Van also assumed that an elastic shakedown in a macroscopic state occurs before the fatigue limit and that both mesoscopic and macroscopic plastic strains and residual stresses are stabilized. Note that a material is said to exhibit elastic shakedown, corresponding to stabilization of elastic responses, if yielding does not occur during unloading and subsequent reloading is wholly elastic.

The theory assumes that for an infinite lifetime, the macroscopic elastic strain tensor ($\underline{\varepsilon}^e(t)$) is the sum of the mesoscopic elastic and plastic strain tensors in the grain ($\underline{\varepsilon}^e_{meso}(t)$, $\underline{\varepsilon}^p_{meso}(t)$). It is expressed as follows:

$$\underline{\varepsilon}^e(t) = \underline{\varepsilon}^e_{meso}(t) + \underline{\varepsilon}^p_{meso}(t) \tag{5.66}$$

By definition,

$$\underline{\varepsilon}^e(t) = \frac{\underline{\sigma}(t)}{E} \tag{5.67}$$

$$\underline{\varepsilon}^e_{meso}(t) = \frac{\underline{\sigma}_{meso}(t)}{E_{meso}} \tag{5.68}$$

where

$$E = \text{the Young's modulus in macroscopic scale}$$
$$E_{meso} = \text{the Young's modulus in mesoscopic level}$$
$$\underline{\sigma}(t) \text{ and } \underline{\sigma}_{meso}(t) = \text{the macroscopic and mesoscopic stress tensors,}$$
$$\text{respectively}$$

Then, by substituting Equations (5.67) and (5.68), Equation (5.66) can be rewritten in term of stresses as

$$\underline{\sigma}_{meso}(t) = \frac{E_{meso}}{E}\underline{\sigma}(t) - E_{meso}\underline{\varepsilon}^p_{meso}. \tag{5.69}$$

Assuming $\frac{E_{meso}}{E} = 1$, Equation (5.69) becomes

$$\underline{\sigma}_{meso}(t) = \underline{\sigma}(t) - E_{meso}\underline{\varepsilon}^p_{meso} = \underline{\sigma}(t) + \underline{\rho}^* \tag{5.70}$$

where a local residual stress tensor in the mesoscopic scale ($\underline{\rho}^*$) is introduced by

$$\underline{\rho}^* = -E_{meso}\underline{\varepsilon}^p_{meso} \tag{5.71}$$

and $\underline{\rho}^*$ is a deviatoric tensor because it is proportional to the mesoscopic plastic strains.

It should be noted that both $\underline{\rho}^*$ and $\underline{\varepsilon}_{meso}^p$ are stabilized and independent of time after a number of loading cycles. Also the mesoscopic hydrostatic pressure $\underline{\sigma}_{meso,h}(t)$ is equal to the macroscopic one:

$$\underline{\sigma}_{meso,h} = \underline{\sigma}_h = \frac{1}{3}(\underline{\sigma}:\underline{I})\underline{I}. \tag{5.72}$$

Thus, the same relation between the instantaneous macroscopic deviatoric stress tensor $\underline{S}(t)$ and mesoscopic tensor $\underline{s}_{meso}(t)$ can be applied:

$$\underline{s}_{meso}(t) = \underline{S}(t) + \underline{\rho}^*. \tag{5.73}$$

Dang Van assumed $\underline{\rho}^* = -\underline{\alpha}^*$ where $\underline{\alpha}^*$ is the center of the smallest von Mises yield surface, σ_{VM}, that completely encloses the path described by the macroscopic deviatoric stress tensor. $\underline{\rho}^*$ is calculated by changing $\underline{\alpha}$, the back stress tensor, to minimize the maximum of the von Mises stresses calculated from the macro deviatoric stress tensor with respect to the updated yield surface center.

An initial guess or a starting point for the yield surface center is calculated as the average of the macro deviatoric stress points. It should be noted that for proportional loading, $\underline{\alpha}^*$ is the average of the macro deviatoric stress points.

The min-max solution process can be expressed as follows:

$$\underline{\alpha}^* = \min_{\underline{\alpha}} \left\{ \max_t \sigma_{VM}(t) \right\} \tag{5.74}$$

$$\sigma_{VM}(t) = \sqrt{\frac{3}{2}\left(\underline{S}(t) - \underline{\alpha}(t)\right):\left(\underline{S}(t) - \underline{\alpha}(t)\right)}$$

$$= \sqrt{\frac{3}{2}\left[(S_{11} - \alpha_{11})^2 + (S_{22} - \alpha_{22})^2 + (S_{33} - \alpha_{33})^2 + 2\left((S_{12} - \alpha_{12})^2 + (S_{23} - \alpha_{23})^2 + (S_{31} - \alpha_{31})^2\right)\right]}. \tag{5.75}$$

This min-max concept is depicted in Figure 5.22. Once $\underline{\alpha}^*$ and $\underline{\rho}^*$ are calculated, the mesoscopic tensor $\underline{s}_{meso}(t)$ can be determined by Equation (5.73) and the instantaneous mesoscopic shear stress can be calculated as

$$\tau_{meso}(t) = \frac{1}{2}\left[s_{meso,1}(t) - s_{meso,3}(t)\right] \tag{5.76}$$

where

$s_{meso,1}(t)$ and $s_{meso,3}(t) =$ the largest and smallest mesoscopic deviatoric principal stresses, respectively

The Dang Van criterion is constructed on a mesoscopic approach to fatigue behavior on a critical shearing plane. It assumes that the mesoscopic shear stress ($\tau_{meso}(t)$) on a grain is responsible for crack nucleation in slip bands within a grain and the mesoscopic hydrostatic stress ($\sigma_{meso,h}(t)$) will influence the opening of these cracks. The formula uses the mesoscopic shear stress and mesoscopic hydrostatic stress to calculate an "equivalent" stress and compare it to a shear fatigue limit (τ_E).

As shown in Figure 5.23, no fatigue damage will occur if

$$\max\{\max_t |\tau_{meso}(t) + \alpha_{DV}\sigma_{meso,h}(t)|\} \leq \tau_E \qquad (5.77)$$

where

$\alpha_{DV} =$ the hydrostatic stress sensitivity

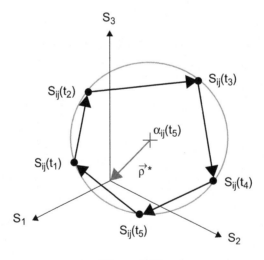

Figure 5.22
The smallest von Mises yield surface enclosing all the macroscopic deviatoric stress tensors.

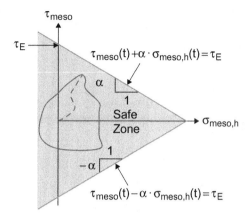

Figure 5.23
The Dang Van criterion for success or failure.

A safety factor for each time instant t can then be defined as follows:

$$SF(t) = \frac{\tau_E}{\tau_{meso}(t) + \alpha_{DV}\sigma_{micro,h}(t)}. \tag{5.78}$$

Example 5.4

The staircase method for fatigue limit testing was conducted on thin-walled tubular specimens subjected to axial sinusoidal loading with R = −1 and R = 0. The two fatigue limits (in amplitude) are found as $\sigma_{E,R=-1}$ of 700 MPa and $\sigma_{E,R=0}$ of 560 MPa.

1. Determine the material parameters used in the Dang Van multiscale approach.

2. The structural component is subjected to complex multiaxial loading resulting in the macroscopic normal stress histories at a critical location as illustrated in Figure 5.24. It is an engineering requirement that this component should have a minimum safety factor of 1.5. Determine whether this component would meet the engineering regulation, using the Dang Van multiscale approach.

3. If the component under the nonproportional macroscopic stressing as shown in Figure 5.10 will survive for an infinite life, please determine whether this component would meet the same engineering regulation, using the Dang Van multiscale approach.

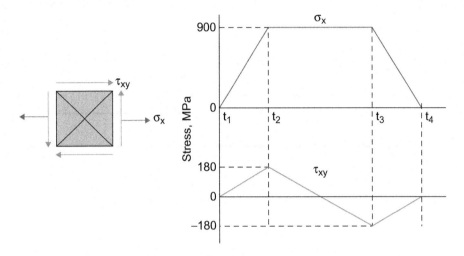

Figure 5.24
Macroscopic plane stress histories at a critical finite element.

Solution

1. For a given fatigue limit of 700 MPa with $R = -1$ loading, at a peak stress point denoted at a time instant t_1, the macroscopic stress, hydrostatic, and deviatoric tensors are

$$\underline{\sigma}(t_1) = \begin{bmatrix} 700 & 0 & 0 \\ 0 & 0 & 0 \\ 0 & 0 & 0 \end{bmatrix} \quad \underline{\sigma}_h(t_1) = \begin{bmatrix} 233.3 & 0 & 0 \\ 0 & 233.3 & 0 \\ 0 & 0 & 233.3 \end{bmatrix}.$$

$$\underline{S}(t_1) = \underline{\sigma}(t_1) - \underline{\sigma}_h(t_1) = \begin{bmatrix} 466.7 & 0 & 0 \\ 0 & -233.3 & 0 \\ 0 & 0 & -233.3 \end{bmatrix}.$$

Similarly, at a valley stress point denoted at a time instant t_2, the macroscopic stress, hydrostatic, and deviatoric tensors are

$$\underline{\sigma}(t_2) = \begin{bmatrix} -700 & 0 & 0 \\ 0 & 0 & 0 \\ 0 & 0 & 0 \end{bmatrix} \quad \underline{\sigma}_h(t_2) = \begin{bmatrix} -233.3 & 0 & 0 \\ 0 & -233.3 & 0 \\ 0 & 0 & -233.3 \end{bmatrix}$$

$$\underline{S}(t_2) = \underline{\sigma}(t_2) - \underline{\sigma}_h(t_2) = \begin{bmatrix} -466.7 & 0 & 0 \\ 0 & 233.3 & 0 \\ 0 & 0 & 233.3 \end{bmatrix}.$$

Due to proportional loading, the center of the final yield surface is equal to the average deviatoric stresses from t_1 to t_2. That means

$$\underline{\alpha}^* = \frac{\underline{S}(t_1) + \underline{S}(t_2)}{2} = \begin{bmatrix} 0 & 0 & 0 \\ 0 & 0 & 0 \\ 0 & 0 & 0 \end{bmatrix}.$$

Thus, the macro residual stress tensor becomes

$$\underline{\rho}^* = -\underline{\alpha}^* = \begin{bmatrix} 0 & 0 & 0 \\ 0 & 0 & 0 \\ 0 & 0 & 0 \end{bmatrix}.$$

The mesoscopic parameters, such as $\underline{s}_{meso}(t_1)$, $\sigma_{meso,h}(t_1)$, and $\tau_{meso}(t_1)$, are determined as follows:

$$\underline{s}_{meso}(t_1) = \underline{S}(t_1) + \underline{\rho}^* \quad \underline{s}_{meso}(t_1) = \begin{bmatrix} 466.7 & 0 & 0 \\ 0 & -233.3 & 0 \\ 0 & 0 & -233.3 \end{bmatrix}$$

$$\sigma_{meso,h}(t_1) = \frac{\sigma_x + \sigma_y + \sigma_z}{3} = \frac{700 + 0 + 0}{3} = 233.3$$

$$\tau_{meso}(t_1) = \frac{1}{2}(466.7 - (-233.3)) = 350.$$

Similarly, $\underline{s}_{meso}(t_2)$, $\sigma_{meso,h}(t_2)$, and $\tau_{meso}(t_2)$ can be obtained as follows:

$$\underline{s}_{meso}(t_2) = \begin{bmatrix} -466.7 & 0 & 0 \\ 0 & 233.3 & 0 \\ 0 & 0 & 233.3 \end{bmatrix}$$

$$\sigma_{meso,h}(t_2) = \frac{-700}{3} = -233.3 \quad \tau_{meso}(t_2) = \frac{1}{2}(233.3 - (-466.7)) = 350.$$

Next, for a given fatigue limit of 560 MPa with R = 0 loading, the macroscopic stress, hydrostatic, and deviatoric tensors at the peak and valley stress points denoted by t_1 and t_2 are

$$\underline{\sigma}(t_1) = \begin{bmatrix} 1120 & 0 & 0 \\ 0 & 0 & 0 \\ 0 & 0 & 0 \end{bmatrix} \quad \underline{\sigma}_h(t_1) = \begin{bmatrix} 373 & 0 & 0 \\ 0 & 373 & 0 \\ 0 & 0 & 373 \end{bmatrix}$$

$$\underline{S}(t_1) = \underline{\sigma}(t_1) - \underline{\sigma}_h(t_1) = \begin{bmatrix} 747 & 0 & 0 \\ 0 & -373 & 0 \\ 0 & 0 & -373 \end{bmatrix}.$$

$$\underline{\sigma}(t_2) = \begin{bmatrix} 0 & 0 & 0 \\ 0 & 0 & 0 \\ 0 & 0 & 0 \end{bmatrix} \quad \underline{\sigma}_h(t_2) = \begin{bmatrix} 0 & 0 & 0 \\ 0 & 0 & 0 \\ 0 & 0 & 0 \end{bmatrix}$$

$$\underline{S}(t_2) = \underline{\sigma}(t_2) - \underline{\sigma}_h(t_2) = \begin{bmatrix} 0 & 0 & 0 \\ 0 & 0 & 0 \\ 0 & 0 & 0 \end{bmatrix}.$$

Due to proportional loading, the center of the final yield surface is equal to the average deviatoric stresses from t_1 to t_2. That means

$$\underline{\alpha}^* = \frac{\underline{S}(t_1) + \underline{S}(t_2)}{2} = \begin{bmatrix} 373.5 & 0 & 0 \\ 0 & -186.5 & 0 \\ 0 & 0 & -186.5 \end{bmatrix}.$$

The macro residual stress tensor $\underline{\rho}^*$ becomes

$$\underline{\rho}^* = -\underline{\alpha}^* = \begin{bmatrix} -373.5 & 0 & 0 \\ 0 & 186.5 & 0 \\ 0 & 0 & 186.5 \end{bmatrix}.$$

The mesoscopic parameters such as $\underline{s}_{meso}(t_1)$, $\sigma_{meso,h}(t_1)$, and $\tau_{meso}(t_1)$ are determined as follows:

$$\underline{s}_{meso}(t_1) = \underline{S}(t_1) + \underline{\rho}^* \quad \underline{s}_{meso}(t_1) = \begin{bmatrix} 406.5 & 0 & 0 \\ 0 & -153.5 & 0 \\ 0 & 0 & -153.5 \end{bmatrix}$$

$$\sigma_{meso,h}(t_1) = \frac{1120}{3} = 373.5 \quad \tau_{meso}(t_1) = \frac{1}{2}(406.5 - (-153.5)) = 280.$$

Similarly, $s_{meso,ij}(t_2)$, $\sigma_{meso,h}(t_2)$, and $\tau_{meso}(t_2)$ can be obtained as follows:

$$\underline{s}_{meso}(t_2) = \underline{S}(t_2) + \underline{\rho}^* \quad \underline{s}_{meso}(t_2) = \begin{bmatrix} -373.5 & 0 & 0 \\ 0 & 186.5 & 0 \\ 0 & 0 & 186.5 \end{bmatrix}$$

$$\sigma_{meso,h}(t_2) = \frac{0+0+0}{3} = 0 \quad \tau_{meso}(t_2) = \frac{1}{2}(186.5 - (-373.5)) = 279.5.$$

The two mesoscopic parameters determined from the two fatigue limits and located in the first quadrant of the $\sigma_{meso,h} - \tau_{meso}$ diagram are used to determine the Dang Van failure criterion. They are ($\sigma_{meso,h}(t_1)$ = 233.3, $\tau_{meso}(t_1)$ = 350) and ($\sigma_{meso,h}(t_2)$ = 373.5, $\tau_{meso}(t_2)$ = 280), from R = −1 and R = 0 loading, respectively. Consequently, the two constants of the fatigue limit criterion (α_{DM} = 0.5 and τ_E = 467 MPa) are obtained.

2. Three macroscopic stress tensors in time are considered to define the complete nonproportional stress history in Figure 5.24. The three points are:

$$\underline{\sigma}(t_1) = \begin{bmatrix} 0 & 0 & 0 \\ 0 & 0 & 0 \\ 0 & 0 & 0 \end{bmatrix} \quad \underline{\sigma}(t_2) = \begin{bmatrix} 900 & 0 & 180 \\ 0 & 0 & 0 \\ 180 & 0 & 0 \end{bmatrix} \quad \underline{\sigma}(t_3) = \begin{bmatrix} 900 & 0 & -180 \\ 0 & 0 & 0 \\ -180 & 0 & 0 \end{bmatrix}$$

$$\text{At } t_1, \underline{\sigma}_h(t_1) = \begin{bmatrix} 0 & 0 & 0 \\ 0 & 0 & 0 \\ 0 & 0 & 0 \end{bmatrix} \quad \underline{S}(t_1) = \begin{bmatrix} 0 & 0 & 0 \\ 0 & 0 & 0 \\ 0 & 0 & 0 \end{bmatrix}.$$

$$\text{At } t_2, \underline{\sigma}_h(t_2) = \begin{bmatrix} 300 & 0 & 0 \\ 0 & 300 & 0 \\ 0 & 0 & 300 \end{bmatrix} \quad \underline{S}(t_2) = \begin{bmatrix} 600 & 0 & 180 \\ 0 & -300 & 0 \\ 180 & 0 & -300 \end{bmatrix}.$$

$$\text{At } t_3, \underline{\sigma}_h(t_3) = \begin{bmatrix} 300 & 0 & 0 \\ 0 & 300 & 0 \\ 0 & 0 & 300 \end{bmatrix} \quad \underline{S}(t_3) = \begin{bmatrix} 600 & 0 & -180 \\ 0 & -300 & 0 \\ -180 & 0 & -300 \end{bmatrix}.$$

The initial guess of a center of the yield surface is equal to the average macroscopic deviatoric stresses from t_1 to t_3.

$$\underline{\alpha}^*_{initial} = \begin{bmatrix} 400 & 0 & 0 \\ 0 & -200 & 0 \\ 0 & 0 & -200 \end{bmatrix}.$$

The von Mises stresses at t_1, t_2, and t_3 are calculated as 600 MPa, 432.7 MPa, and 432.7 MPa, respectively. The maximum von Mises stress is 600 MPa calculated on the macro deviatoric stresses with respect to the initial guessed yield surface center. By minimizing the maximum von Mises stress during several iterations, the final yield center solution can be obtained as follows:

$$\underline{\alpha}^* = \begin{bmatrix} 336 & 0 & 0 \\ 0 & -168 & 0 \\ 0 & 0 & -168 \end{bmatrix}.$$

The final corresponding von Mises stresses at t_1, t_2, and t_3 are 504 MPa, 504 MPa, and 504 MPa. Thus, the macro residual stress tensor is

$$\underline{\rho}^* = -\underline{\alpha}^* = \begin{bmatrix} -336 & 0 & 0 \\ 0 & 168 & 0 \\ 0 & 0 & 168 \end{bmatrix}.$$

Finally, the mesoscopic stress tensors at t_1 are

$$\underline{s}_{meso}(t) = \underline{S}(t) + \underline{\rho}^* \qquad \underline{s}_{meso}(t_1) = \begin{bmatrix} -336 & 0 & 0 \\ 0 & 168 & 0 \\ 0 & 0 & 168 \end{bmatrix}$$

$$\sigma_{meso,h}(t_1) = \frac{1}{3}(0) = 0 \qquad \tau_{meso}(t_1) = \frac{1}{2}\left(168 - (-336)\right) = 252$$

$$252 + 0.5 \cdot 0 = 252 < 467$$

$$SF(t1) = \frac{467}{252} = 1.85 > 1.5 \,(\text{meeting the engineering requirement}).$$

At t_2,

$$\underline{s}_{meso}(t_2) = \begin{bmatrix} 600 & 0 & 180 \\ 0 & -300 & 0 \\ 180 & 0 & -300 \end{bmatrix} + \begin{bmatrix} -336 & 0 & 0 \\ 0 & 168 & 0 \\ 0 & 0 & 168 \end{bmatrix}$$

$$= \begin{bmatrix} 264 & 0 & 180 \\ 0 & -132 & 0 \\ 180 & 0 & -132 \end{bmatrix}$$

$$s_{meso,1}(t_2) = 333.6 \quad s_{meso,2}(t_2) = -132 \quad s_{meso,3}(t_2) = -201.6$$

$$\sigma_{meso,h}(t_2) = \frac{1}{3}(900) = 300 \qquad \tau_{meso}(t_2) = \frac{1}{2}(333.6 - (-201.6)) = 267.6$$

$$267.6 + 0.5 \times (300) = 417.6 < 467$$

$$SF(t_2) = \frac{467}{417.6} = 1.12 < 1.5 \,(\text{NOT meeting the engineering requirement}).$$

At t_3,

$$\underline{s}_{meso}(t_3) = \begin{bmatrix} 600 & 0 & -180 \\ 0 & -300 & 0 \\ -180 & 0 & -300 \end{bmatrix} + \begin{bmatrix} -336 & 0 & 0 \\ 0 & 168 & 0 \\ 0 & 0 & 168 \end{bmatrix}$$

$$= \begin{bmatrix} 264 & 0 & -180 \\ 0 & -132 & 0 \\ -180 & 0 & -132 \end{bmatrix}$$

$$s_{meso,1}(t_3) = 333.6 \quad s_{meso,2}(t_3) = -132 \quad s_{meso,3}(t_3) = -201.6$$

$$\sigma_{meso,h}(t_3) = \frac{1}{3}(900) = 300 \qquad \tau_{meso}(t_3) = \frac{1}{2}(333.6 - (-201.6)) = 267.6$$

$$267.6 + 0.5 \times (300) = 417.6 < 467$$

$$SF(t_3) = \frac{467}{417.6} = 1.12 < 1.5 \text{ (NOT meeting the engineering requirement)}.$$

In conclusion, this component subjected to the nonproportional loading will not meet the engineering requirement because the minimum safety factor of 1.12 found at both time instant t_2 and t_3 is less than the required 1.5.

3. Four macroscopic plane stress tensors in time are considered to define the complete nonproportional stress history in Figure 5.10. Following the same calculation procedures as stated in step 2, the mesoscopic stresses and the factors of safety are tabulated in Table 5.5. It therefore is concluded that this

Table 5.5: Output of Mesoscopic Stresses Calculation Procedures (Stress Unit in MPa)

	Macroscopic Stresses				
	σ_{xx}	σ_{yy}	τ_{xy}	σ_{zz}	σ_h
1	−300	−90	−50	0	−130
2	300	90	−50	0	130
3	300	90	50	0	130
4	−300	−90	50	0	−130

	Macroscopic Deviatoric Stresses and von Mises Stress						
	S_{xx}	S_{yy}	S_{zz}	S_{xy}	S_{xz}	S_{yz}	σ_{VM}
1	−170	40	130	−50	0	0	280
2	170	−40	−130	−50	0	0	280
3	170	−40	−130	50	0	0	280
4	−170	40	130	50	0	0	280
α^*_{ini}	0	0	0	0	0	0	
α^*	0	0	0	0	0	0	280

	Mesoscopic Deviatoric Stresses					
	$S_{meso,xx}$	$S_{meso,yy}$	$S_{meso,zz}$	$S_{meso,xy}$	$S_{meso,xz}$	$S_{meso,yz}$
1	−170	40	130	−50	0	0
2	170	−40	−130	−50	0	0
3	170	−40	−130	50	0	0
4	−170	40	130	50	0	0

	Principal Stresses, Shear Stresses, Hydrostatic Stresses, and SF					
	$S_{meso,1}$	$S_{meso,2}$	$S_{meso,3}$	τ_{meso}	$\sigma_{meso,h}$	SF
1	130	51.3	−181.3	155.6	−130	5.15
2	181.3	−51.3	−130.0	155.6	130	2.11
3	181.3	−51.3	−130.0	155.6	130	2.11
4	130	51.3	−181.3	155.6	−130	5.15

component under the nonproportional loading would survive for infinite life because the minimum safety factor of 2.11 exceeds the engineering requirement (1.5).

Summary

Based on the assumption there is very little plastic deformation in the HCF regime, the pseudo or fictitious stress components calculated from a linear elastic FEA can be used to estimate the fatigue damage parameter and predict life from the synthesized pseudo stress-life curve.

Multiaxial cyclic loading can be classified into two categories: nonproportional and proportional loading. Nonproportional loading is the loading path that causes the principal stress axis or the maximum shear stress axis of a local element to rotate with time and with respect to a local coordinate system. On the other hand, proportional loading will result in a stationary principal stress axis or a maximum shear stress axis.

Nonproportional loading causes equal or more damage than proportional loading, based on the same von Mises stress range. The cause of this phenomenon has been explained by an additional nonproportional strain hardening due to slip behavior of the material. The severity of nonproportional hardening is dependent on the ease on which slip systems interact and the type of loading path.

Numerous stress-based multiaxial fatigue damage models have been developed and can be categorized into four different classifications: empirical formula approach, equivalent stress approach, critical plane approach, and Dang Van multiscale approach.

The empirical formula approach is referred to as the best-fit equation to the experimental fatigue limit data tested under combined normal and shear stresses, and it is only applicable to the biaxial stress state in a fully reversed loading condition.

The equivalent stress approach, based on either the von Mises or Tresca yield theory, was originally developed to define fatigue limit criteria of various materials under proportional loading. All the equivalent stress formulations are limited to fatigue assessment under proportional loading, except for the two methods introduced by Sonsino (1995) and Lee et al. (2007) where the von Mises equivalent stress formula has been modified to account for the nonproportional hardening effect.

The critical plane approach provides a physical interpretation of the damage initiation process. The crack orientation can be identified by searching for the most damaging plane among numerous potential crack initiation planes, where the damage parameter is usually defined as a function of the shear and normal stresses on such a plane. However, based on the pseudo stress input excluding strain hardening in the material modeling, it is very challenging to account for the nonproportional hardening effect on the fatigue damage.

The Dang Van multiscale approach is a popular approach for assessing mesoscopic fatigue damage at the fatigue limit. It is assumed that an elastic shakedown in a macroscopic state occurs before the fatigue limit and that both mesoscopic and macroscopic plastic strains and residual stresses are stabilized.

References

Archer, R. (1987). Fatigue of welded steel attachments under combined direct stress and shear stress. Paper No. 50, International Conference on Fatigue of Welded Constructions. Brighton, UK: The Welding Institute.

Borodii, M. V., & Strizhalo, V. O. (2004). Hardening and lifetime predicting under biaxial low-cycle fatigue. In *Proceedings of the seventh international conference on biaxial/ multiaxial fatigue and fracture* (pp. 279–284). Berlin, Germany.

Borodii, M. V., & Shukaev, S. M. (2007). Additional cyclic strain hardening and its relation to material structure, mechanical characteristics, and lifetime. *International Journal of Fatigue, 29,* 1184–1191.

Brown, M. W., & Miller, K. J. (1973). A theory for fatigue failure under multiaxial stress-strain conditions. *Applied Mechanics Group, 187,* 745–755.

Dang Van, K., Griveau, B., & Message, O. (1982). On a new multiaxial fatigue limit criterion: Theory and application. In M. W. Brown & K. Miller (Eds.), *Biaxial and Multiaxial Fatigue* (pp. 479–496). EGF Publication 3. London: Mechanical Engineering Publications.

Dang Van, K. (1993). Macro-micro approach in high-cycle multiaxial fatigue. In D. L. McDowell & R. Ellis (Eds.), *Advances in Multiaxial Fatigue,* ASTM STP 1991 (pp. 120–130). Philadelphia: ASTM.

Dietmann, H. (1973a). Werkstoffverhalten unter mehrachsiger schwingender Beanspruchung, Teil 1: Berechnungsmöglichkeiten. *Materialwissenschaft und Werkstofftechnik (Journal of Materials Science and Engineering Technology), 5,* 255–263.

Dietmann, H. (1973b). Werkstoffverhalten unter mehrachsiger schwingender Beanspruchung, Teil 2: Experimentelle Untersuchungen. *Materialwissenschaft und Werkstofftechnik (Journal of Materials Science and Engineering Technology), 6*, 322–333.

Doong, S. H., Socie, D. F., & Robertson, I. M. (1990). Dislocation substructures and nonproportional hardening. *Journal of Engineering Materials and Technology, 112*(4), 456–465.

Findley, W. N. (1959). A theory for the effect of mean stress on fatigue of metal under combined torsion and axial load or bending. *Journal of Engineering for Industry, 81*, 301–306.

Gaier, C., & Dannbauer, H. (2008). A multiaxial fatigue analysis method for ductile, semi-ductile, and brittle materials. *Arabian Journal for Science and Engineering, 33*(1B), 224–235.

Gough, H. J., & Pollard, H. V. (1935). The strength of metals under combined alternating stresses. *Proceedings of the Institution of Mechanical Engineers, 131*, 3–103.

Gough, H. J. (1950). Engineering steels under combined cyclic and static stresses. *Journal of Applied Mechanics, 50*, 113–125.

Hashin, Z. (1981). Fatigue failure criteria for combined cyclic stress. *International Journal of Fracture, 17*(2), 101–109.

Itoh, T., Sakane, M., Ohnami, M., & Socie, D. F. (1995). Nonproportional low cycle fatigue criterion for type 304 stainless steel. *Journal of Engineering Materials and Technology, 117*, 285–292.

Itoh, T., Sakane, M., Ohnami, M., & Socie, D. F. (1997). Nonproportional low cycle fatigue of 6061 aluminum alloy under 14 strain paths. *Proceedings of 5th International Conference on Biaxial/Multiaxial Fatigue and Fracture* (pp. 173–187). Cracow, Poland.

Krempl, E., & Lu, H. (1983). Comparison of the stress responses of an aluminum alloy tube to proportional and alternate axial and shear strain paths at room temperature. *Mechanics of Materials, 2*, 183–192.

Lee, S.B. (1985). A criterion for fully reversed out-of-phase torsion and bending. In K. J. Miller & M. W. Brown, (Eds.), *Multiaxial Fatigue*, ASTM STP 853 (pp. 553–568). Philadelphia: ASTM.

Lee, Y. L., & Chiang, Y. J. (1991). Fatigue predictions for components under biaxial reversed loading. *Journal of Testing and Evaluation, JTEVA, 19*(5), 359–367.

Lee, Y. L., Tjhung, T., & Jordan, A. (2007). A life prediction model for welded joints under multiaxial variable amplitude loading histories. *International Journal of Fatigue, 29*, 1162–1173.

McDiarmid, D. L. (1974). A new analysis of fatigue under combined bending and twisting. *The Aeronautical Journal of the Royal Aeronautical Society, 78*, 325–329.

McDiarmid, D. L. (1987). Fatigue under out-of-phase bending and torsion. *Fatigue and fracture of engineering materials and structures, 9*(6), 457–475.

Miner, M. A. (1945). Cumulative damage in fatigue. *Journal of Applied Mechanics, 67*, A159–A164.

Palmgren, A. (1924). Die Lebensdauer von Kugellagern, (The service life of ball bearings). *Zeitschrift des Vereinesdeutscher Ingenierure, 68*(14), 339–341.

Rotvel, F. (1970). Biaxial fatigue tests with zero mean stresses using tubular specimens. *International Journal of Mechanics and Science, 12*, 597–613.

Shamsaei, N., & Fatemi, A. (2010). Effect of microstructure and hardness on non-proportional cyclic hardening coefficient and predictions. *Material Science and Engineering, A527*, 3015–3024.

Sines, G. (1959). Behavior of metals under complex static and alternating stresses. In G. Sines & J. L. Waisman (Eds.), *Metal Fatigue* (pp. 145–169). New York: McGraw-Hill.

Siljander, A., Kurath, P., & Lawrence, F. V. Jr. (1992). Nonproportional fatigue of welded structures. In M. R. Mitchell & R. W. Langraph (Eds.), *Advanced in Fatigue Lifetime Predictive Technique*, ASTM STP 1122 (pp. 319–338). Philadelphia: ASTM.

Sonsino, C. M. (1995). Multiaxial fatigue of welded joints under in-phase and out-of-phase local strains and stresses. *International Journal of Fatigue, 17*(1), 55–70.

Socie, D. F., & Marquis, G.B. (2000). *Multiaxial Fatigue* (pp. 285–294). Warrendale, PA: SAE International.

Wang, C. H., & Brown, M. W. (1996). Life prediction techniques for variable amplitude multiaxial fatigue—Part 1: Theories. *Journal of Engineering Materials and Technology, 118*, 367–374.

CHAPTER 6

Strain-Based Uniaxial Fatigue Analysis

Yung-Li Lee
Chrysler Group LLC

Mark E. Barkey
The University of Alabama

Chapter Outline

Introduction

The local strain–life method is based on the assumption that the life spent on crack nucleation and small crack growth of a notched component is identical to that of a smooth laboratory specimen under the same cyclic deformation (i.e., strain controlled material behavior at the local crack initiation site).

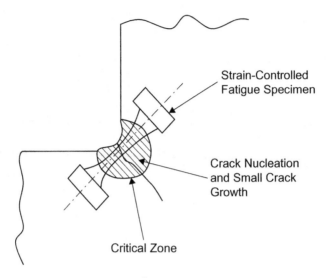

Figure 6.1
Concept of the local strain–life approach.

Using this concept as illustrated in Figure 6.1, it is possible to determine the fatigue initiation life of a cyclically loaded component if the relationship between the localized strain in the specimen and fatigue life is known. This local strain versus life relationship, typically represented as a log-log plot of strain amplitude versus fatigue life in reversals, is generated by conducting cyclic axial strain-controlled tests on smooth, polished material specimens.

Figure 6.2 shows configurations and dimensions of the commonly used material specimens (round bar and flat plate). Cyclic strain-controlled testing is recommended because the material at the stress concentration area in a component may be subjected to cyclic plastic deformation even when the bulk of the component behaves elastically during cyclic loading. The experimental test programs and data reduction technique for cyclic and fatigue material properties can be found elsewhere (Lee et al., 2005) and are excluded from this chapter.

This chapter presents the local strain–life method in a uniaxial state of stress in the following sequences: estimates of cyclic stress–strain and fatigue properties, mean stress correction models, and notch stress and strain analysis.

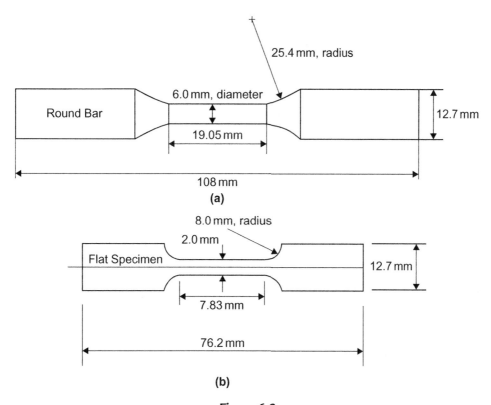

Figure 6.2
Fatigue test specimen (a) configurations and (b) dimensions (provided by
Dr. Benda Yan of ArcelorMittal R&D).

Steady State Cyclic Stress–Strain Relation

For most metals, fatigue life can be characterized by steady-state behavior because for the constant strain-amplitude controlled tests, the stress–strain relationship becomes stable after rapid hardening or softening in the initial cycles, that is, about the first several percent of the total fatigue life. The cyclically stable stress–strain response is termed as the hysteresis loop, as illustrated in Figure 6.3.

The inside of the hysteresis loop defined by the total strain range ($\Delta\varepsilon$) and total stress range ($\Delta\sigma$) represents the elastic plus plastic work on a material undergoing loading and unloading. Usually, the stabilized hysteresis loop is taken at

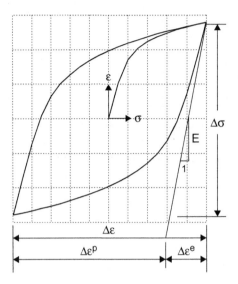

Figure 6.3
Nomenclature for a hysteresis loop.

half of the total fatigue life. It is assumed that the total strain range can be decomposed into elastic and plastic strain components ($\Delta\varepsilon^{e}$, $\Delta\varepsilon^{p}$), which can be expressed as follows:

$$\Delta\varepsilon = \Delta\varepsilon^{e} + \Delta\varepsilon^{p} \tag{6.1}$$

where

E = modulus of elasticity

$$\Delta\varepsilon^{e} = \frac{\Delta\sigma}{E}. \tag{6.2}$$

When a family of stabilized hysteresis loops created by various strain amplitude levels is plotted on the same σ-ε coordinate, a cyclic stress–strain curve can be constructed by the locus of the loop tips, as shown in Figure 6.4, and expressed using the familiar Ramberg–Osgood equation:

$$\varepsilon = \varepsilon^{e} + \varepsilon^{p} = \frac{\sigma}{E} + \left(\frac{\sigma}{K'}\right)^{1/n'} \tag{6.3}$$

where

K' = the cyclic strength coefficient
n' = the cyclic strain hardening exponent
$'$ (superscript) = the parameters associated with "cyclic behavior" to differentiate them from monotonic behavior parameters

The cyclic yield stress (σ'_y) can be defined as the stress at 0.2% of plastic strain on a cyclic stress–strain curve.

Masing (1926) proposed that the stress amplitude (σ_a) versus strain amplitude (ε_a) curve follows the same expression as described by the cyclic stress–strain curve:

$$\varepsilon_a = \varepsilon_a^e + \varepsilon_a^p = \frac{\sigma_a}{E} + \left(\frac{\sigma_a}{K'}\right)^{1/n'} \tag{6.4}$$

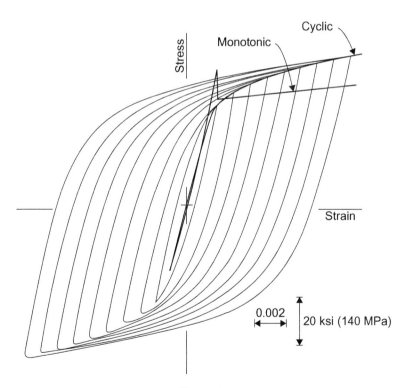

Figure 6.4
Construction of a cyclic stress–strain curve.

where

ε_a^e and ε_a^p = the elastic and plastic strain amplitudes, respectively

The Masing assumption is valid for homogeneous materials that have symmetric behavior in tension and compression. Equation (6.4) can be rewritten in terms of strain range ($\Delta\varepsilon$) and stress range ($\Delta\sigma$) as follows:

$$\frac{\Delta\varepsilon}{2} = \frac{\Delta\varepsilon^e}{2} + \frac{\Delta\varepsilon^p}{2} = \frac{\Delta\sigma}{2E} + \left(\frac{\Delta\sigma}{2K'}\right)^{1/n'}. \tag{6.5}$$

Equation (6.5) can be further reduced to the following equation:

$$\Delta\varepsilon = \frac{\Delta\sigma}{E} + 2\left(\frac{\Delta\sigma}{2K'}\right)^{1/n'}. \tag{6.6}$$

Note that for a given strain increment, the stress increment with respect to a reference turning point can be calculated from Equation (6.6), and vice versa. Equation (6.6) has been widely used for describing and tracking the hysteresis behavior under variable amplitude loading conditions.

When a material response returns back to its previously experienced deformation, it will remember the past path to reach such a state and will follow this path with additional increase in deformation. This is the so-called "memory" effect observed in materials undergoing complex loading histories, which should be considered and can be accounted for by the proper choice of a reference turning point. For example, as can be seen in Figure 6.5, in the third reversal (from

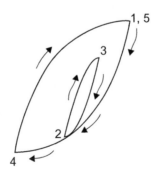

Figure 6.5
Illustration of material memory.

Point 3 to Point 4), Point 1, instead of Point 3, should be used as a reference point after the smaller reversal from Points 3 to 2 is closed, and this loading reversal will follow the path from Points 1 to 4.

Fully Reversed, Constant Amplitude Strain–Life Relation

Based on Morrow's proposal (1965), the relation of the total strain amplitude (ε_a) and fatigue life to failure in reversals ($2N_f$) can be expressed in the following form:

$$\varepsilon_a = \varepsilon_a^e + \varepsilon_a^p = \frac{\sigma_f'}{E}(2N_f)^b + \varepsilon_f'(2N_f)^c \qquad (6.7)$$

where

    σ_f' = the fatigue strength coefficient
    ε_f' = the fatigue ductility coefficient
    b = the fatigue strength exponent
    c = the fatigue ductility exponent

The previous equation is called the strain–life equation for the zero mean stress case, and is the foundation of the local strain–life approach for fatigue. This equation is a summation of two separate curves for elastic strain amplitude versus life ($\varepsilon_a^e - 2N_f$) and for plastic strain amplitude versus life ($\varepsilon_a^p - 2N_f$).

Dividing the strain–life equation (Basquin, 1910) by E, the elastic strain amplitude versus life curve can be obtained as follows:

$$\varepsilon_a^e = \frac{\Delta\varepsilon^e}{2} = \frac{\sigma_a}{E} = \frac{\sigma_f'}{E}(2N_f)^b \qquad (6.8)$$

and the plastic strain amplitude versus life curve, simultaneously developed by Manson (1953) and Coffin (1954), is expressed as

$$\varepsilon_a^p = \frac{\Delta\varepsilon^p}{2} = \varepsilon_f'(2N_f)^c. \qquad (6.9)$$

Both Equations (6.8) and (6.9) are fitted to the experimental data for stress versus life and plastic strain versus life, in which the fatigue life $2N_f$ is chosen as the independent variable. Eliminating $2N_f$ between Equations (6.8) and (6.9) and comparison with the plastic strain amplitude in Equation (6.4) leads to the following estimates:

$$n' = \frac{b}{c} \tag{6.10}$$

$$K' = \frac{\sigma'_f}{(\varepsilon'_f)^{n'}}. \tag{6.11}$$

When plotted on log-log coordinates, both curves become straight lines as shown in Figure 6.6, where the transition fatigue life in reversals ($2N_T$) is defined as the intersection of the elastic and the plastic straight lines. A transition fatigue life occurs when the magnitude of plastic strain amplitude is equal to that of elastic strain amplitude.

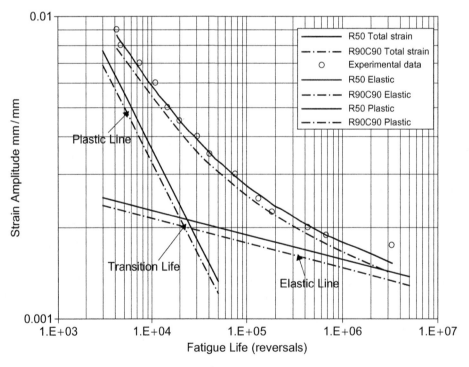

Figure 6.6
Schematic of a total strain–life curve.

The regime to the left of this point where fatigue life is less than the transition fatigue life is considered the plastic strain dominant regime, the so-called low cycle fatigue (LCF) regime. The regime to the right where fatigue life is higher than the transition fatigue life is the elastic strain dominant regime, the high cycle fatigue (HCF) regime.

Overall, steels with high hardness and ultimate tensile strength have lower transition fatigue life. Equating Equations (6.8) and (6.9) leads to the estimate of the transition fatigue life as follows:

$$2N_T = \left(\frac{\varepsilon_f' E}{\sigma_f'}\right)^{1/(b-c)}. \tag{6.12}$$

Estimate of Cyclic Material and Fatigue Properties

If there are no experimental strain–life fatigue data available, an estimate of cyclic and fatigue behavior of a material can be helpful in the design stage. However, the parameter estimation should not eliminate the need for real data. Lee and Song (2006) reviewed and evaluated the existing estimation techniques (Muralidharan & Manson, 1988; Baumel & Seeger, 1990; Roessle & Fatemi, 2000; Meggiolaro & Castro, 2004) for cyclic and fatigue properties. They concluded that for a given ultimate tensile strength ($S_{t,u}$), the uniform material law (Baumel & Seeger, 1990) is recommended for both steels and titanium alloys and the medians method (Meggiolaro & Castro, 2004), for aluminum alloys. Both the uniform material law and the medians method are summarized in Table 6.1.

Several models have been proposed to estimate ultimate tensile strength from hardness because there has been a strong correlation between hardness and ultimate tensile strength. Again, Lee and Song (2006) reviewed most of them and found that Mitchell's equation (Mitchell, 1979) provides the best results for both steels and aluminum alloys. His equation is defined as follows:

$$S_{t,u}(\text{MPa}) = 3.45 \text{HB} \tag{6.13}$$

where

HB = the Brinell hardness

Table 6.1: Estimated Cyclic Material and Fatigue Properties Based
on the Uniform Material Law and the Medians Method

Material Properties	Steels[1]	Titanium and Aluminum Alloys[1]	Aluminum Alloys[2]
σ_f'	$1.5S_{t,u}$	$1.67S_{t,u}$	$1.9S_{t,u}$
b	−0.087	−0.095	−0.11
ε_f'	0.59ψ	0.35	0.28
c	−0.58	−0.069	−0.066
K′	$1.65S_{t,u}$	$1.61S_{t,u}$	—
n′	0.15	0.11	—

Notes: $\psi = 1$ if $\frac{S_{t,u}}{E} \leq 3 \times 10^{-3}$ or $\psi = 1.375 - 125.0\frac{S_{t,u}}{E}$ if $\frac{S_{t,u}}{E} > 3 \times 10^{-3}$.

[1] Uniform material law
[2] Medians method

Strain–Life Equations with Mean Stress

Most experimental data support the fact that compressive mean normal stresses are beneficial, and tensile mean normal stresses are detrimental to fatigue life. However, it actually depends on the damage mechanism in the material; a shear parameter sensitive material would not necessarily have a longer fatigue life if the compressive stress is not aligned with the shear plane. This has been observed under the condition when the fatigue behavior falls in the high cycle fatigue regime where elastic strain is dominant.

In conjunction with the local strain–life approach, many mean stress correction models have been proposed to quantify the effect of mean stresses on fatigue behavior. The modified Morrow equation (Morrow, 1968) and the Smith–Watson–Topper model (Smith et al., 1970) are commonly used and described in the following sections.

Morrow

Morrow (1968) originally presented his mean stress correction model in the stress-life equation. By postulating the mean stress effect is negligible in the LCF regime and can be modeled by the Morrow equation for its noticeable effect in the HCF regime, the strain–life equation (so-called the modified Morrow equation) is then modified as follows:

$$\varepsilon_a = \frac{\sigma_f' - \sigma_m}{E}(2N_f)^b + \varepsilon_f'(2N_f)^c \tag{6.14}$$

where

σ_m = a mean stress

This equation has been extensively used for steels and used with considerable success in the HCF regime. Walcher, Gray, and Manson (1979) have noted that for other materials, such as Ti-6Al-4V, σ_f' is too high a value for the mean stress correction, and an intermediate value of $k_m\sigma_f'$ is introduced. Thus, a generic formula was proposed:

$$\varepsilon_a = \frac{k_m\sigma_f' - \sigma_m}{E}(2N_f)^b + \varepsilon_f'(2N_f)^c. \tag{6.15}$$

This equation requires additional test data to determine $k_m\sigma_f'$.

Smith, Watson, and Topper

Smith, Watson, and Topper (1970) developed another mean stress correction model, by postulating the fatigue damage in a cycle is determined by the product of $\sigma_{max}\varepsilon_a$, where σ_{max} is the maximum stress. They stated that $\sigma_a\varepsilon_a$ for a fully reversed test is equal to $\sigma_{max}\varepsilon_a$ for a mean stress test.

Later in 1995, Langlais and Vogel (1995) expressed this concept in the following form:

$$\sigma_{max}\varepsilon_a = \sigma_{a,rev}\,\varepsilon_{a,rev} \quad \text{for} \quad \sigma_{max} > 0 \tag{6.16}$$

where $\sigma_{a,rev}$ and $\varepsilon_{a,rev}$ are the fully reversed stress and strain amplitudes, respectively, that produce an equivalent fatigue damage due to the SWT parameter.

The value of $\varepsilon_{a,rev}$ should be obtained from the fully reversed, constant amplitude strain–life curve, Equation (6.7), and the value of $\sigma_{a,rev}$, from the cyclic stress–strain curve, Equation (6.3). The SWT parameter predicts no fatigue damage if the maximum tensile stress becomes zero and negative. The solutions to Equation (6.16) can be obtained by using the Newton–Raphson iterative procedure.

For a special case of Equation (6.16), where a material satisfies the compatibility condition among fatigue and strain properties (i.e., $n' = b/c$ and $K' = \sigma_f'/(\varepsilon_f')^{n'}$), the maximum tensile stress for fully-reversed loading is then given by

$$\sigma_{max} = \sigma_a = \sigma_f'(2N_f)^b. \tag{6.17}$$

Also by multiplying the fully reversed, constant amplitude strain–life equation, the SWT mean stress correction formula becomes

$$\sigma_{max}\varepsilon_a = \frac{(\sigma_f')^2}{E}(2N_f)^{2b} + \sigma_f'\varepsilon_f'(2N_f)^{b+c} \quad \sigma_{max} > 0. \tag{6.18}$$

Equation (6.18) has been the widely adopted SWT equation that has been successfully applied to grey cast iron (Fash & Socie, 1982), hardened carbon steels (Koh & Stephens, 1991), microalloyed steels (Forsetti & Blasarin, 1988), and precipitation-hardened aluminum alloys in the 2000 and 7000 series (Dowling, 2009).

Example 6.1

A single active strain gage was placed at the notch root of a notched plate in the loading axis of the plate. The notched component made of SAE 1137 carbon steel has the following material properties:

E = 209,000 MPa, K' = 1230 MPa, n' = 0.161, σ_f' = 1006 MPa,

b = −0.0809, ε_f' = 1.104, c = −0.6207.

The recorded strain time history due to the applied load is repetitive; it is shown in Figure 6.7. Determine the fatigue life of the SAE 1137 notched plate by the following procedures:

1. Plot the cyclic stress–strain response (hysteresis loop).

2. Estimate the fatigue life of the notched plate with the SWT formula.

3. Estimate the fatigue life of the notched plate with the modified Morrow equation.

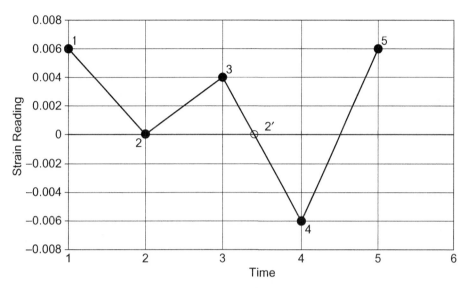

Figure 6.7
Variable amplitude strain-time history.

Solution

During cyclic loading, it is assumed that the material follows the cyclic stress–strain curve for the initial loading and the hysteresis stress–strain behavior for the subsequent loading reversals. The stresses at strain increments of 0.0006 are calculated for the entire block. The stress–strain curve up to point 1 is calculated using a cyclic stress–strain curve equation (the Ramberg–Osgood equation):

$$\varepsilon_1 = \frac{\sigma_1}{E} + \left(\frac{\sigma_1}{K'}\right)^{1/n'}.$$

From then on all incremental reversals with respect to a reference turning point are calculated based on Masing's model:

$$\Delta\varepsilon = \frac{\Delta\sigma}{E} + 2\left(\frac{\Delta\sigma}{2K'}\right)^{1/n'}.$$

Due to the difficulty of algebraically solving for σ and $\Delta\sigma$ in these equations, they are solved using the iterative solver tool in MS Excel. The resulting data are tabulated in Table 6.2, where the reference turning points are listed to

Table 6.2: Calculation Summary of True Stresses and Strains

Reversal	ε	Δε	Δσ (MPa)	σ (MPa)	Ref Point
0 to 1	0.0000	*Reference Point*		0.0	0
	0.0006	0.0006	122.7	122.7	0
	0.0012	0.0012	195.1	195.1	0
	0.0018	0.0018	226.4	226.4	0
	0.0024	0.0024	245.3	245.3	0
	0.0030	0.003	259.0	259.0	0
	0.0036	0.0036	269.7	269.7	0
	0.0042	0.0042	278.6	278.6	0
	0.0048	0.0048	286.2	286.2	0
	0.0054	0.0054	292.9	292.9	0
	0.0060	0.006	298.8	298.8	0
1 to 2	0.0060	*Reference Point*		298.8	1
	0.0054	−0.0006	125.3	173.5	1
	0.0048	−0.0012	245.7	53.1	1
	0.0042	−0.0018	335.4	−36.6	1
	0.0036	−0.0024	390.2	−91.4	1
	0.0030	−0.003	426.2	−127.4	1
	0.0024	−0.0036	452.7	−153.9	1
	0.0018	−0.0042	473.5	−174.6	1
	0.0012	−0.0048	490.6	−191.8	1
	0.0006	−0.0054	505.2	−206.4	1
	0.0000	−0.006	518.0	−219.2	1
2 to 3	0.0000	*Reference Point*		−219.2	2
	0.0004	0.0004	83.8	−135.4	2
	0.0008	0.0008	166.8	−52.3	2
	0.0012	0.0012	245.7	26.6	2
	0.0016	0.0016	310.1	91.0	2
	0.0020	0.002	356.6	137.5	2
	0.0024	0.0024	390.2	171.0	2
	0.0028	0.0028	415.6	196.5	2
	0.0032	0.0032	435.8	216.7	2
	0.0036	0.0036	452.7	233.5	2
	0.0040	0.004	467.0	247.8	2
3 to 4	0.0040	*Reference Point*		247.8	3
	0.0030	−0.001	207.4	40.4	3
	0.0020	−0.002	356.6	−108.8	3
	0.0010	−0.003	426.2	−178.4	3
	0.0000	−0.004	467.0	−219.2	3
	0.0060	*Reference Point*		298.8	1
	−0.0010	−0.007	536.2	−237.3	1
	−0.0020	−0.008	551.6	−252.8	1

Table 6.2: Cont'd

Reversal	ε	$\Delta\varepsilon$	$\Delta\sigma$ (MPa)	σ (MPa)	Ref Point
3 to 4	−0.0030	−0.009	565.1	−266.3	1
	−0.0040	−0.01	577.1	−278.3	1
	−0.0050	−0.011	587.8	−289.0	1
	−0.0060	−0.012	597.6	−298.8	1
4 to 5	−0.0060	*Reference Point*		−298.8	4
	−0.0048	0.0012	245.7	−53.1	4
	−0.0036	0.0024	390.2	91.4	4
	−0.0024	0.0036	452.7	153.9	4
	−0.0012	0.0048	490.6	191.8	4
	0.0000	0.006	518.0	219.2	4
	0.0012	0.0072	539.4	240.6	4
	0.0024	0.0084	557.2	258.4	4
	0.0036	0.0096	572.4	273.6	4
	0.0048	0.0108	585.7	286.9	4
	0.0060	0.012	597.6	298.8	4

properly describe the material memory effect. From Point 1 to Point 2 (the first reversal), Point 1 is the reference point. From Point 2 to Point 3 (the second reversal), Point 2 is the new reference point. The next reversal from Point 3 to Point 4 requires two parts: the first part from Point 3 to Point 2' and the remaining part from Point 2' to Point 4.

Point 2' indicates the return path from Point 3 (a reference point), and is equivalent to Point 2 on the hysteresis due to the material memory effect. After closing the hysteresis loop from Point 3 to Point 2, Point 1 becomes the reference point for the remaining reversal from Point 2' to Point 4. Finally, Point 4 is the reference point for the last reversal Point 4 to Point 5. The calculated stresses and strains for all the reversals are tabulated in Table 6.2, which are also used to plot the hysteresis loop of the single block cycle, shown in Figure 6.8.

The next step is to extract the hysteresis loops and identify the maximum and minimum stress and strain points for each loop so these data could be entered into a damage calculation. Please note that the three-point rainflow cycle counting technique can be used to check the identified hysteresis loops. Using this information, the Smith–Watson–Topper and the modified Morrow formulas are used to evaluate the number of block cycles under which this component will survive.

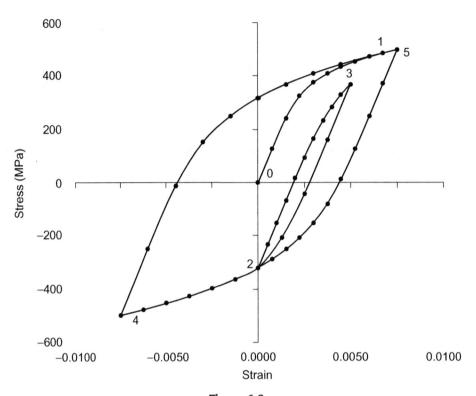

Figure 6.8
Simulated hysteresis loops for the given strain time history.

For each counted cycle, an individual damage number is calculated by employing the linear damage accumulation rule, which is also known as the Palmgren–Miner linear damage rule (Palmgren, 1924; Miner, 1945). The total damage D is defined as:

$$D = \sum_{i=1}^{kn} \frac{n_i}{N_{f,i}} \qquad (6.19)$$

where

    n_i = the number of applied cycles to a constant stress amplitude
    kn = the total number of the stress blocks
    $N_{f,i}$ = the so-called fatigue life as the number of cycles to failure calculated from either the modified Morrow equation or the SWT model

Table 6.3: Summary of Damage Calculation Based on the SWT Parameter

n_i	ε_{max}	ε_{min}	σ_{max} (MPa)	σ_{min} (MPa)	ε_a	N_f (cycles)	d_i
1	0.004	0.000	368.3	−317.7	0.002	194303	5.15E-6
1	0.006	−0.006	497.6	−497.5	0.006	4476	2.23E-4
						Σd_i per block = 2.28E-4	
						# of blocks to failure = 4400	

Table 6.4: Summary of Damage Calculation Based on the Modified Morrow Parameter

n_i	ε_{max}	ε_{min}	σ_{max} (MPa)	σ_{min} (MPa)	σ_m (MPa)	ε_a	N_f (cycles)	d_i
1	0.004	0.000	368.3	−317.7	25.3	0.002	211905	4.72E-6
1	0.006	−0.006	497.6	−497.5	0.005	0.006	4835	2.07E-4
							Σd_i per block = 2.12E-4	
							# of blocks to failure = 4700	

The number of repeats of the given load time history (the number of blocks) can be estimated by assuming failure occurs when D = 1. The results of the SWT and modified Morrow damage calculations and life predictions are tabulated in Tables 6.3 and 6.4, respectively.

Notch Analysis

Notch analysis is referred to as a numerical analysis procedure to estimate the local stress–strain response at a notch root (a stress concentration site) based on the pseudo (fictitious) stress time history from a linear elastic finite element analysis (FEA). For a component with a well-defined notch geometry and configuration, the pseudo stress (σ^e) can also be obtained by the product of the elastic stress concentration factor (K_t) and nominal stress (S).

There have been numerous efforts devoted to the development of notch analyses for expeditious stress–strain calculations. Among these, Neuber's rule (Neuber, 1961) and Molsky–Glinka's energy density method (Molsky & Glinka, 1981) have been widely used and will be discussed in the following sections.

Neuber

Neuber (1961) analyzed a grooved body subjected to monotonic torsional loading and derived a rule for nonlinear material behavior at the notch root. Neuber's paper was written in terms of shear parameters only. Others subsequently extended its meaning for normal stress and strain terms. It is observed, as shown in Figure 6.9, that after local yielding occurs, the local true notch stress (σ) is less than the pseudo stress predicted by the theory of elasticity and the local true notch strain (ε) is greater than that estimated by the theory of elasticity.

Normalizing the local true notch stress with respect to the nominal stress (S) and the true notch strain to the nominal strain (e) leads to the true stress concentration (K_σ) factor and the true strain concentration (K_ε) factor, respectively. Neuber then proposed a hypothesis that the elastic stress concentration factor is the geometric mean of the true stress and strain concentration factors; that is,

$$K_t = \sqrt{K_\sigma \cdot K_\varepsilon}. \tag{6.20}$$

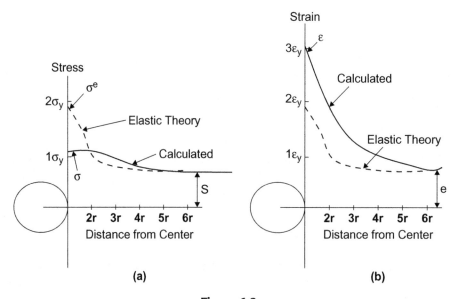

Figure 6.9
Stresses (a) and strains (b) at a notch.

Squaring both sides of Equation (6.20) leads to the following famous Neuber expression:

$$\sigma\varepsilon = K_t^2 Se. \tag{6.21}$$

The physical interpretation of the Neuber rule is shown in Figure 6.10 (note that for the graphical illustration, both sides of Equation (6.21) were divided by 2).

During cyclic loading, it is assumed that the material follows the cyclic stress–strain curve for the initial loading and the hysteresis stress–strain behavior for the subsequent loading reversals. Therefore, in terms of the initial cyclic stress–strain curve, the Neuber equation can be written as

$$\sigma_1\varepsilon_1 = K_t^2 S_1 e_1 \tag{6.22}$$

and, in terms of the hysteresis stress–strain curve, as

$$\Delta\sigma\Delta\varepsilon = K_t^2 \Delta S \Delta e \tag{6.23}$$

where the subscript 1 in Equation (6.22) refers to the initial cyclic loading condition.

Equations (6.22) and (6.23) represent the equations of a hyperbola for given K_t and nominal stress–strain data. The right side of the equation is often referred to

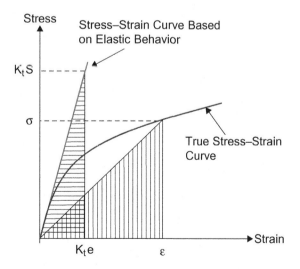

Figure 6.10
Interpretation of the Neuber model.

as Neuber's constant. To solve for the two unknowns σ_1 and ε_1 or $\Delta\sigma$ and $\Delta\varepsilon$, an additional equation for cyclic material behavior is required. Depending on the nominal stress–strain behavior, the Neuber equation will be discussed.

Nominally Elastic Behavior

When the bulk of a notched component behaves elastically and plasticity takes place locally at the notch root (called nominally elastic behavior), the following Neuber equations hold:

For the initial cyclic stress–strain curve

$$e_1 = \frac{S_1}{E} \tag{6.24}$$

$$\varepsilon_1 = \frac{\sigma_1}{E} + \left(\frac{\sigma_1}{K'}\right)^{1/n'}. \tag{6.25}$$

For the hysteresis stress–strain curve

$$\Delta e = \frac{\Delta S}{E} \tag{6.26}$$

$$\Delta\varepsilon = \frac{\Delta\sigma}{E} + 2\left(\frac{\Delta\sigma}{2K'}\right)^{1/n'}. \tag{6.27}$$

where

   S_1 = the nominal stress for initial loading
   e_1 = the nominal strain for initial loading
   σ_1 = the local true notch stress on the initial cyclic loading curve
   ε_1 = the local true notch strain on the initial cyclic loading curve
   Δe = the nominal strain range
   ΔS = the nominal stress range

Substituting the elastic nominal stress–strain and the local cyclic stress–strain relations—Equations (6.24) through (6.27)—into the Neuber equations (Equations (6.22) and (6.23)) results in the following equations:

$$\frac{\sigma_1^2}{E} + \sigma_1\left(\frac{\sigma_1}{K'}\right)^{1/n'} = \frac{(K_t S_1)^2}{E} = \frac{(\sigma_1^e)^2}{E} \tag{6.28}$$

and

$$\frac{(\Delta\sigma)^2}{E} + 2 \cdot \Delta\sigma \left(\frac{\Delta\sigma}{2K'}\right)^{1/n'} = \frac{(K_t \Delta S)^2}{E} = \frac{(\Delta\sigma^e)^2}{E}. \tag{6.29}$$

Given K_t and nominal stress or the pseudo stress data, these equations for the local stress can be solved by using the Newton–Raphson iteration technique. Once the local stress is determined, Equation (6.25) or (6.27) will be used to obtain the corresponding local strain value.

As a rule of thumb, the assumption of the nominally elastic material behavior works well when nominal stress is below 30% of the cyclic yield stress.

Nominally Gross Yielding of a Net Section

When nonlinear net section behavior is considered, the nominal stress (S) and nominal strain (e) need to follow a nonlinear material relationship and Neuber's rule has to be modified too. Seeger and Heuler (1980) proposed the modified version of a nominal stress S^M to account for the general yielding:

$$S^M = S\left(\frac{K_t}{K_p}\right) \tag{6.30}$$

where K_p is known as the limit load factor or the plastic notch factor, and is defined as

$$K_p = \frac{L_p}{L_y} \tag{6.31}$$

where

L_y and L_p = the loads producing first yielding and gross yielding of a net
             section, respectively

A finite element analysis with an elastic-perfectly plastic material can be utilized to determine L_p and L_y. The modified stress and strain (S^M and e^M) follow the cyclic stress–strain equation:

$$e^M = \frac{S^M}{E} + \left(\frac{S^M}{K'}\right)^{1/n'}. \tag{6.32}$$

Since the pseudo notch stress is independent of the definition of nominal stress, a modified elastic stress concentration K^M associated with S^M is introduced in the following form:

$$\sigma^e = S^M K_t^M = SK_t \qquad (6.33)$$

and the Neuber rule can be rewritten

$$\sigma\varepsilon = (K_t^M)^2 S^M e^M. \qquad (6.34)$$

The new form of the Neuber rule can be rearranged and extended to

$$\sigma\varepsilon = \frac{(SK_t)^2}{E}\left(\frac{e^M E}{S^M}\right). \qquad (6.35)$$

Equation (6.35) is the generalized Neuber rule for nonlinear net section behavior. Note that if the $S^M - e^M$ curve remains in the elastic range, the factor $\left(e^M E / S^M\right)$ becomes unity and the generalized Neuber equation reduces to the known Neuber equation.

Modified Neuber Rule

To account for the cyclic material behavior in the local strain–life approach, Topper et al. (1969) proposed to modify the Neuber rule by replacing K_t with the fatigue notch factor (K_f). This has been criticized for both accounting twice the notch root plasticity effects and incorporating the S-N empiricism into the "more fundamentally satisfying" theory. However, by knowing these conflicting results, we recommend that the modified Neuber rule be used for local notch stress–strain estimates and fatigue life predictions.

The use of the K_f factor in Neuber's rule would experience a problem with a notched component where the notch geometry and configuration is complex because the nominal stress S and the K_t factor are difficult to quantify. However, the product of $K_t S$ (the pseudo stress) is obtained from a linear elastic finite element analysis. Therefore, there is a need to convert $K_t S$ to $K_f S$ in the modified Neuber rule.

The K_t–K_f relationship can only be obtained experimentally. In the past empirical average stress models (Peterson, 1959; Neuber, 1946; Heywood, 1962) have been developed to estimate the K_t–K_f factor with the reference to a notch radius (r) and the ultimate tensile strength $S_{t,u}$. Moreover, the following expression for the K_t/K_f ratio was developed by Siebel and Stieler (1955):

$$\frac{K_t}{K_f} = 1 + \sqrt{C_{ss}\overline{G}}$$
(6.36)

where

  C_{ss} = a material constant dependent on yield strength (σ_y)
  $\overline{G}$ = the relative stress gradient, defined as

$$\overline{G} = \frac{1}{\sigma^e_{max}}\left(\frac{d\sigma^e(x)}{dx}\right)_{x=0}$$
(6.37)

where

  σ^e_{max} = the maximum local pseudo stress
  $\sigma^e(x)$ = the theoretically calculated pseudo stress distribution near a notch
      root
  x = the normal distance from the notch root (as shown in Figure 6.11)

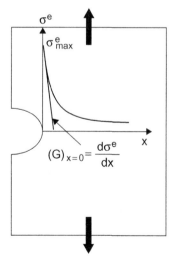

Figure 6.11
Definition of relative stress gradient.

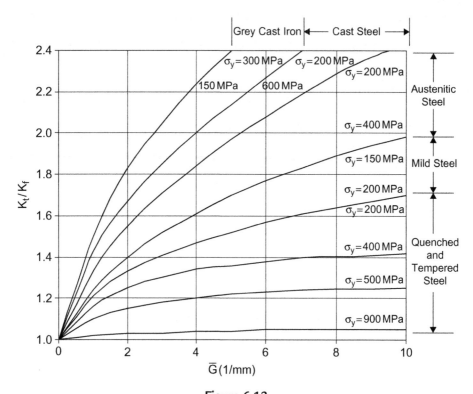

Figure 6.12
Empirical relative stress gradient correction factor for steel, aluminum and cast iron (Siebel and Stieler, 1955).

Equation (6.36) is the generalized formula for the K_t/K_f ratio for various materials with different yield strengths as illustrated in Figure 6.12.

This unique relationship between the K_t/K_f ratio and $\overline{G}$ provides a promising opportunity for use in the modified Neuber model. The K_t/K_f ratio can be obtained from Equation (6.36) for the given $\overline{G}$ and material yield strength. And the pseudo stress at a notch ($\sigma_1^e = K_t S_1$) or the pseudo stress range at a notch ($\Delta\sigma^e = K_t \Delta S$) can be obtained from the elastic finite element analysis.

Therefore, letting $K_t/K_f = A_{SS}$, the modified Neuber equation with the assumption of nominally elastic behavior can be rewritten as follows:

$$\frac{\sigma_1^2}{E} + \sigma_1 \left(\frac{\sigma_1}{K'}\right)^{1/n'} = \frac{(\sigma_1^e)^2}{A_{SS}^2 E} \qquad (6.38)$$

and

$$\frac{(\Delta\sigma)^2}{E} + 2 \cdot \Delta\sigma \left(\frac{\Delta\sigma}{2K'}\right)^{1/n'} = \frac{(\Delta\sigma^e)^2}{A_{ss}^2 E}. \tag{6.39}$$

These two equations containing the stress gradient effect for the local stress or stress range can be solved using the Newton–Raphson iteration technique.

Molsky and Glinka

Molsky and Glinka (1981) proposed another notch analysis method, which assumes the strain energy density at the notch root (W_e) is related to the energy density due to nominal stress and strain (W_S) by a factor of K_t^2. That means

$$W_e = K_t^2 W_S. \tag{6.40}$$

Figure 6.13 illustrates the physical interpretation of the strain energy density method. If nominally elastic behavior of the notched specimen is assumed, the following strain energy equations can be obtained:

$$W_S = \frac{1}{2}\frac{S_a^2}{E} \tag{6.41}$$

and

$$W_e = \frac{\sigma_a^2}{2 \cdot E} + \frac{\sigma_a}{1 + n'}\left(\frac{\sigma_a}{K'}\right)^{1/n'}. \tag{6.42}$$

Substituting Equations (6.41) and (6.42) into Equation (6.40) leads to the well-known energy density formula:

$$\frac{\sigma_a^2}{E} + \frac{2\sigma_a}{1 + n'}\left(\frac{\sigma_a}{K'}\right)^{1/n'} = \frac{(K_t S_a)^2}{E}. \tag{6.43}$$

For initial loading, Equation (6.43) can be reduced to

$$\frac{\sigma_1^2}{E} + \frac{2\sigma_1}{1 + n'}\left(\frac{\sigma_1}{K'}\right)^{1/n'} = \frac{(K_t S_1)^2}{E} = \frac{(\sigma_1^e)^2}{E}. \tag{6.44}$$

For stabilized hysteresis behavior, Equation (6.43) can be reduced to

$$\frac{(\Delta\sigma)^2}{E} + \frac{4 \cdot \Delta\sigma}{1 + n'}\left(\frac{\Delta\sigma}{2K'}\right)^{1/n'} = \frac{(K_t \Delta S)^2}{E} = \frac{(\Delta\sigma^e)^2}{E} \tag{6.45}$$

where σ_1^e and $\Delta\sigma^e$ are the local pseudo stress and the pseudo stress change at a notch root based on a linear elastic finite element analysis.

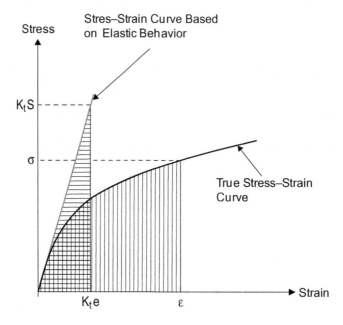

Figure 6.13
Interpretation of the strain energy density method.

Example 6.2

A notched thin plate subjected to a load-time history was analyzed by a linear elastic finite element analysis. The linear elastic finite element analysis indicates that the pseudo stress at the notch root is 0.035 MPa as a result of a unit applied load (P = 1 Newton). This plate is made of SAE 1005 steel that has the following cyclic and fatigue properties:

E = 207,000 MPa, K′ = 1240 MPa, n′ = 0.27, σ'_f = 886 MPa,

b = −0.14, ε'_f = 0.28, c = −0.5.

It is assumed that the notched plate follows nominally elastic behavior. Estimate the fatigue life of the notched thin plate made of SAE 1005 steel, with the following, subjected to the variable amplitude loading condition shown in Figure 6.14.

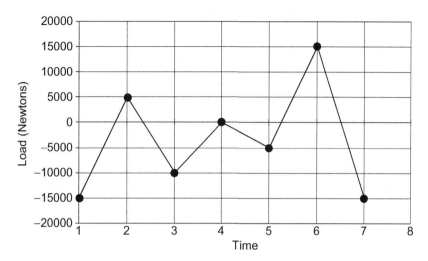

Figure 6.14
Variable amplitude load-time history.

1. Use the Molsky–Glinka energy density method to simulate the local hysteresis behavior.

2. Predict the fatigue life of the notched plate using the SWT mean stress correction formula.

3. Predict the fatigue life of the notched plate using the modified Morrow mean stress method.

Solution

A notched thin plate subjected to a load-time history is analyzed by a linear elastic finite element analysis. The finite element results indicate that the pseudo stress at the notch root is 0.035 MPa due to a unit applied load (P = 1 N). This gives a linear relationship between all further applied loads and localized pseudo stress.

$$\frac{\sigma^e}{P} = \frac{0.035\,\text{MPa}}{1\,\text{Newton}}$$

From this loading history and the linear relationship developed between input load and elastic stress, the local pseudo stresses (σ^e) can be calculated. For example, at the first load point:

$$\sigma^e = \left(\frac{0.035\,\text{MPa}}{1\,\text{Newton}}\right) \cdot (-15{,}000\,\text{Newton}) = -525\,\text{MPa}$$

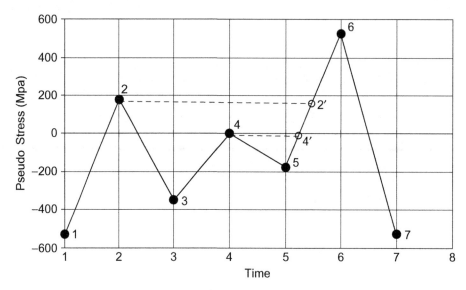

Figure 6.15
Variable amplitude σ^e-time history.

The complete pseudo stress versus time history is illustrated in Figure 6.15.

The hysteresis loops can be constructed through all seven data points. Each reversal is divided into 10 parts in order to get a well-defined graph. The Molsky–Glinka energy density method is employed to convert the pseudo stress into true stress and strain data. Taking the absolute value of the input stress or stress change, the initial loading curve up to point 1 is calculated using Equation (6.44) and from then on all the incremental reversals with respect to a reference turning point are calculated based on Equation (6.45). Since σ and $\Delta\sigma$ are difficult to solve for algebraically, the iterative solver tool in MS Excel was used.

The resulting data for local true stresses and strains are tabulated in Table 6.5 and the resulting hysteresis loops are shown in Figure 6.16. Please note that to describe the material memory effect, a reference turning point needs to be chosen correctly when calculating the incremental reversals.

For example, there will be three reference points involved in simulating the reversal from Point 5 to Point 6. Points 5, 3, and 1 are the reference points for the points in reversals: Point 5 to Point 4', Point 4' to Point 2',

Table 6.5: Calculation Summary Based on Local Pseudo Stresses
for True Stresses and Strains

Reversal	σ^e (MPa)	$\Delta\sigma^e$ (MPa)	$\Delta\sigma$ (MPa)	$\Delta\varepsilon$	σ (MPa)	ε	Ref Point
	0	*Reference Point*			0.0	0.0000	0
	−53	−53	−51.3	−0.0003	−51.3	−0.0003	0
	−105	−105	−94.1	−0.0005	−94.1	−0.0005	0
	−158	−158	−126.5	−0.0008	−126.5	−0.0008	0
	−210	−210	−152.2	−0.0012	−152.2	−0.0012	0
0 to 1	−263	−263	−173.5	−0.0015	−173.5	−0.0015	0
	−315	−315	−191.9	−0.0019	−191.9	−0.0019	0
	−368	−368	−208.3	−0.0024	−208.3	−0.0024	0
	−420	−420	−223.1	−0.0028	−223.1	−0.0028	0
	−473	−473	−236.7	−0.0033	−236.7	−0.0033	0
	−525	−525	−249.3	−0.0038	−249.3	−0.0038	0
	−525	*Reference Point*			−249.3	−0.0038	1
	−455	70	69.4	0.0003	−179.9	−0.0035	1
	−385	140	133.6	0.0007	−115.7	−0.0031	1
	−315	210	188.1	0.0011	−61.2	−0.0028	1
	−245	280	233.3	0.0014	−16.0	−0.0024	1
1 to 2	−175	350	271.4	0.0019	22.1	−0.0020	1
	−105	420	304.3	0.0023	55.1	−0.0015	1
	−35	490	333.5	0.0028	84.2	−0.0010	1
	35	560	359.8	0.0033	110.5	−0.0005	1
	105	630	383.8	0.0038	134.6	0.0000	1
	175	700	406.0	0.0044	156.8	0.0006	1
	175	*Reference Point*			156.8	0.0006	2
	123	−53	52.3	−0.0003	104.5	0.0003	2
	70	−105	102.6	−0.0005	54.2	0.0001	2
	18	−158	148.2	−0.0008	8.5	−0.0002	2
	−35	−210	188.1	−0.0011	−31.4	−0.0005	2
2 to 3	−88	−263	222.7	−0.0013	−66.0	−0.0008	2
	−140	−315	253.1	−0.0016	−96.3	−0.0011	2
	−193	−368	280.0	−0.0020	−123.3	−0.0014	2
	−245	−420	304.3	−0.0023	−147.6	−0.0017	2
	−298	−473	326.5	−0.0027	−169.8	−0.0021	2
	−350	−525	347.0	−0.0030	−190.2	−0.0025	2
	−350	*Reference Point*			−190.2	−0.0025	3
	−315	0	0.2	0.0000	−190.0	−0.0025	3
3 to 4	−280	35	35.0	0.0002	−155.3	−0.0023	3
	−245	70	69.4	0.0003	−120.8	−0.0021	3
	−210	105	102.6	0.0005	−87.6	−0.0020	3
	−175	140	133.6	0.0007	−56.6	−0.0018	3

(Continued)

Table 6.5: Calculation Summary Based on Local Pseudo Stresses
for True Stresses and Strains—cont'd

Reversal	σ^e (MPa)	$\Delta\sigma^e$ (MPa)	$\Delta\sigma$ (MPa)	$\Delta\varepsilon$	σ (MPa)	ε	Ref Point
	−140	175	162.1	0.0009	−28.1	−0.0016	3
	−105	210	188.1	0.0011	−2.1	−0.0014	3
	−70	245	211.7	0.0012	21.5	−0.0012	3
	−35	280	233.3	0.0014	43.1	−0.0010	3
	0	315	253.1	0.0016	62.8	−0.0008	3
	0		*Reference Point*		62.8	−0.0008	4
	−18	−18	17.5	−0.0001	45.4	−0.0009	4
	−35	−35	35.0	−0.0002	27.9	−0.0010	4
	−53	−53	52.3	−0.0003	10.6	−0.0011	4
	−70	−70	69.4	−0.0003	−6.6	−0.0012	4
4 to 5	−88	−88	86.2	−0.0004	−23.4	−0.0012	4
	−105	−105	102.6	−0.0005	−39.7	−0.0013	4
	−123	−123	118.4	−0.0006	−55.6	−0.0014	4
	−140	−140	133.6	−0.0007	−70.8	−0.0015	4
	−158	−158	148.2	−0.0008	−85.4	−0.0016	4
	−175	−175	162.1	−0.0009	−99.3	−0.0017	4
	−175		*Reference Point*		−99.3	−0.0017	5
	−105	70	69.4	0.0003	−29.9	−0.0013	5
	−35	140	133.6	0.0007	34.3	−0.0010	5
	−350		*Reference Point*		−190.2	−0.0025	3
	35	385	288.4	0.0021	98.2	−0.0004	3
	105	455	319.3	0.0026	129.1	0.0001	3
5 to 6	175	525	347.0	0.0030	156.8	0.0006	3
	−525		*Reference Point*		−249.3	−0.0038	1
	245	770	426.7	0.0050	177.5	0.0012	1
	315	840	446.2	0.0056	196.9	0.0018	1
	385	910	464.5	0.0063	215.2	0.0025	1
	455	980	481.9	0.0070	232.7	0.0031	1
	525	1050	498.6	0.0077	249.3	0.0038	1
	525		*Reference Point*		249.3	0.0038	6
	420	−105	102.6	−0.0005	146.7	0.0033	6
	315	−210	188.1	−0.0011	61.2	0.0028	6
	210	−315	253.1	−0.0016	−3.8	0.0022	6
	105	−420	304.3	−0.0023	−55.1	0.0015	6
6 to 7	0	−525	347.0	−0.0030	−97.7	0.0008	6
	−105	−630	383.8	−0.0038	−134.6	0.0000	6
	−210	−735	416.6	−0.0047	−167.3	−0.0009	6
	−315	−840	446.2	−0.0056	−196.9	−0.0018	6
	−420	−945	473.3	−0.0066	−224.1	−0.0028	6
	−525	−1050	498.6	−0.0077	−249.3	−0.0038	6

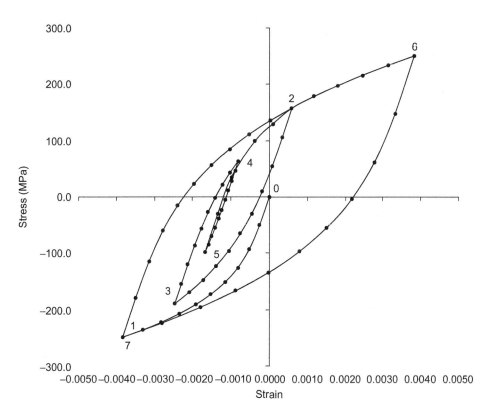

Figure 6.16
Simulation of hysteresis loops based on the block cycle loading and the
Molsky–Glinka energy density method.

and Point 2' to Point 6. Point 4' and 2' are the return points for the
memory effect, and are identical to Points 4 and 2 on the hysteresis
loops. These reference points can be easily identified with the application
of the three-point rainflow cycle counting technique.

The next step is to extract the hysteresis loops and identify the maximum
and minimum stress and strain points for each loop so these data could
be used as input into a damage calculation. Using this information, the
SWT and the modified Morrow mean stress correction models are used
to evaluate the number of block cycles.

For each counted cycle, an individual damage number is calculated
according to the linear damage rule. The sum of these yields the total
damage for one block cycle. Since failure is defined as a damage value of 1,

Table 6.6: Summary of Damage Calculation Based on the SWT Parameter

n_i	ε_{max}	ε_{min}	σ_{max} (MPa)	σ_{min} (MPa)	ε_a	N_f (Cycles)	d_i
1	−0.0008	−0.0017	62.8	−99.3	0.00043	3,215,072	3.11E-8
1	0.000587	−0.00246	156.8	−190.2	0.00152	87,124	1.15E-5
1	0.003831	−0.00383	249.3	−249.3	0.00383	5157	1.94E-4
					Σd_i per block = 2.06E-4		
					# of blocks to failure = 4900		

Table 6.7: Summary of Damage Calculation Based on the Modified Morrow Parameter

n_i	ε_{max}	ε_{min}	σ_{max} (MPa)	σ_{min} (MPa)	σ_m (MPa)	ε_a	N_f (Cycles)	d_i
1	−0.0008	−0.0017	62.8	−99.3	−18.2	0.00043	1,700,984	5.88E-8
1	0.00058	−0.0024	156.8	−190.2	−16.7	0.00152	78,444	1.27E-5
1	0.00383	−0.00383	349.3	−249.3	0	0.00383	5508	1.82E-4
						Σd_i per block = 1.95E-4		
						# of blocks to failure = 5100		

the inverse of the total damage per block cycle will yield the number of block cycles until failure.

The results of these calculations are tabulated in Tables 6.6 and 6.7 where the SWT and the modified Morrow methods predict that the component will fail after 4900 blocks and 5100 blocks of variable amplitude loading shown earlier in Figure 6.14, respectively.

Applications

As a result of various manufacturing and processing conditions, the surface of a real structure will be different from the smooth, polished material specimens used to determine the strain–life fatigue properties. Thus, there is a need to account for the effects of residual stress, surface treatment, and finish on fatigue in the high cycle fatigure regime. The following present some practical notch analysis models to account for these effects.

Residual Stress Effect

If the residual stress (σ_r) or residual strain (ε_r) at a notch root due to a manufacturing process is obtained prior to any operating load reversals, both of the modified Neuber rule and the Molsky–Glinka energy density method for initial loading needs to be modified. The following variants have been proposed based on the modified Neuber rule:

1. Lawrence et al. (1982):

$$\frac{\sigma_1^2}{E} + \sigma_1 \left(\frac{\sigma_1}{K'}\right)^{1/n'} = \frac{(K_f S_1 + \sigma_r)^2}{E}. \tag{6.46}$$

2. Reemsnyder (1981):

$$\left(1 - \frac{\sigma_r}{\sigma_1}\right)^2 \left[\frac{\sigma_1^2}{E} + \sigma_1 \left(\frac{\sigma_1}{K'}\right)^{1/n'}\right] = \frac{(K_t S_1)^2}{E}. \tag{6.47}$$

3. Baumel and Seeger (1989):

$$\sigma_1 (\varepsilon_1 - \varepsilon_r) = \frac{(K_t S_1)^2}{E}. \tag{6.48}$$

Surface Finish Effect

If the surface treatment factor (C_S) and the roughness correction factor ($C_{\sigma,R}$) are known from FKM-Guideline (Haibach, 2003), the Lawrence model can be further revised to account for the surface finish effect:

$$\frac{\sigma_1^2}{E} + \sigma_1 \left(\frac{\sigma_1}{K'}\right)^{1/n'} = \frac{(K_f^* S_1 + \sigma_r)^2}{E} \tag{6.49}$$

where

$$K_f^* = \frac{C_S}{K_f + \dfrac{1}{C_{\sigma,R}} - 1}. \tag{6.50}$$

Per FKM-Guideline, the surface treatment factor for steel and cast iron materials are tabulated in Table 6.8. The values in the table are applicable to components of 30 to 40 mm diameter, while the values in the parenthesis are for 8 to 15 mm diameter.

Table 6.8: Surface Treatment Factors for Various Materials

Surface Treatment	Unnotched Components	Notched Components
Steel		
Chemo-Thermal Treatment		
Nitriding		
Depth of case 0.1–0.4 mm	1.10–1.15	1.30–2.00
Surface hardness 700–1000 HV10	(1.15–1.25)	(1.90–3.00)
Case hardening		
Depth of case 0.2–0.8 mm	1.10–1.50	1.20–2.00
Surface hardness 670–750 HV10	(1.20–2.00)	(1.50–2.50)
Carbo-nitriding		
Depth of case 0.2–0.8 mm	(1.80)	
Surface hardness 670–750 HV10		
Mechanical Treatment		
Cold rolling	1.10–1.25	1.30–1.80
	(1.20–1.40)	(1.50–2.20)
Shot peening	1.10–1.20	1.10–1.50
	(1.10–1.30)	(1.40–2.50)
Thermal Treatment		
Inductive hardening		
Flame-hardening		
Depth of case 0.9–1.5 mm	1.20–1.50	1.50–2.50
Surface hardness 51–64 HRC	(1.30–1.60)	(1.60–2.8)
Cast Iron Materials		
Nitriding	1.10 (1.15)	1.3 (1.9)
Case hardening	1.1 (1.2)	1.2 (1.5)
Cold rolling	1.1 (1.2)	1.3 (1.5)
Shot peening	1.1 (1.1)	1.1 (1.4)
Inductive hardening, flame-hardening	1.2 (1.3)	1.5 (1.6)

Source: Adapted from FKM-Guideline, published by Forschungskuratorium Maschinenebau, 2003.

Per the FKM-Guideline, the roughness factor under normal stress is:

$$C_{\sigma,R} = 1 - a_R \log(R_Z) \log\left(\frac{2S_{t,u}}{S_{t,min,u}}\right) \tag{6.51}$$

where

a_R = a roughness constant listed in Table 6.9

R_Z = the average roughness value of the surface in μm

$S_{t,u}$ = the ultimate tensile strength in MPa

$S_{t,min,u}$ = the minimum ultimate tensile strength in MPa in Table 6.9

Table 6.9: a_R and $S_{t,min,u}$ for Various Materials

Materials	a_R	$S_{t,min,u}$ MPa
Steel	0.22	400
Steel castings	0.20	400
Ductile iron	0.16	400
Malleable cast iron	0.12	350
Grey cast iron	0.06	100
Wrought aluminum alloys	0.22	133
Cast aluminum alloys	0.20	133

Source: Adapted from FKM-Guideline, published by Forschungskuratorium
Maschinenebau, 2003.

An average roughness value, $R_Z = 200$ µm, applies for a rolling skin, a forging skin, and the skin of cast irons. For steels, the roughness value of a ground surface varies from 1 µm to 12 µm, and the value of a finished surface ranges from 6.3 µm to 100 µm.

Summary

The Masing equation for describing and tracking the hysteresis behavior of a homogeneous material under variable amplitude load time history has been introduced. For any given strain increment, the stress increment with respect to a reference turning point can be easily calculated by the equation, and vice versa. The proper choice of a reference turning point to account for the memory effect of a material undergoing complex loading condition is important in the material simulation process.

The Morrow strain–life equation for zero mean stress has been described. In conjunction with the local strain–life approach, the modified Morrow equation and the SWT model have been proposed to quantify the effect of mean stresses on fatigue behavior.

If there are no experimental strain–life fatigue data available, an estimate of cyclic and fatigue behavior of a material can be helpful in the design stage. For a given ultimate tensile strength ($S_{t,u}$), the uniform material law by Baumel and Seeger (1990) is recommended for both steels and titanium alloys, and the medians method by Meggiolaro and Castro (2004), for aluminum alloys.

The modified Neuber rule and the Molsky–Glinka energy density method are the two popular techniques to estimate the true stress–strain behavior at a stress concentration area based on the pseudo stress time history obtained from the linear elastic FEA. Their limitations and applications have been addressed. Moreover, the practical notch analysis models to account for the effects of residual stress, surface treatment, and finish on fatigue in the HCF regime have been presented.

References

Basquin, O. H. (1910). The exponential law of endurance tests. *Proceedings of American Society for Testing and Materials, 10*, 625–630.

Baumel, A., Jr., & Seeger, T. (1989). Thick surface layer model—life calculations for specimens with residual stress distribution and different material zones. In *Proceedings of international conference on residual stress (ICRS2)* (pp. 809–914), London.

Baumel, A., Jr., & Seeger, T. (1990). *Material data for cyclic loading: Supplemental Volume*. Amsterdam: Elsevier Science.

Coffin, L. F., Jr. (1954). A study of the effect of cyclic thermal stresses on a ductile metal. *Transactions of ASME, 76*, 931–950.

Dowling, N. E. (2009). Mean stress effects in strain–life fatigue. *Fatigue and Fracture of Engineering Materials and Structures, 32*, 1004–1019.

Fash, J., & Socie, D. F. (1982). Fatigue behavior and mean effects in grey cast iron. *International Journal of Fatigue, 4*(3), 137–142.

Forsetti, P., & Blasarin, A. (1988). Fatigue behavior of microalloyed steels for hot forged mechanical components. *International Journal of Fatigue, 10*(3), 153–161.

Haibach, E. (2003). *FKM-guideline: Analytical strength assessment of components in mechanical engineering* (5th rev. ed., English version). Berlin: Forschungskuratorium Maschinenebau.

Heywood, R. B. (1962). *Designing against failure*. London: Chapman & Hall.

Koh, S. K., & Stephens, R. I. (1991). Mean stress effects on low cycle fatigue for a high strength steel. *Fatigue and Fracture Engineering Materials and Structures, 14*(4), 413–428.

Langlais, T. E., & Vogel, J. H. (1995). Overcoming limitations of the conventional strain–life fatigue damage model. *Journal of Engineering Materials and Technology, 117*, 103–108.

Lawrence, F. V., Burk, J. V., & Yung, J. Y. (1982). Influence of residual stress on the predicted fatigue life of weldments. *Residual Stress in Fatigue, ASTM STP 776*, 33–43.

Lee, P. S., & Song, J. H. (2006). Estimation methods for strain–life fatigue properties from hardness. *International Journal of Fatigue, 28*, 386–400.

Lee, Y. L., Pan, J., Hathaway, R., & Barkey, M. (2005). *Fatigue testing and analysis: Theory and practice*. Boston: Elsevier/Butterworth-Heinemann.

Manson, S. S. (1953). Behavior of materials under conditions of thermal stress. In *Heat transfer symposium* (pp. 9–75). The University of Michigan Engineering Research Institute, Ann Arbor.

Masing, G. (1926). Eigenspannungen and Verfestigung beim Messing. In *Proceedings of the 2nd international congress for applied mechanics* (pp. 332–335), Zurich.

Meggiolaro, M. M., & Castro, T. P., Jr. (2004). Statistical evaluation of strain–life fatigue crack initiation prediction. *International Journal of Fatigue, 26*, 463–476.

Miner, M. A. (1945). Cumulative damage in fatigue. *Journal of Applied Mechanics, 67*, A159–A164.

Mitchell, M. R. (1979). Fundamentals of modern fatigue analysis for design. In *ASM symposium on fatigue and microstructure*, St. Louis, MO, October 14–15.

Molsky, K., & Glinka, G. (1981). A method of elastic-plastic stress and strain calculation at a notch root. *Materials Science and Engineering, 50*, 93–100.

Morrow, J. D. (1965). Cyclic plastic strain energy and fatigue of metals. *Internal Friction, Damping, and Cyclic Plasticity, ASTM STP 378*, 45–86.

Morrow, J. D. (1968). Fatigue design handbook, section 3.2. *SAE Advanced in Engineering, 4*, 21–29.

Muralidharan, U., & Manson, S. S. (1988). A modified universal slopes equation for estimating of fatigue characteristics of metals. *Journal of Engineering Materials and Technology, ASME, 110*, 55–58.

Neuber, H. (1946). *Theory of notch stress*. Ann Arbor, MI: J. W. Edwards.

Neuber, H. (1961). Theory of stress concentration for shear–strained prismatical bodies with arbitrary nonlinear stress-strain law. *Journal of Applied Mechanics, Transaction of ASME, 28*, 544–550.

Palmgren, A. (1924). Die lebensdauer von kugellagern. *Zeitschrift des Vereinesdeutscher Ingenierure, 68*(14), 339–341

Peterson, R. E. (1959). *Analytical approach to stress concentration effects in aircraft materials* (Technical Report 59-507). Dayton, OH: U.S. Air Force—WADC Symp. Fatigue Metals.

Reemsnyder, H. S. (1981). Evaluating the effect of residual stresses on notched fatigue resistance. In *Materials, experimentation, and design in fatigue—Proceedings of Fatigue '81* (pp. 273–295). Guilford, England: Westbury Press, distributed in the United States by Ann Arbor Science Publishers, Woborn, MA.

Roessle, M. L., & Fatemi, A. (2000). Strain-controlled fatigue properties of steels and some simple approximations. *International Journal of Fatigue, 22*, 495–511.

Seeger, T., & Heuler, P. (1980). Generalized application of Neuber's rule. *Journal of Testing and Evaluation, 8*(4), 199–204.

Siebel, E., & Stieler, M. (1955). Significance of dissimilar stress distributions for cycling loading. *VDI-Zeitschrift, 97*(5), 121–126 (in German).

Smith, K. N., Watson, P., & Topper, T. H. (1970). A stress-strain function for the fatigue of metals. *Journal of Materials, 5*(4), 767–778.

Topper, T. H., Wetzel, R. M., & Morrow, J. (1969). Neuber's rule applied to fatigue of notched specimens. *Journal of Materials, 4*(1), 200–209.

Walcher, J., Gray, D., & Manson, S. S. (1979). Aspects of cumulative fatigue damage analysis of cold end rotating structures. AIAA/SAE/ASME 15th Joint Propulsion Conference, Las Vegas, NV, Paper 79-1190.

CHAPTER 7

Fundamentals of Cyclic Plasticity Theories

Yung-Li Lee
Chrysler Group LLC

Mark E. Barkey
The University of Alabama

Chapter Outline

Introduction

Early studies of cyclic plasticity concentrated mainly on the monotonic and uni-axial loading conditions. But for the past four decades, efforts have been directed toward cyclic multiaxial plasticity for both proportional and nonproportional loading. Numerous new theories have been developed since, however, many of these theories have complex mathematical formulations and require a large number of material constants to achieve reasonable correlation between simulation and experimental data.

Consequently the simplicity and the physical clarity desired may be lost. In this chapter, the historical and recent development of "simple" plasticity theories will be presented with emphasis on both physical interpretation of all the formulations and determination of the material parameters required. Please note that the plasticity theories reviewed herein do not apply to rate-dependent yielding, materials at elevated temperature, and anisotropic materials.

The applications of the existing plasticity theories to multiaxial notch analysis based on pseudo stresses are also reviewed. Pseudo stresses (also called ficti-tious stresses) are the stresses calculated from a linear elastic finite element ana-lysis. True stress and strain components at a notch are the essential parameters for fatigue life predictions. Nonlinear finite element analysis could be the perfect solution to calculate the notch stresses and strains, but its usage may be very limited due to intensive CPU time consumption. Therefore, estimation techni-ques for the multiaxial notch stresses and strains based on the pseudo stresses are important and needed.

Tensor Notations

Stress and strain are considered as symmetric, second-order tensors, and they are represented by $\underline{\sigma}$ and $\underline{\varepsilon}$, respectively, with components that are listed by

$$\underline{\sigma} = \sigma_{ij} = \begin{bmatrix} \sigma_{11} & \sigma_{12} & \sigma_{13} \\ \sigma_{21} & \sigma_{22} & \sigma_{23} \\ \sigma_{31} & \sigma_{32} & \sigma_{33} \end{bmatrix} = \begin{bmatrix} \sigma_x & \tau_{xy} & \tau_{xz} \\ \tau_{yx} & \sigma_y & \tau_{yz} \\ \tau_{zx} & \tau_{zy} & \sigma_z \end{bmatrix}$$

and

$$\underline{\varepsilon} = \varepsilon_{ij} = \begin{bmatrix} \varepsilon_{11} & \varepsilon_{12} & \varepsilon_{13} \\ \varepsilon_{21} & \varepsilon_{22} & \varepsilon_{23} \\ \varepsilon_{31} & \varepsilon_{32} & \varepsilon_{33} \end{bmatrix} = \begin{bmatrix} \varepsilon_x & \varepsilon_{xy} & \varepsilon_{xz} \\ \varepsilon_{yx} & \varepsilon_y & \varepsilon_{yz} \\ \varepsilon_{zx} & \varepsilon_{zy} & \varepsilon_z \end{bmatrix}.$$

The unit tensor is

$$\underline{I} = I_{ij} = \delta_{ij} = \begin{bmatrix} 1 & 0 & 0 \\ 0 & 1 & 0 \\ 0 & 0 & 1 \end{bmatrix}$$

where

δ_{ij} is the Kronecker delta

A colon between two tensors denotes their inner product, dot, or scalar. For example,

$$\underline{\sigma} : \underline{\varepsilon} = \sigma_{ij}\varepsilon_{ij} = \sigma_{11}\varepsilon_{11} + \sigma_{12}\varepsilon_{12} + \sigma_{13}\varepsilon_{13} + \sigma_{21}\varepsilon_{21} + \sigma_{22}\varepsilon_{22} + \sigma_{23}\varepsilon_{23}$$
$$+ \sigma_{31}\varepsilon_{31} + \sigma_{32}\varepsilon_{32} + \sigma_{33}\varepsilon_{33}$$

and

$$\underline{\sigma} : \underline{I} = \sigma_{kk} = \sigma_{11} + \sigma_{22} + \sigma_{33} = \sigma_x + \sigma_y + \sigma_z.$$

Also the norm of a tensor is defined as

$$|\underline{\sigma}| = \sqrt{\underline{\sigma} : \underline{\sigma}}.$$

It is a common practice to refer to some of the tensors as vectors, particularly in describing a hardening rule, even though a tensor and a vector are not physically equivalent.

Theory of Elasticity

An additive decomposition of the elastic ($d\underline{\varepsilon}^e$) and plastic ($d\underline{\varepsilon}^p$) components of the total strain increment ($d\underline{\varepsilon}$) is assumed and expressed as

$$d\underline{\varepsilon} = d\underline{\varepsilon}^e + d\underline{\varepsilon}^p. \tag{7.1}$$

This equation is often used to define the plastic strain, and is valid for small strains compared to unity. The volume change due to plastic straining is also assumed to be zero, and is the plastic incompressibility condition, which can be represented by

$$d\underline{\varepsilon}^p : \underline{I} = 0. \tag{7.2}$$

For the elastic part, assuming Hooke's law is applicable, the elastic strain versus stress relationship can be expressed by

$$d\underline{\varepsilon}^e = \frac{1+v}{E}\left[d\underline{\sigma} - \frac{v}{1+v}(d\underline{\sigma}:\underline{I})\underline{I}\right] \tag{7.3}$$

or

$$d\underline{\varepsilon}^e = \frac{d\underline{\sigma}}{2G} - \frac{v}{E}(d\underline{\sigma}:\underline{I})\underline{I} \tag{7.4}$$

where

$\quad d\underline{\sigma} =$ the stress increment tensor
$\quad E =$ Young's modulus of elasticity
$\quad G =$ the elastic shear modulus
$\quad v =$ Poisson's ratio

The elastic shear modulus can be represented by

$$G = \frac{E}{2(1+v)}. \tag{7.5}$$

The stress tensor can be assumed to be decomposed into a deviatoric ($\underline{S}$) stress tensor and a hydrostatic ($\underline{\sigma}_h$) stress tensor, respectively, and can be written by

$$\underline{\sigma} = \underline{S} + \underline{\sigma}_h \tag{7.6}$$

where hydrostatic stress is the average normal stress at a point defined as

$$\underline{\sigma}_h = \frac{1}{3}(\underline{\sigma}:\underline{I})\underline{I}. \tag{7.7}$$

It should be noted that hydrostatic stress does not influence plastic yielding and the principal directions. Therefore, when yielding, the deviatoric stress tensor is responsible for plastic deformation and can be expressed as follows:

$$\underline{S} = \underline{\sigma} - \frac{1}{3}(\underline{\sigma}:\underline{I})\underline{I}. \tag{7.8}$$

According to Hooke's law, the elastic deviatoric strain increment tensor $(d\underline{e}^e)$ is assumed to be linearly related to the deviatoric stress increment tensor as

$$d\underline{S} = 2G d\underline{e}^e. \tag{7.9}$$

Based on its definition, the elastic deviatoric strain increment tensor can be expressed by

$$d\underline{e}^e = d\underline{e} - d\underline{e}^p = \left[d\underline{\varepsilon} - \frac{1}{3}(d\underline{\varepsilon} : \underline{I})\underline{I} \right] - d\underline{\varepsilon}^p. \tag{7.10}$$

Using Equation (7.10), the deviatoric stress increment tensor can be rewritten,

$$\frac{d\underline{S}}{2G} = d\underline{\varepsilon} - d\underline{\varepsilon}^p - \frac{1}{3}(d\underline{\varepsilon} : \underline{I})\underline{I}. \tag{7.11}$$

The relationship between plastic strains versus deviatoric stresses can be described in the following sections on theories of plasticity.

Monotonic Plasticity Theories

Monotonic plasticity theories describe plastic deformation behavior of a material under monotonic loading. The following key elements are required to develop an algorithm that will simulate the nonlinear monotonic stress–strain behavior of a material:

- Definition of a yield surface function
- Application of an isotropic hardening rule
- Introduction of a flow rule
- Calculation of plastic strains

Yield Surface Function

A yield surface function or yield criterion $f(\underline{S})$ defines the region of purely elastic behavior under any multiaxial state of stress, and is represented as a surface in stress space. The yield criterion is given by

$f(\underline{S}) < 0$: elastic deformation
$f(\underline{S}) = 0$: plastic deformation

The commonly accepted yield criterion for metals is the von Mises yield criterion, which assumes the yield surface function $f(J_2)$ is a function of the second invariant of the deviatoric stress tensor (J_2) and has the form

$$f(J_2) = 3J_2 - 3k^2 \qquad (7.12)$$

where k is the yield strength in shear. The second deviatoric stress invariant can be written in terms of deviatoric principal stresses as

$$J_2 = \frac{1}{2}(\underline{S}:\underline{S}) = \frac{1}{2}(S_1^2 + S_2^2 + S_3^2). \qquad (7.13)$$

Also, in terms of principal stresses, it is given that

$$J_2 = \frac{1}{6}\left[(\sigma_1 - \sigma_2)^2 + (\sigma_2 - \sigma_3)^2 + (\sigma_3 - \sigma_1)^2\right]. \qquad (7.14)$$

The square root of $3J_2$ is defined as the von Mises or equivalent stress, and can be written in terms of deviatoric stresses and principal stresses as

$$\sigma_{eq} = \sqrt{3J_2} = \sqrt{\frac{3}{2}\underline{S}:\underline{S}} = \frac{1}{\sqrt{2}}\sqrt{(\sigma_1 - \sigma_2)^2 + (\sigma_2 - \sigma_3)^2 + (\sigma_3 - \sigma_1)^2}. \qquad (7.15)$$

In this case of $f(J_2) = 0$, the equivalent yield stress σ_{eq} under uniaxial loading is identical to the uniaxial yield stress σ_y, which implies $\sigma_y = \sqrt{3}k$. Equivalent to Equation (7.12), the yield surface function can have the following two variants:

$$f(\underline{S}) = \underline{S}:\underline{S} - 2k^2 = 0 \qquad (7.16)$$

or

$$f(\underline{S}) = \frac{3}{2}\underline{S}:\underline{S} - \sigma_y^2 = 0. \qquad (7.17)$$

Equations (7.16) and (7.17) are also the commonly used von Mises yield criteria where both k and σ_y are dependent on the accumulated plastic strain, p.

Similarly, the increment of an equivalent plastic strain dp is defined as

$$dp = \sqrt{\frac{2}{3}d\underline{\varepsilon}^p : d\underline{\varepsilon}^p}. \qquad (7.18)$$

This choice of the coefficient in Equation (7.18) will also result in an identical uniaxial plastic strain increment. The accumulated plastic strain is then obtained by

$$p = \int dp. \tag{7.19}$$

Isotropic Hardening

Isotropic hardening is used to describe the deformation hardening behavior for a material under monotonic loading. Isotropic hardening, as illustrated in Figure 7.1, assumes that the initial yield surface expands uniformly without translation and distortion as plasticity occurs. The size increase in the yield surface depends on the stress, hardening property, and temperature.

The isotropic hardening model does not take into account the Bauschinger effect, which means cyclic yield strength reduction under a loading reversal. If temperature is excluded here, the size of the yield surface is governed by the accumulated plastic work or accumulated plastic strain. For example, the yield surface function is $f = \text{func}(\underline{S} \text{ or } \underline{\sigma}, H)$ where H is a hardening function.

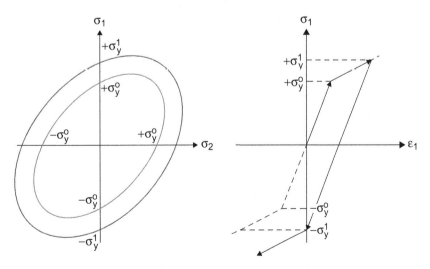

Figure 7.1
Isotropic hardening.

The hardening function can be either work or strain hardening. If the size increase of a yield surface depends on plastic work ($W^P = \int \sigma_{ij} d\varepsilon_{ij}^P$), then the material behavior is called work hardening and the yield surface can be written as

$$f = \text{func}(\underline{S} \text{ or } \underline{\sigma}, W^P). \tag{7.20}$$

If the size increase of a yield surface depends on the accumulated equivalent plastic strain, then it is named strain hardening whose yield surface can be written as

$$f = \text{func}(\underline{S} \text{ or } \underline{\sigma}, p). \tag{7.21}$$

Strain hardening is often assumed and employed in the plasticity theories. Thus, the yield criterion can be written as

$$f(\underline{S}) = \frac{3}{2}\underline{S} : \underline{S} - \sigma_y^2(p) = 0 \tag{7.22}$$

and

$$\sigma_y(p) = \sigma_{yo} + R(p) \tag{7.23}$$

where

    σ_{yo} = the initial yield stress
    $R(p)$ = the isotropic strain hardening function

$R(p)$ has the following formulation with an initial condition $R(0) = 0$:

$$dR(p) = b(R^L - R)dp. \tag{7.24}$$

in which R^L is the saturated value with increasing plastic strain and b is the material parameter determining the rate at which saturation is reached. Both R^L and b can be obtained from the uniaxial stress–strain curve. Provided that σ_y^L is the limit or saturated yield stress, then $R^L = \sigma_y^L - \sigma_{yo}$. The solution to the differential equation, Equation (7.24), is

$$R(p) = R^L(1 - e^{-bp}). \tag{7.25}$$

Flow Rules

A flow rule defines the plastic deformation vector by determining the orientation and magnitude of plastic deformation of a material as plasticity occurs. There have been many flow rules proposed over the years, some of which will be reviewed in the following two sections.

Classical Flow Rules

Levy (1870) and von Mises (1913) postulated that with the negligible elastic strain the total strain increment is proportional to the deviatoric stress and has the following relation:

$$d\underline{\varepsilon} = d\lambda_1 \underline{S} \tag{7.26}$$

where

$d\lambda_1 =$ a multiplier

Later Prandtl (1925) and Reuss (1930) postulated that the plastic strain increment is proportional to the deviatoric stress, which has the following relation:

$$d\underline{\varepsilon}^p = d\lambda_2 \underline{S} \tag{7.27}$$

where

$d\lambda_2 =$ a multiplier that can be determined from an equivalent stress versus accumulated plastic strain $(\sigma_{eq} - p)$ curve

It is assumed that if σ_{eq} is a function of p, then the material would have strain hardening behavior defined as

$$\sigma_{eq} = H(p) \tag{7.28}$$

where

$H() =$ a hardening function

The multiplier $d\lambda_2$ can be determined by squaring Equation (7.27):

$$d\underline{\varepsilon}^p : d\underline{\varepsilon}^p = d\lambda_2^2 \underline{S} : \underline{S}. \tag{7.29}$$

Therefore, Equation (7.29) can be written as

$$\frac{3}{2}(dp)^2 = \frac{2}{3}d\lambda_2^2(\sigma_{eq})^2, \tag{7.30}$$

which results in

$$d\lambda_2 = \frac{3}{2}\frac{dp}{\sigma_{eq}}. \tag{7.31}$$

Now differentiating the strain hardening function (Equation 7.28) with respect to dp has

$$d\sigma_{eq} = \frac{\partial H(p)}{\partial p} dp. \tag{7.32}$$

Equation (7.32) can be expressed as

$$dp = \frac{d\sigma_{eq}}{\dfrac{\partial H(p)}{\partial p}} = \frac{d\sigma_{eq}}{h}. \tag{7.33}$$

Here h is termed as the plastic modulus, the tangent modulus of a uniaxial stress–plastic strain curve, as depicted in Figure 7.2. Substituting Equation (7.33) back to (7.31) yields

$$d\lambda_2 = \frac{3}{2h} \frac{d\sigma_{eq}}{\sigma_{eq}}. \tag{7.34}$$

Finally the Prandtl–Reuss flow equation can be expressed as

$$d\underline{\varepsilon}^p = \frac{3}{2h} \frac{d\sigma_{eq}}{\sigma_{eq}} \underline{S}. \tag{7.35}$$

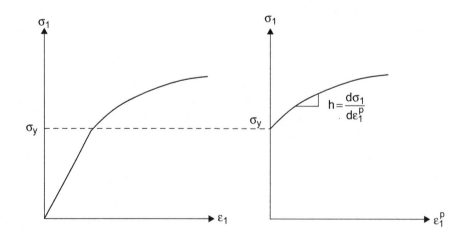

Figure 7.2
Uniaxial stress versus strain and plastic strain curves.

Postulating that the total plastic strain would have the same type of relationship as the Prandtl–Reuss incremental one, Hencky (1924) developed the following flow equation:

$$\underline{\varepsilon}^p = \frac{3}{2} \frac{p}{\sigma_{eq}} \underline{S}. \tag{7.36}$$

Both flow rules imply that the present state of plastic strain is independent of its loading path. Also both flow rules would be identical under the special circumstances of linear strain hardening and a monotonically increasing load.

Associated Flow Rule

Von Mises (1928) assumed that there exists a plastic potential function $Q(\underline{S})$ that can be related to the plastic flow as

$$d\underline{\varepsilon}^p = d\lambda_3 \frac{\partial Q(\underline{S})}{\partial \underline{S}} \tag{7.37}$$

where

   $d\lambda_3 = $ a multiplier

The plastic potential function represents a surface in the stress space and the plastic strain increment is a vector normal to the surface.

Therefore, Equation (7.37) is referred to as the normality flow rule. For most materials, a yield surface function $f(\underline{S})$ is the criterion for development of plastic deformation. So it is a common approach in the theory of plasticity to assume that the plastic potential function is identical to the yield surface function ($Q(\underline{S}) = f(\underline{S})$). Thus, the plastic flow can be written as

$$d\underline{\varepsilon}^p = d\lambda_3 \frac{\partial f(\underline{S})}{\partial \underline{S}}. \tag{7.38}$$

This indicates the plastic strain increment vector is normal to the yield surface, and is known as the associated flow rule. But if $Q(\underline{S}) \neq f(\underline{S})$, Equation (7.37) would be used and termed the nonassociated flow rule.

Plastic Strains

$d\lambda_3$ in the associated flow rule (Equation 7.38) can be determined by setting up the dot product of an exterior normal and a stress vector tangent to the yield surface equal to zero,

$$(d\underline{S} - hd\underline{\varepsilon}^p) : \frac{\partial f}{\partial \underline{S}} = 0 \qquad (7.39)$$

where

$\frac{\partial f}{\partial \underline{S}}$ = the gradient of the yield surface with respect to $\underline{S}$
h = the generalized plastic modulus that has related the plastic flow to the stress increment component normal to the yield surface

This concept of normality is illustrated in Figure 7.3. By substituting Equation (7.38) into Equation (7.39), $d\lambda_3$ is obtained as follows:

$$d\lambda_3 = \frac{1}{h} \frac{d\underline{S} \frac{\partial f}{\partial \underline{S}}}{\frac{\partial f}{\partial \underline{S}} \frac{\partial f}{\partial \underline{S}}}. \qquad (7.40)$$

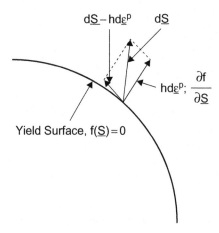

Figure 7.3
Normality condition of stress states.

Finally, after substituting Equation (7.40) into Equation (7.38) and with an introduction of $\underline{n}$, the unit normal to a yield surface, the associated flow rule can be simplified in the following:

$$d\underline{\varepsilon}^p = \frac{1}{h}(\underline{n}:d\underline{S})\underline{n} \tag{7.41}$$

and

$$\underline{n} = \frac{\partial f}{\partial \underline{S}} \bigg/ \left| \frac{\partial f}{\partial \underline{S}} \right|. \tag{7.42}$$

Cyclic Plasticity Theories

A material under cyclic reversed loading will usually exhibit the Bauschinger effect, meaning during a reversed loading the elastic stress range is unchanged and the yield strength is reduced, as shown in Figure 7.4. The kinematic hardening rule can simulate the Bauschinger effect for material behavior under cyclic loading. Kinematic hardening assumes that during plastic deformation, the subsequent yield surface translates as a rigid body in the stress space, maintaining the size, shape, and orientation of the initial yield surface.

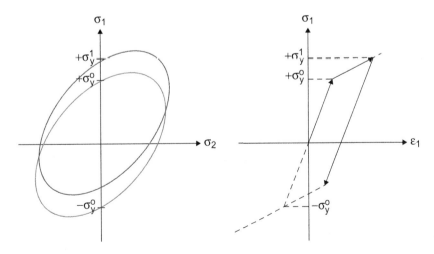

Figure 7.4
Illustration of kinematic hardening.

The incremental plasticity theory based on the kinematic hardening model has advanced significantly in the past decades. When the von Mises yield criterion is assumed, the kinematic hardening can be expressed as

$$f = \frac{3}{2}(\underline{S} - \underline{\alpha}) : (\underline{S} - \underline{\alpha}) - \sigma_y^2 = 0 \qquad (7.43)$$

where

$\underline{\alpha}$ = the back stress tensor, which defines the translational center of the yield surface

σ_y = the yield stress for the size of the yield surface

The yield surface is supposed to be a function of accumulated plastic strain p, but is assumed to be a constant here. This type of kinematic hardening is schematically illustrated in Figure 7.4.

To describe nonlinear material hardening behavior, kinematic hardening models can be classified into (1) multiple-surface (or nested-surface) models, (2) two-surface models, and (3) single-surface models. The first multiple-surface model was proposed by Mroz (1967). In this approach, the nonlinear stress–strain relation is represented by a number of straight-line segments. The linearization results in a series of yield surfaces, each with its own center and size.

As illustrated in Figure 7.5, Mroz's model uses a finite number of yield surfaces in the stress space to approximate the nonlinear hardening behavior of materials, where the flow of each yield surface represents a constant plastic modulus. Mroz specifies that during loading, these individual surfaces do not intersect but consecutively contact and push each other. To meet this criterion, the active yield surface must translate in the direction that connects the current stress point ($\underline{S}$) on the surface with the conjugate point ($\underline{S}^L$) on the next inactive yield or limit surface, which has the same outward normal ($\underline{n}$) as the normal to the active surface at $\underline{S}$.

The active yield surface (f) is defined by the von Mises yield criterion, Equation (7.43), while the next inactive yield or limit surface (f^L) can be expressed by

$$f^L = \frac{3}{2}\underline{S}^L : \underline{S}^L - (\sigma_y^L)^2 = 0 \qquad (7.44)$$

where

$\sqrt{2/3}\sigma_y^L$ = the radius of the inactive yield or limit surface in the deviatoric stress space

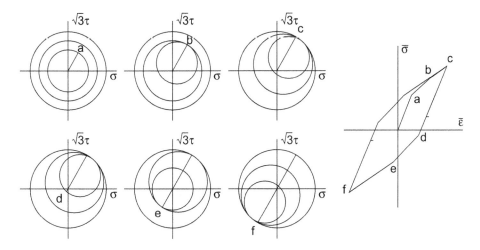

Figure 7.5
The multiple-surface model in a biaxial stress state
Source: Adapted from Socie and Marquis, 2000.

Then, based on Mroz's translational rule, $\underline{S}^L$ is defined as

$$\underline{S}^L = \frac{\sigma_y^L}{\sigma_y}(\underline{S} - \underline{\alpha}).\qquad(7.45)$$

The incremental translation of a loading surface can be expressed in the following expression:

$$d\underline{\alpha} = d\mu_4(\underline{S}^L - \underline{S})\qquad(7.46)$$

where

$\quad d\mu_4 = $ a multiplier

The two-surface models were introduced by Dafalias and Popov (1975, 1976) and Krieg (1975). Figure 7.6 illustrates the concept of the model with two surfaces, a stationary limit surface and a loading surface that may translate in the stress space inside the limit surface. The material nonlinearity is described based on the relative position of the loading surface with respect to the limit surface. The Mroz translational rule is adopted to avoid any overlapping of both surfaces.

The single-surface model was originally developed by Armstrong and Frederick (1966). The model was later modified by many researchers (Chaboche, 1977; Chaboche & Rousselier, 1983; Chen & Keer, 1991; Ohno & Wang, 1993a,

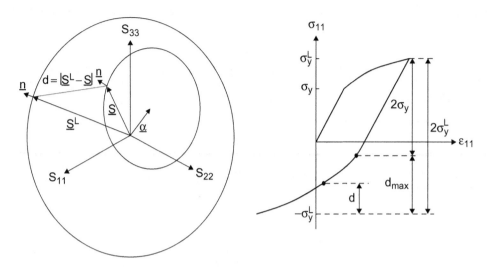

Figure 7.6
A two-surface model with the Mroz translational rule.

1993b; Jiang & Sehitoglu, 1996a,b). Essentially it is a single yield surface with an introduction of a nonlinear kinematic hardening rule to describe the nonlinear cyclic material behavior, as shown in Figure 7.7. The original Armstrong–Frederick model has the following translational rule for the back stress:

$$d\underline{\alpha} = \mu_1 d\underline{\varepsilon}^p - \mu_2 dp\underline{\alpha} \tag{7.47}$$

where

μ_1 and μ_2 = material constants

Others proposed to use multiple back stress increments in order to achieve good correlations with experimental data in "ratcheting." It is assumed that the total back stresses are composed of additive parts,

$$\underline{\alpha} = \sum_{i=1}^{M} \underline{\alpha}^{(i)} \tag{7.48}$$

where

$\underline{\alpha}$ = the total back stress
$\underline{\alpha}^{(i)}$ = a part of the total back stress, i = 1, 2, ..., M
M = the number of back stress parts considered

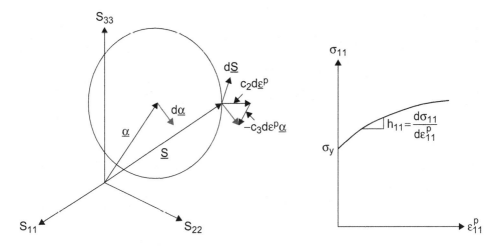

Figure 7.7

Single-surface model with a nonlinear hardening rule.

Each back stress part follows the Armstrong–Frederick relation,

$$d\underline{\alpha}^{(i)} = \mu_3^{(i)} \left(\mu_4^{(i)} d\underline{\varepsilon}^p - W^{(i)} dp\underline{\alpha}^{(i)} \right)$$

$$(i = 1, 2, \ldots, M)$$

(7.49)

where

$\mu_3^{(i)}$ and $\mu_4^{(i)}$ = material constants associated with the i-th part of back stress $\underline{\alpha}^{(i)}$

$W^{(i)}$ = a function of back stress range, which determines the nonlinear recovery behavior of this model

A number of efforts (Chaboche, 1977; Chaboche & Rousselier, 1983; Ohno & Wang, 1993a, 1993b; Jiang & Sehitoglu, 1996a, 1996b) have been focused on introducing a proper $W^{(i)}$ to improve the performance of their models. In addition to the choice for one of these kinematic hardening models, the following key elements need to be included to develop an algorithm that will simulate the nonlinear cyclic stress–strain behavior of a material:

• The elastic and plastic process

• An associated flow rule

- The consistency condition

- A kinematic hardening rule

- The generalized plastic modulus

- Formulation of stresses and strains

Elastic and Plastic Processes

A process is said to be elastic if the current stress state is interior to the yield surface. If the current stress is on the yield surface, and the stress points toward the interior side of the tangent plane of a yield surface, then the process is also elastic. Otherwise the process is said to be plastic. The fundamental quantity that distinguishes elastic and plastic processes is the sign of the normal component of the trial stress $(\underline{n} : d\underline{S})$.

As depicted in Figure 7.8, the elastic or unloading process takes place if

$$f < 0 \quad \text{and} \quad \underline{n} : d\underline{S} < 0. \tag{7.50}$$

A plastic process is defined if

$$f = 0 \quad \text{and} \quad \underline{n} : d\underline{S} > 0. \tag{7.51}$$

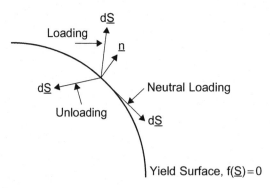

Figure 7.8
Loading criteria.

A neutral process occurs if

$$f = 0 \quad \text{and} \quad \underline{n} : d\underline{S} = 0. \tag{7.52}$$

For neutral or nearly neutral loading, the stresses remain on the yield surface but the center of the yield surface does not move. To ensure the stresses remain on the yield surface, the value of the current yield function is calculated after the stress increment has been added to obtain the current stress state. But, the addition of an elastic stress increment $(A_1 \Delta \underline{S})$ would result in stresses that are not on the yield surface. A consistency condition in a finite difference form is needed to correct this situation by using either the radial return technique developed by Hughes (1984) or the interactive calculation procedure.

This consistency condition, which is based on the interactive calculation procedure yields

$$f = \frac{3}{2}(\underline{S} + A_1 \Delta \underline{S} - \underline{\alpha}) : (\underline{S} + A_1 \Delta \underline{S} - \underline{\alpha}) - \sigma_y^2 = 0 \tag{7.53}$$

$$f = \frac{2}{3}[(\Delta \underline{S} : \Delta \underline{S})A_1^2 + 2\Delta \underline{S} : (\underline{S} - \underline{\alpha})A_1 + (\underline{S} - \underline{\alpha}) : (\underline{S} - \underline{\alpha})] - \sigma_y^2 = 0. \tag{7.54}$$

A_1 should take the smaller root of the two solutions of Equation (7.54). This furnishes a simple method for the numerical computation to correct the stress increment for neutral loading.

Associated Flow Rule

When a material point is yielding under the action of a total strain or stress increment, a flow rule is needed to determine the plastic strain increment. By substituting the von Mises yield criterion, the unit normal to the yield surface becomes

$$\underline{n} = \sqrt{\frac{3}{2}} \frac{\underline{S} - \underline{\alpha}}{\sigma_y}. \tag{7.55}$$

Thus, the associated flow rule, Equation (7.41), becomes

$$d\underline{\varepsilon}^p = \frac{3}{2h\sigma_y^2}[(\underline{S} - \underline{\alpha}) : d\underline{S}](\underline{S} - \underline{\alpha}). \tag{7.56}$$

Consistency Condition

During loading, the new (or updated) stress state following the application of stress increment must lie on the new (or subsequent) yield surface. That means that the yield surface must move or change to accommodate the change in the stress increment. Suppose that the yield surface is given as

$$f(\underline{S}, \underline{\alpha}, \sigma_y) = 0. \tag{7.57}$$

The equation of the new surface due to stress increments is given by

$$f(\underline{S} + d\underline{S}, \underline{\alpha} + d\underline{\alpha}, \sigma_y + d\sigma_y) = 0. \tag{7.58}$$

Since both incremental values are infinitesimal, this equation can be expanded as

$$f(\underline{S} + d\underline{S}, \underline{\alpha} + d\underline{\alpha}, \sigma_y + d\sigma_y) = f(\underline{S}, \underline{\alpha}, \sigma_y) + \frac{\partial f}{\partial \underline{S}} : d\underline{S} + \frac{\partial f}{\partial \underline{\alpha}} : d\underline{\alpha} + \frac{\partial f}{\partial \sigma_y} d\sigma_y = 0. \tag{7.59}$$

The consistency condition can be written as

$$df = \frac{\partial f}{\partial \underline{S}} : d\underline{S} + \frac{\partial f}{\partial \underline{\alpha}} : d\underline{\alpha} + \frac{\partial f}{\partial \sigma_y} d\sigma_y = 0. \tag{7.60}$$

For the von Mises yield criterion with the kinematic hardening rule, where σ_y is assumed to be a constant, the consistency condition yields the following equation:

$$d\overline{S} : \overline{n} = d\overline{\alpha} : \overline{n}. \tag{7.61}$$

Kinematic Hardening Models

When kinematic hardening is assumed, the translation of the yield surface must be specified. This can be done by prescribing the change of the location of the center of the current yield surface in the stress space, meaning the evolution of the back stress tensor ($d\underline{\alpha}$).

Several kinematic hardening models have been proposed to describe the change of the back stress tensor. A generic kinematic hardening rule that is associated with $d\underline{\varepsilon}^p$ and $\underline{\alpha}$ can be expressed as follows:

$$d\underline{\alpha} = d\lambda_5 \underline{A} \tag{7.62}$$

where

$\underline{A}$ = the translational direction of a yield surface
$d\lambda_5$ = the scalar that can be determined from the consistence condition

There are three kinematic hardening rules that fall into this category and can be expressed as

Prager's rule (1955):

$$\underline{A} = d\underline{\varepsilon}^p \tag{7.63}$$

Mroz's rule (1967):

$$\underline{A} = S^L - S = \sqrt{\frac{2}{3}} \frac{h(\sigma_y^L - \sigma_y)}{d\underline{S}:\underline{n}} d\underline{\varepsilon}^p - \underline{\alpha} \tag{7.64}$$

Armstrong–Frederick's rule (1966):

$$\underline{A} = d\underline{\varepsilon}^p - c_1 d\varepsilon_{eq}^p \underline{\alpha}. \tag{7.65}$$

Another general kinematic hardening rule that is dependent on $d\underline{S}$ and $\underline{\alpha}$ or $\underline{n}$ can be represented by

$$d\underline{\alpha} = d\lambda_6 \underline{B} \tag{7.66}$$

where

$\underline{B}$ = the translational direction of the yield surface
$d\lambda_6$ = the scalar that can be determined from the consistence condition

There are three kinematic hardening rules that fall into this category and are expressed as

Phillips–Lee's rule (1979):

$$\underline{B} = d\underline{S} \tag{7.67}$$

Ziegler's rule (1959):

$$\underline{B} = \underline{S} - \underline{\alpha} \tag{7.68}$$

Rolovic–Tipton's rule (2000a, 2000b):

$$\underline{B} = \frac{d\underline{S}}{|d\underline{S}|} + c_2\underline{n}. \tag{7.69}$$

The Mroz translational rule is chosen in the two-surface model for its simplicity and geometrical consistency, while the original Armstrong–Frederick rule is recommended for use in the single-surface model for fewer material parameters required than the modified ones. The two kinematic models are discussed in the following sections.

Determination of a Generalized Plastic Modulus

The following sections provide examples that illustrate determination of the plastic modulus used in the incremental plasticity model, which is based on the uniaxial cyclic stress–strain curve of a material.

Two-Surface Model with the Mroz Kinematic Rule

The Mroz incremental movement of the yield surface center can be described in a deviatoric stress space as

$$d\underline{\alpha} = d\mu_4(\underline{S}^L - \underline{S}) \tag{7.70}$$

where

$d\mu_4 = $ a positive scalar to be determined by the consistency condition

The consistency condition yields the following equation:

$$d\underline{S} : \underline{n} = d\underline{\alpha} : \underline{n}. \tag{7.71}$$

Therefore, Equation (7.70) becomes

$$d\mu_4 = \frac{d\underline{S} : \underline{n}}{(\underline{S}^L - \underline{S}) : \underline{n}} \tag{7.72}$$

and the associated flow rule is then obtained by

$$d\underline{\varepsilon}^p = \frac{3}{2h\sigma_y^2}[(\underline{S} - \underline{\alpha}) : d\underline{S}](\underline{S} - \underline{\alpha}). \tag{7.73}$$

Considering the uniaxial loading case, $S_{22} = S_{33} = -1/2S_{11} = -1/3\sigma_{11}$ and $\alpha_{22} = \alpha_{33} = -1/2\alpha_{11}$. Hence, the generalized hardening modulus (h) can be reduced to

$$h = \frac{2}{3}\frac{d\sigma_{11}}{d\varepsilon_{11}^p}. \tag{7.74}$$

Assuming the uniaxial stress–plastic strain curve can be represented by the Ramberg–Osgood equation, the tangent modulus of the curve is

$$\frac{d\sigma_{11}}{d\varepsilon_{11}^p} = K'n'\left(\frac{\sigma}{K'}\right)^{\frac{n'-1}{n'}} \tag{7.75}$$

where

  K' = the cyclic strength coefficient
  n' = the cyclic hardening exponent

Substituting Equation (7.75) into Equation (7.74) yields the following equation:

$$h = \frac{2}{3}K'n'\left(\frac{\sigma}{K'}\right)^{\frac{n'-1}{n'}}. \tag{7.76}$$

The distance between the loading and the limit surfaces is a variable to describe the generalized plastic modulus. In the two-surface model (Bannantine, 1989; Lee et al., 1995), it is assumed the hardening modulus depends on the distance (d) from the current stress point ($\underline{S}$) of a yield surface to the conjugate stress point ($\underline{S}^L$) on the limit surface with the same outward normal.

Thus, d can be expressed as

$$d = \sqrt{\frac{3}{2}(\underline{S}^L - \underline{S}) : (\underline{S}^L - \underline{S})}. \tag{7.77}$$

The maximum distance in the uniaxial stress–strain curve is

$$d_{max} = 2(\sigma_y^L - \sigma_y). \tag{7.78}$$

Figure 7.6 earlier in the chapter depicts the distance between the loading and the limit surfaces and its relationship in the uniaxial stress–strain curve. Let D be the normalized difference between d and d_{max}

$$D = \frac{d_{max} - d}{d_{max}}. \tag{7.79}$$

The normalized D varies between zero and one. When initial yielding occurs after a reversed unloading, D is zero and the corresponding uniaxial stress is

$$\sigma = \pm (2\sigma_y - \sigma_y^L). \tag{7.80}$$

When the limit surface is reached, D is equal to one and the uniaxial stress is

$$\sigma = \pm \sigma_y^L. \tag{7.81}$$

Thus, the uniaxial stress can be related to the normalized D in the following form:

$$\sigma = \pm [2(\sigma_y^L - \sigma_y)(D - 1) + \sigma_y^L]. \tag{7.82}$$

Substituting this uniaxial stress into the generalized hardening modulus becomes

$$h = \frac{2}{3} K'n' \left[\frac{|2(\sigma_y^L - \sigma_y)(D - 1) + \sigma_y^L|}{K'} \right]^{\frac{n'-1}{n'}}. \tag{7.83}$$

Also in another two-surface model (Wang & Brown, 1993), a variable d_1 is introduced and defined as the projection of the vector connecting the loading stress point and its conjugate point on the same normal to the loading surface,

$$d_1 = \sqrt{\frac{3}{2}(\underline{S}^L - \underline{S}) : \underline{n}} = \sigma_y^L - \sqrt{\frac{3}{2}\underline{S} : \underline{n}}. \tag{7.84}$$

It can be shown that for proportional loading, the equivalent uniaxial stress is

$$\sigma = 2\sigma_y^L - d_1 - d^m \tag{7.85}$$

where

d^m = a discrete memory variable to account for the memory effect

This was described in Wang and Brown (1993). So the generalized hardening modulus becomes:

$$h = \frac{2}{3} K' n' \left(\frac{2\sigma_y^L - d_1 - d^m}{K'} \right)^{\frac{n'-1}{n'}}. \tag{7.86}$$

Single-Surface Model with the Armstrong–Frederick Kinematic Rule

The von Mises yield criterion is expressed in the form

$$f = \frac{3}{2} (\underline{S} - \underline{\alpha}) : (\underline{S} - \underline{\alpha}) - \sigma_y^2 = 0. \tag{7.87}$$

Plus, the original Armstrong–Frederick kinematic model is used:

$$d\underline{\alpha} = \mu_1 d\underline{\varepsilon}^p - \mu_2 dp\underline{\alpha} \tag{7.88}$$

where

μ_1 and μ_2 = constant material parameters

This will be determined from the uniaxial cyclic stress–strain curve. The associated flow rule is also given as follows:

$$d\underline{\varepsilon}^p = \frac{3}{2h\sigma_y^2} [(\underline{S} - \underline{\alpha}) : d\underline{S}](\underline{S} - \underline{\alpha}). \tag{7.89}$$

For the von Mises yield criterion with the kinematic hardening rule, the consistency condition yields the equation,

$$d\underline{S} : \underline{n} = d\underline{\alpha} : \underline{n} \tag{7.90}$$

where

$$\underline{n} = \sqrt{\frac{3}{2}} \frac{\underline{S} - \underline{\alpha}}{\sigma_y}. \tag{7.91}$$

Substituting the kinematic model and the flow rule into the previous consistency equation has

$$h = \mu_1 - \mu_2 \frac{(\underline{S} - \underline{\alpha}) : \underline{\alpha}}{\sigma_y}. \tag{7.92}$$

The material parameters μ_1 and μ_2 can be obtained from the uniaxial cyclic stress–strain curve. Considering the uniaxial loading case, $S_{22} = S_{33} = -1/2 S_{11} = -1/3\sigma_{11}$ and $\alpha_{22} = \alpha_{33} = -1/2\alpha_{11}$. Hence, the generalized plastic modulus (h) can be reduced to

$$h = \mu_1 - \frac{3}{2}\mu_2 \frac{(S_{11} - \alpha_{11})\alpha_{11}}{\sigma_y} \tag{7.93}$$

and the von Mises yield criterion can be expressed as follows:

$$\frac{3}{2}(\underline{S} - \underline{\alpha}) : (\underline{S} - \underline{\alpha}) = \sigma_y^2. \tag{7.94}$$

Expanding Equation (7.94) results in

$$\frac{9}{4}(S_{11} - \alpha_{11})^2 = \sigma_y^2. \tag{7.95}$$

Finally squaring the root of Equation (7.95) has

$$\frac{3}{2}|S_{11} - \alpha_{11}| = \sigma_y. \tag{7.96}$$

Therefore, after substituting Equation (7.96) into Equation (7.93), the generalized plastic modulus can be rewritten as

$$h = \mu_1 - \mu_2\alpha_{11} \quad S_{11} > \alpha_{11} \tag{7.97}$$

$$h = \mu_1 + \mu_2\alpha_{11} \quad S_{11} < \alpha_{11} \tag{7.98}$$

where minus and plus refer to loading and reversed loading, respectively. Also, the flow rule becomes

$$d\varepsilon_{11}^p = \frac{2}{3h} d\sigma_{11}. \tag{7.99}$$

By considering only the loading condition, the tangent modulus of the uniaxial cyclic stress–strain curve can be expressed in terms of the generalized plastic modulus in the following:

$$h_{11} = \frac{d\sigma_{11}}{d\varepsilon_{11}^p} = \frac{3}{2}h = \frac{3}{2}(\mu_1 - \mu_2\alpha_{11}) \tag{7.100}$$

and the yield criterion has

$$\alpha_{11} = S_{11} - \frac{2}{3}\sigma_y = \frac{2}{3}\sigma_{11} - \frac{2}{3}\sigma_y. \tag{7.101}$$

Substituting Equation (7.101) into Equation (7.100) yields

$$\frac{d\sigma_{11}}{d\varepsilon_{11}^p} + \mu_2\sigma_{11} = \frac{3}{2}\mu_1 + \mu_2\sigma_y \quad \text{or} \quad d\sigma_{11}$$

$$= \left(\sigma_y - \sigma_{11} + \frac{3}{2}\frac{\mu_1}{\mu_2}\right)(\mu_2 d\varepsilon_{11}^p). \tag{7.102}$$

Imposing the initial condition, $\sigma_{11} = \sigma_y$ at $\varepsilon_{11}^p = 0$, the solution of the preceding linear differential equation is

$$\sigma_{11} = \sigma_y + \frac{3}{2}\frac{\mu_1}{\mu_2}\left(1 - e^{-\mu_2\varepsilon_{11}^p}\right). \tag{7.103}$$

Therefore, the material parameters, μ_1 and μ_2, can be determined by fitting the uniaxial cyclic stress-plastic strain data into Equation (7.103).

Figure 7.9 depicts the correlation between the Ramberg–Osgood equation and Equation (7.103) with the fitted material parameters, μ_1 and μ_2. Equation (7.103)

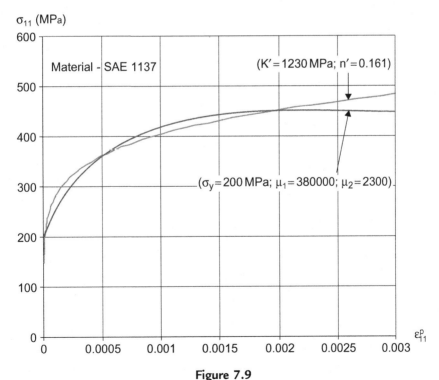

Figure 7.9

Comparison between the original and fitted cyclic stress-plastic strain curves.

gives an exponential shape to the uniaxial stress–strain curve that saturates with increasing plastic strain, where the value of $\sigma_y + \frac{3}{2}\frac{\mu_1}{\mu_2}$ is the maximum saturated stress and μ_2 determines the rate at which saturation is achieved.

Advanced Cyclic Plasticity Models for Nonproportional Hardening

Nonproportional hardening is an additional strain hardening of some materials under nonproportional loading where the principal axes change with time. A special material modeling technique is required to predict this type of material behavior.

The phenomenon of this additional strain hardening as a result of the slip behavior of the material has been explained (Tanaka, 1994; Itoh et al., 1995). As shown in Figure 7.10, the continuous change of the principal stress plane or the maximum shear stress plane due to nonproportional loading increases the interaction between slip systems resulting in plastic deformation along the different slip systems. The cross slip interaction due to plastic deformation can induce an additional strain hardening as compared to that observed in proportional loading.

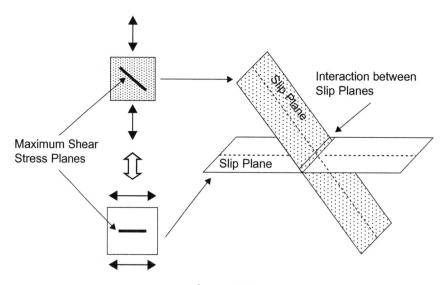

Figure 7.10
Interaction between slip planes due to nonproportional loading.

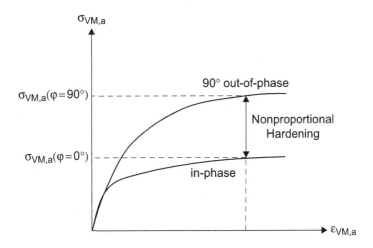

Figure 7.11

Comparison of equivalent stress–strain curves for the same material under in-phase and 90° out-of-phase loadings.

This strain hardening phenomenon is illustrated in Figure 7.11, which shows the effective stress–strain curves for the same material under proportional (in-phase) and nonproportional (90° out-of-phase) loadings. The additional strain hardening due to nonproportional loading can be described by the nonproportional hardening coefficient, α_{NP}, defined by

$$\alpha_{NP} = \frac{\sigma_{VM,a}(\varphi = 90^{\circ})}{\sigma_{VM,a}(\varphi = 0^{\circ})} - 1 \qquad (7.104)$$

where

$$\varphi = \text{the phase angle between two loads}$$
$$\sigma_{VM,a}(\varphi = 90^{\circ}) = \text{the 90° out-of-phase von Mises stress amplitude}$$
$$\sigma_{VM,a}(\varphi = 0^{\circ}) = \text{the in-phase von Mises stress amplitude at the same strain amplitude}$$

The severity of nonproportional hardening is dependent on the ease on which slip systems interact and the type of loading path. Materials such as aluminum alloys have weak interactions and show wavy slips because dislocations can easily change their slip planes as the maximum shear stress plane rotates, resulting in no or small additional strain hardening.

A typical material, such as Type 304 stainless steel (Doong et al., 1990; Itoh et al., 1995), shows higher strain hardening effects than aluminum alloys (Krempl & Lu, 1983; Doong et al., 1990; Itoh et al., 1997) based on the identical nonproportional loading tests. Also the study (Tanaka et al., 1985; Itoh et al., 1995) showed that different nonproportional loading paths produce different degrees of nonproportional hardening, among which the 90° out-of-phase loading path has the largest degree of nonproportional hardening.

The aforementioned cyclic plasticity models in previous sections fail to account for nonproportional hardening and these models need to be modified with a special parameter to capture this phenomenon. There have been many attempts to describe nonproportional hardening, among which the fourth rank tensor developed by Tanaka (1994) is the promising method to account for the nonproportional hardening, based on the investigation by Zhang and Jiang (2008).

Tanaka introduced a nonproportional parameter A and an internal state variable $\underline{C}$ describing the internal dislocation structure, both of which are defined as follows:

$$A = \sqrt{1 - \frac{\underline{n}(\underline{C} : \underline{C})\underline{n}}{\underline{C} : \underline{C}}} \tag{7.105}$$

and

$$d\underline{C} = C_c(\underline{n} : \underline{n} - \underline{C})d\varepsilon_{eq}^p \tag{7.106}$$

where

C_c = a material constant

Based on the framework described by Zhang and Jiang (2008), an advanced plasticity model with Tanaka's nonproportional parameter A is introduced to account for nonproportional hardening. This advanced model considers a combined isotropic and kinematic hardening. Therefore, the yield criterion can be written

$$f(\underline{S}, \underline{\alpha}, \sigma_y(p)) = \sqrt{\frac{3}{2}(\underline{S} - \underline{\alpha}) : (\underline{S} - \underline{\alpha})} - \sigma_y(p) = 0. \tag{7.107}$$

For the evolution of the isotropic hardening, $\sigma_y(p)$ follows the expression

$$\sigma_y(p) = \sigma_{yo} + R(p) \tag{7.108}$$

where

σ_{yo} = the initial yield stress
$R(p)$ = the isotropic strain hardening function

$R(p)$ has the following exponential formulation:

$$R(p) = R_{NP}(1 - e^{-b_{NP} \cdot p}) \tag{7.109}$$

in which b_{NP} is a material parameter and R_{NP} is the saturated value with increasing plastic strain determined. R_{NP} is proposed as follows:

$$R_{NP} = Aq_N + (1 - A)q_P \tag{7.110}$$

where q_N and q_P are the material parameters for nonproportional hardening ($A = 1$) and proportional hardening ($A = 0$), respectively.

A simplified version of the evolution of the back stress tensor in the Zhang and Jiang model (2008) is given by

$$d\underline{\alpha} = cr\left[\underline{n} - \left(\frac{\sqrt{\underline{\alpha} : \underline{\alpha}}}{r}\right)^X \frac{\underline{\alpha}}{\sqrt{\underline{\alpha} : \underline{\alpha}}}\right]d\varepsilon_{eq}^p + \frac{\underline{\alpha}}{r}dr \tag{7.111}$$

where c and X are material constants, and the evolution of r has the following form:

$$dr = b_r[1 + (m_1 - m_2q)A](R_{NP} - r)d\varepsilon_{eq}^p \tag{7.112}$$

in which b_r, m_1, and m_2 are material constants and q is the value of the memory size.

Stresses and Strains

For the strain-controlled test, the deviatoric stress tensor in terms of total strain and plastic strain increment tensors can be rewritten,

$$\frac{d\underline{S}}{2G} = d\underline{\varepsilon} - d\underline{\varepsilon}^p - \frac{1}{3}(d\underline{\varepsilon} : \underline{I})\underline{I}. \tag{7.113}$$

With the introduction of the flow rule, this equation reduces as

$$\frac{d\underline{S}}{2G} = d\underline{\varepsilon} - \frac{1}{h}(\underline{n}:d\underline{S})\underline{n} - \frac{1}{3}(d\underline{\varepsilon}:\underline{I})\underline{I}. \tag{7.114}$$

Dotting both sides of Equation (7.114) by $\underline{n}$ yields

$$\frac{d\underline{S}}{2G}:\underline{n} = d\underline{\varepsilon}:\underline{n} - \frac{1}{h}(\underline{n}:d\underline{S})\underline{n}:\underline{n} - \frac{1}{3}(d\underline{\varepsilon}:\underline{I})\underline{I}:\underline{n}. \tag{7.115}$$

Since the volume change due to plastic strain change is zero,

$$d\underline{\varepsilon}^p:\underline{I} = 0; \quad \underline{I}:\underline{n} = 0. \tag{7.116}$$

Thus, Equation (7.115) reduces to

$$\frac{d\underline{S}}{2G}:\underline{n} = d\underline{\varepsilon}:\underline{n} - \frac{1}{h}(\underline{n}:d\underline{S}) \tag{7.117}$$

or

$$\underline{n}:d\underline{S} = \frac{2Gh}{2G+h}(\underline{n}:d\underline{\varepsilon}). \tag{7.118}$$

Finally, substituting Equation (7.118) into Equation (7.114) results in

$$d\underline{S} = 2G\left[d\underline{\varepsilon} - \frac{2G}{2G+h}(\underline{n}:d\underline{\varepsilon})\underline{n} - \frac{1}{3}(d\underline{\varepsilon}:\underline{I})\underline{I}\right]. \tag{7.119}$$

For the stress-controlled test, the elastic strain increment can be rewritten:

$$d\bar{\varepsilon}^e = \frac{1+v}{E}\left[d\bar{\sigma} - \frac{v}{1+v}(d\bar{\sigma}:\bar{I})\bar{I}\right]. \tag{7.120}$$

And with the application of the flow rule, the total strain increment is

$$d\bar{\varepsilon} = \frac{1+v}{E}\left[d\bar{\sigma} - \frac{v}{1+v}(d\bar{\sigma}:\bar{I})\bar{I}\right] + \frac{1}{h}(\bar{n}:d\bar{\sigma})\bar{n}. \tag{7.121}$$

From both strain-controlled and stress-controlled tests, the inelastic stress–strain response is extremely dependent on the generalized plastic modulus (h), which is closely related to the active yield surface position and the uniaxial plastic modulus.

Applications: Notch Analyses Based on Pseudo Stresses

The following sections present three popular multiaxial notch analyses—the Hoffman–Seeger method, the Buczynski–Glinka method, and the Lee–Chiang–Wong method—in detail.

Hoffmann and Seeger

Hoffmann and Seeger (1989) developed an approach to estimate notch stresses and strains based on the pseudo stresses calculated from a linear elastic finite element analysis (FEA). The assumptions are:

- The surface stress and strain components at a notch root are of interest

- The principal stress and strain axes are fixed in orientation

- The ratio of the in-plane principal strains is constant

- The out-of-plane principal stress is zero, $\sigma_3 = 0$

- The uniaxial stress–strain curve can be extended for use with suitable equivalent stress and strain parameters such as the von Mises parameter

- Hencky's flow rule is adopted

The elastic values of the von Mises stress σ_{eq}^e and strain ε_{eq}^e from a linear elastic FEA are computed:

$$\sigma_{eq}^e = \frac{\sigma_1^e}{|\sigma_1^e|}\sqrt{(\sigma_1^e)^2 + (\sigma_2^e)^2 - \sigma_1^e\sigma_2^e} \tag{7.122}$$

$$\varepsilon_{eq}^e = \frac{\sigma_{eq}^e}{E}. \tag{7.123}$$

The equivalent stress σ_{eq} and strain ε_{eq} can be obtained by solving the following modified Neuber rule (1961):

$$\sigma_{eq}\varepsilon_{eq} = \sigma_{eq}^e\varepsilon_{eq}^e = \frac{(\sigma_{eq}^e)^2}{E} \tag{7.124}$$

and the Ramberg–Osgood equation:

$$\varepsilon_{eq} = \frac{\sigma_{eq}}{E} + \left(\frac{\sigma_{eq}}{K'}\right)^{\frac{1}{n'}} \tag{7.125}$$

where

 E = Young's modulus
 K' = cyclic strength coefficient
 n' = the cyclic hardening exponent

It is assumed Hencky's flow rule is valid and has the following expression:

$$\underline{\varepsilon} = \frac{1+v}{E}\underline{\sigma} - \frac{v}{E}(\underline{\sigma}:\underline{I})\underline{I} + \frac{3}{2}\frac{\varepsilon_{eq}^p}{\sigma_{eq}}\left[\underline{\sigma} - \frac{1}{3}(\underline{\sigma}:\underline{I})\underline{I}\right] \qquad (7.126)$$

or

$$\underline{\varepsilon} = \left(\frac{1+v}{E} + \frac{3}{2}\frac{\varepsilon_{eq}^p}{\sigma_{eq}}\right)\underline{\sigma} - \left[\frac{v}{E} + \frac{1}{2}\frac{\varepsilon_{eq}^p}{\sigma_{eq}}\right](\underline{\sigma}:\underline{I})\underline{I}. \qquad (7.127)$$

It is also assumed proportional principal stresses and $\sigma_3 = 0$. Then the maximum principal strain ε_1 can be expressed as follows:

$$\varepsilon_1 = \left(\frac{1+v}{E} + \frac{3}{2}\frac{\varepsilon_{eq}^p}{\sigma_{eq}}\right)\cdot\sigma_1 - \left(\frac{v}{E} + \frac{1}{2}\frac{\varepsilon_{eq}^p}{\sigma_{eq}}\right)\cdot(\sigma_1 + \sigma_2) \qquad (7.128)$$

or

$$\varepsilon_1 = \left(\frac{1}{E} + \frac{\varepsilon_{eq}^p}{\sigma_{eq}}\right)\cdot\sigma_1 - \left(\frac{v}{E} + \frac{1}{2}\frac{\varepsilon_{eq}^p}{\sigma_{eq}}\right)\cdot\sigma_2. \qquad (7.129)$$

But the two terms on the right side of Equation (7.129) can be further expanded:

$$\left(\frac{1}{E} + \frac{\varepsilon_{eq}^p}{\sigma_{eq}}\right)\cdot\sigma_1 = \left(\frac{\sigma_{eq} + E\varepsilon_{eq}^p}{E}\frac{1}{\sigma_{eq}}\right)\cdot\sigma_1 = \frac{\varepsilon_{eq}}{\sigma_{eq}}\cdot\sigma_1 \qquad (7.130)$$

or

$$\left(\frac{v}{E} + \frac{1}{2}\frac{\varepsilon_{eq}^p}{\sigma_{eq}}\right)\cdot\sigma_2 = \left(\frac{2\sigma_{eq}\cdot v + E\varepsilon_{eq}^p}{2E\sigma_{eq}}\right)\cdot\sigma_2 = \left[\frac{1}{2} - \left(\frac{1}{2} - v\right)\frac{\sigma_{eq}}{E\varepsilon_{eq}}\right]\frac{\varepsilon_{eq}}{\sigma_{eq}}\sigma_2. \quad (7.131)$$

Substituting Equations (7.130) and (7.131) into Equation (7.129) yields

$$\varepsilon_1 = \frac{\varepsilon_{eq}}{\sigma_{eq}}\cdot(\sigma_1 - v'\cdot\sigma_2) \qquad (7.132)$$

where

$v' = $ the generalized Poisson ratio

This is defined as

$$v' = \frac{1}{2} - \left(\frac{1}{2} - v\right)\cdot\frac{\sigma_{eq}}{E\cdot\varepsilon_{eq}}. \qquad (7.133)$$

Similarly, ε_2 and ε_3 can be determined by

$$\varepsilon_2 = \frac{\varepsilon_{eq}}{\sigma_{eq}} \cdot (\sigma_2 - v'\sigma_1) \tag{7.134}$$

$$\varepsilon_3 = -v' \cdot \frac{\varepsilon_{eq}}{\sigma_{eq}} \cdot (\sigma_1 + \sigma_2). \tag{7.135}$$

Combining Equations (7.132) and (7.134) has

$$a = \frac{\sigma_2}{\sigma_1} = \frac{\dfrac{\varepsilon_2}{\varepsilon_1} + v'}{1 + v' \cdot \dfrac{\varepsilon_2}{\varepsilon_1}}. \tag{7.136}$$

Provided that ε_1 and ε_2 denote surface strains and ε_3 is normal to the surface, the assumption of a constant surface strain ratio has

$$\frac{\varepsilon_2}{\varepsilon_1} = \frac{\varepsilon_2^e}{\varepsilon_1^e} = \frac{\dfrac{1}{E} \cdot (\sigma_2^e - v \cdot \sigma_1^e)}{\dfrac{1}{E} \cdot (\sigma_1^e - v \cdot \sigma_2^e)}. \tag{7.137}$$

The principal stresses and strains can then be calculated:

$$\sigma_1 = \sigma_{eq} \cdot \frac{1}{\sqrt{1 - a + a^2}} \tag{7.138}$$

$$\sigma_2 = a \cdot \sigma_1. \tag{7.139}$$

Finally, the steps to calculate the local notch surface stresses and strains based on pseudo stresses are summarized in the following:

1. The elastic values of the von Mises stress and strain from a linear elastic FEA are computed:

$$\sigma_{eq}^e = \frac{\sigma_1^e}{|\sigma_1^e|} \sqrt{(\sigma_1^e)^2 + (\sigma_2^e)^2 - \sigma_1^e \sigma_2^e}$$

$$\varepsilon_{eq}^e = \frac{\sigma_{eq}^e}{E}.$$

2. The equivalent stress and strain can be obtained by solving the following modified Neuber rule:

$$\sigma_{eq}\varepsilon_{eq} = \sigma_{eq}^{e}\varepsilon_{eq}^{e} = \frac{\left(\sigma_{eq}^{e}\right)^{2}}{E}$$

$$\varepsilon_{eq} = \frac{\sigma_{eq}}{E} + \left(\frac{\sigma_{eq}}{K'}\right)^{\frac{1}{n'}}.$$

3. The assumption of a constant surface strain ratio leads to

$$\frac{\varepsilon_{2}}{\varepsilon_{1}} = \frac{\varepsilon_{2}^{e}}{\varepsilon_{1}^{e}} = \frac{\frac{1}{E}\cdot\left(\sigma_{2}^{e} - v\cdot\sigma_{1}^{e}\right)}{\frac{1}{E}\cdot\left(\sigma_{1}^{e} - v\cdot\sigma_{2}^{e}\right)}.$$

4. Based on Hencky's flow rule, the generalized Poisson ratio and the principal stress ratio can be obtained as

$$v' = \frac{1}{2} - \left(\frac{1}{2} - v\right)\cdot\frac{\sigma_{eq}}{E\cdot\varepsilon_{eq}}$$

$$a = \frac{\sigma_{2}}{\sigma_{1}} = \frac{\frac{\varepsilon_{2}}{\varepsilon_{1}} + v'}{1 + v'\cdot\frac{\varepsilon_{2}}{\varepsilon_{1}}}.$$

5. The principal stresses and strains can be calculated as follows:

$$\sigma_{1} = \sigma_{eq}\cdot\frac{1}{\sqrt{1 - a + a^{2}}}$$

$$\sigma_{2} = a\cdot\sigma_{1}$$

$$\varepsilon_{1} = \frac{\varepsilon_{eq}}{\sigma_{eq}}\cdot\left(\sigma_{1} - v'\cdot\sigma_{2}\right)$$

$$\varepsilon_{2} = \frac{\varepsilon_{eq}}{\sigma_{eq}}\cdot\left(\sigma_{2} - v'\sigma_{1}\right)$$

$$\varepsilon_{3} = -v'\cdot\frac{\varepsilon_{eq}}{\sigma_{eq}}\cdot\left(\sigma_{1} + \sigma_{2}\right).$$

Buczynski and Glinka

Moftakhar et al. (1995) proposed the concept of equality of the total strain energy at the notch root by modifying the Neuber rule for the multiaxial notch stress–strain analysis:

$$\underline{\sigma}^e : \underline{\varepsilon}^e = \underline{\sigma} : \underline{\varepsilon} \tag{7.140}$$

where the product of $\underline{\sigma}^e$ and $\underline{\varepsilon}^e$ is the total strain energy due to pseudo stress and strain tensors at a notch and the product of $\underline{\sigma}$ and $\underline{\varepsilon}$ is the total strain energy due to true stress and strain tensors at the same notch.

This is depicted in Figure 7.12. With the Hencky flow rule, the following additional stress–strain equation is introduced:

$$\underline{\varepsilon} = \frac{1+v}{E}\underline{\sigma} - \frac{v}{E}(\underline{\sigma}:\underline{I})\underline{I} + \frac{3}{2}\frac{\varepsilon^p_{eq}}{\sigma_{eq}}\left[\underline{\sigma} - \frac{1}{3}(\underline{\sigma}:\underline{I})\underline{I}\right] \tag{7.141}$$

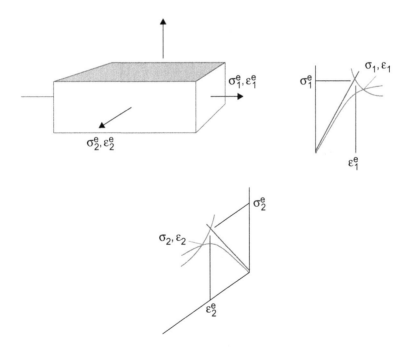

Figure 7.12
Generalized Neuber rule.

where $\varepsilon_{eq}^p = \frac{\sigma_{eq}}{E_{eq,t}}$. In this case, $\underline{\sigma}$ and $\underline{\varepsilon}$ can be obtained by solving the simultaneous Equations (7.140) and (7.141).

However, after reviewing this total strain-based approach, Chu (1995) found that the accuracy of the estimated notch stresses and strains depends on the selected coordinate system and the local notch constraint conditions and that the conflict between the flow rule and the modified Neuber rule may result in singularity in the solutions at some specific stress ratios. Therefore, it is recommended to use the incremental format of the strain energy to improve the accuracy.

Buczynski and Glinka (2000) developed a method analogous to the original Neuber rule by assuming that the elastic incremental strain energy density equals the true incremental strain energy density. It is graphically shown in Figure 7.13 and mathematically expressed in terms of deviatoric stress and strain spaces:

$$\underline{S}_k^e : \Delta\underline{e}^e + \Delta\underline{S}^e : \underline{e}_k^e = \underline{S}_k : \Delta\underline{e} + \Delta\underline{S} : \underline{e}_k \qquad (7.142)$$

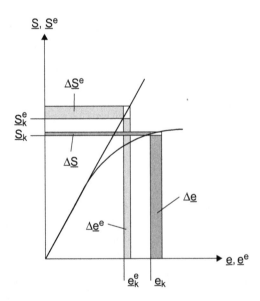

Figure 7.13
Generalized Neuber rule in deviatoric stress
and strain increments.

where

$\underline{S}_k^e, \underline{e}_k^e, \Delta\underline{S}^e$, and $\Delta\underline{e}^e =$ the deviatoric pseudo stress and strain tensors at a load step k, and the incremental deviatoric pseudo stress and strain tensors, respectively

$\underline{S}_k, \underline{e}_k, \Delta\underline{S}$, and $\Delta\underline{e} =$ the true deviatoric stress and strain tensors at a load step k, and the true incremental deviatoric stress and strain tensors, respectively

The quantities on the left side of Equation (7.142) are known from the outputs of a linear elastic FEA due to the incremental applied loads. The two variables ($\Delta\underline{S}$ and $\Delta\underline{e}$) on the right side of the equation are unknown and can be determined by introducing an additional equation based on the Prandtl–Reuss flow rule as:

$$\Delta\underline{e} = \frac{\Delta\underline{S}}{2G} + \frac{3}{2}\frac{\Delta\varepsilon_{eq}^p}{\sigma_{eq}}\underline{S}_k. \qquad (7.143)$$

Consequently, the pseudo and true deviatoric stress and strain tensors at the next load step k+1 can be updated as

$$\underline{S}_{k+1}^e = \underline{S}_k^e + \Delta\underline{S}^e \qquad (7.144)$$

$$\underline{e}_{k+1}^e = \underline{e}_k^e + \Delta\underline{e}^e \qquad (7.145)$$

$$\underline{S}_{k+1} = \underline{S}_k + \Delta\underline{S} \qquad (7.146)$$

$$\underline{e}_{k+1} = \underline{e}_k + \Delta\underline{e} \qquad (7.147)$$

where

$$\Delta\underline{e}^e = \frac{\Delta\underline{S}^e}{2G}. \qquad (7.148)$$

In terms of true stress and strain spaces, the equality of total strain energy at a notch root can be expressed as follows:

$$\underline{\sigma}_k^e : \Delta\underline{\varepsilon}^e + \Delta\underline{\sigma}^e : \underline{\varepsilon}_k^e = \underline{\sigma}_k : \Delta\underline{\varepsilon} + \Delta\underline{\sigma} : \underline{\varepsilon}_k. \qquad (7.149)$$

Assuming the nominally elastic behavior, Equation (7.149) can be written as

$$\frac{2}{E}(\underline{\sigma}_k^e : \Delta\underline{\sigma}^e) = \underline{\sigma}_k : \Delta\underline{\varepsilon} + \Delta\underline{\sigma} : \underline{\varepsilon}_k. \qquad (7.150)$$

Similarly, the Prandtl–Reuss stress–strain incremental relation for isotropic hardening is introduced as

$$\Delta\underline{\varepsilon} = \frac{1+v}{E}\Delta\underline{\sigma} - \frac{v}{E}(\Delta\underline{\sigma}:\underline{I})\underline{I} + \frac{3}{2}\frac{\varepsilon^p_{eq}}{\sigma_{eq}}\left[\Delta\underline{\sigma} - \frac{1}{3}(\Delta\underline{\sigma}:\underline{I})\underline{I}\right]. \qquad (7.151)$$

As a result, the true stress and strain incremental tensors ($\Delta\underline{\sigma}$, $\Delta\underline{\varepsilon}$) can be calculated by solving the previous simultaneous equations–(7.150) and (7.151).

Lee, Chiang, and Wong

Lee, Chiang, and Wong (1995) developed a two-step calculation procedure to estimate multiaxial notch stresses and strains using pseudo stresses as an input to a two-surface plasticity model. At the same time, Barkey, Socie, and Hsia (1994) proposed a similar approach by using an anisotropic plasticity model with the structural stress versus strain relation. Later, Gu and Lee (1997) and Lee and Gu (1999) extended the two-step calculation procedure by using an endochronic approach to calculate notch stresses and strains based on pseudo stress solutions.

The complete two-step solution procedure is summarized here. The first step of this approach is to create a uniaxial pseudo stress versus true strain relation ($\sigma^e_1 - \varepsilon_1$ curve) by using either Neuber's rule (Neuber, 1961) or Molsky–Glinka's energy density method (Molsky & Glinka, 1981). The new cyclic strength coefficient and strain hardening exponent (K^* and n^*) can then be estimated by fitting the previous $\sigma^e_1 - \varepsilon_1$ curve. Subsequently, the local true strains can be calculated by using the stress control plasticity model with the new material properties (K^* and n^*). The second step is to obtain the local true stresses by using the strain control plasticity model with the cyclic material properties (K' and n').

Figure 7.14 illustrates the concept of this approach. The $\sigma_1 - \varepsilon_1$ curve is the uniaxial material stress–strain behavior. The linear elastic finite element solution follows the $\sigma^e_1 - \varepsilon^e_1$ curve and can be converted into the $\sigma^e_1 - \varepsilon_1$ curve by using Neuber's or Molsky–Glinka's method. With any given value of σ^e_1, the local strain ε_1 can be obtained from the $\sigma^e_1 - \varepsilon_1$ curve. And, after ε_1 is calculated, the local stress σ_1 can be easily determined from the $\sigma_1 - \varepsilon_1$ curve. It is assumed that

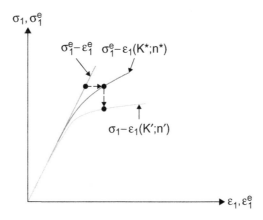

Figure 7.14
Illustration of the two-step calculation concept in uniaxial
stress versus strain curves.

the $\sigma_1^e - \varepsilon_1$ and $\sigma_1 - \varepsilon_1$ behavior in multiaxial state of stresses can be described by any cyclic plasticity model.

Summary

An additive decomposition of the elastic and plastic components of the total strain is always assumed throughout all the plasticity theories. The elastic strain tensor can be calculated based on the theory of elasticity; the plastic strain tensor is based on the theory of plasticity.

The theory of elasticity has been reviewed. The formulas based on Hooke's law for calculation of elastic strain tensor due to stress and deviatoric stress tensors were presented.

There are many classical plasticity theories for plastic strain tensor calculation for materials undergoing monotonic loading. They are common in adopting the isotropic hardening model where plastic deformation of the material under loading will result in even size expansion of the yield surface with respect to its center, but different in defining the flow rule specifying the orientation and magnitude of the plastic deformation. Some of the popular flow rules (such as Levy, Mises,

Prandtl, Reuss, Hencky, and the associated flow rule) are discussed. Their equations for calculation of plastic strain tensor are also documented.

There are numerous theories of plasticity to simulate the hysteresis behavior for materials subjected to cyclic loading. They are common in adopting the associated flow rule for the plastic deformation tensor, but different in using various kinematic hardening models (such as nested-surface model, two-surface model, and single-surface model) to account for the Bauschinger effect in cyclic reversed loading.

In each model, there are various translational rules (for example, Parger's, Mroz's, Armstrong–Frederick's, Jiang–Sehitoglu's, Phillips–Lee's, Ziegler's, and Rolovic–Tipton's) developed based on experimental observation to update the back stress tensor of the yield surface during a plastic deformation increment. These kinematic hardening models with the popular translational rules are discussed. Two examples for calculation of plastic strain tensor are illustrated.

An advanced plasticity model, a combined isotropic and kinematic hardening model developed by Zhang and Jiang which adopts the fourth-rank tensor developed by Tanaka to account for nonproportional hardening has been presented.

The application of the existing plasticity theories to multiaxial notch analysis based on pseudo stresses is also reviewed. Three popular multiaxial notch analyses (the Hoffman–Seeger, Buczynski–Glinka, and Lee–Chiang–Wong methods) are also presented in detail.

References

Armstrong, P. J., & Frederick, C. O. (1966). *A mathematical representation of the multiaxial Bauschinger effect*, CEGB Report RD/B/N 731, Berkeley Nuclear Laboratories.

Bannantine, J. A. (1989). *A variable amplitude multiaxial life prediction method*, (UILU-ENG 89-3605), A report of the material engineering-mechanical behavior, College of Engineering, University of Illinois at Urbana–Champaign, October.

Barkey, M. E., Socie, D. F., & Hsia, K. J. A. (1994). A yield surface approach to the estimation of notch strains for proportional and nonproportional cyclic loading. *Journal of Engineering Materials and Technology, 116*, 173–180.

Buczynski, A., & Glinka, G. (2000). Multiaxial stress-strain notch analysis. In S. Kalluri & P. J. Bonacause (Eds.), *Multiaxial fatigue and deformation: testing and prediction*, ASTM 1387 (pp. 82–98). West Conshohocken, PA: ASTM.

Chaboche, J. L. (1977). Viscoplastic constitutive equations for the description of cyclic and anisotropic behavior of metals. *Bulletin de l'Acad Polonaise des Sciences, Series Sc. Et Techn., 25*(1), 33–42.

Chaboche, J. L., & Rousselier, G. (1983). On the plastic and viscoplastic constitutive equations, part I: Rules developed with internal variable concept. *ASME Journal of Pressure Vessel Technology, 105*, 153–158.

Chen, W. R., & Keer, L. M. (1991). An application of incremental plasticity theory to fatigue life prediction of steels. *ASME Journal of Engineering Materials and Technology, 113*, 404–410.

Chu, C.-C. (1995). *Incremental multiaxial Neuber correction for fatigue analysis.* SAE Paper No. 950705, SAE World Congress, Detroit.

Dafalias, Y. F., & Popov, E. P. (1975). A model of nonlinearly hardening materials for complex loading. *Acta Mechanica, 21*, 173–192.

Dafalias, Y. F., & Popov, E. P. (1976). Plastic internal variables formalism of cyclic plasticity. *ASME Journal of Applied Mechanics, 43*(4), 645–651.

Doong, S. H., Socie, D. F., & Robertson, I. M. (1990). Dislocation substructures and nonproportional hardening. *Journal of Engineering Materials and Technology, 112*(4), 456–465.

Gu, R., & Lee, Y. L. (1997). A new method for estimating nonproportional notch-root stresses and strains. *ASME Journal of Engineering Materials and Technology, 119*, 40–45.

Hencky, H. (1924). Zur theorie plastischer deformationen und der hierdurch hervorgerufenen nachspannunger. *Zeits. Angew. Math. u. Mech., 4*, 323–334.

Hoffmann, M., & Seeger, T. (1989). Stress-strain analysis and life predictions of a notched shaft under multiaxial loading. In *Multiaxial Fatigue: Analysis and Experiments*, Chapter 6, SAE AE-14, 81–96. Warrendale, PA: SAE International.

Hughes, T. J. R. (1984). Numerical implementation of constitutive model: Rate-independent deviatoric plasticity. In S. Nemat-Nasser et al. (Eds.), *Theoretical foundation for large-scale computations for nonlinear material behavior* (pp. 24–57). Boston: Martinus Nijhoff Publishers.

Itoh, T., Sakane, M., Ohnami, M., & Socie, D. F. (1995). Nonproportional low cycle fatigue criterion for type 304 stainless steel. *Journal of Engineering Materials and Technology, 117*, 285–292.

Itoh, T., Sakane, M., Ohnami, M., & Socie, D. F. (1997). Nonproportional low cycle fatigue of 6061 aluminum alloy under 14 strain paths. *Proceedings of 5th International Conference on Biaxial/Multiaxial Fatigue and Fracture, 1* (pp. 173–187). Cracow, Poland.

Jiang, Y., & Sehitoglu, H. (1996a). Modeling of cyclic ratcheting plasticity, Part I: Development of constitutive relations. *ASME Journal of Applied Mechanics, 63*, 720–725.

Jiang, Y., & Sehitoglu, H. (1996b). Modeling of cyclic ratcheting plasticity, Part II: Comparisons of model simulations with experiments. *ASME Journal of Applied Mechanics, 63*, 726–733.

Krempl, E., & Lu, H. (1983). Comparison of the stress responses of an aluminum alloy tube to proportional and alternate axial and shear strain paths at room temperature. *Mechanics of Materials, 2*, 183–192.

Krieg, R. D. (1975). A practical two surface plasticity theory. *ASME Journal of Applied Mechanics, 42*(3), 641–646.

Lee, Y. L., Chiang, Y. J., & Wong, H. H. (1995). A constitutive model for estimating multiaxial notch strains. *ASME Journal of Engineering Materials and Technology, 117*, 33–40.

Lee, Y. L., & Gu, R. (1999). Multiaxial notch stress-strain analysis and its application to component life predictions, Chap. 8. In T. Cordes, & K. Lease (Eds.), *Multiaxial fatigue of an inducted hardened shaft* (pp. 71–78). SAE AE-28. Warrendale, PA: SAE International.

Levy, M. (1870). Memoire sur les equations generales des mouvements interieurs des corps solides ductiles au dela des limitesou l'elasticite pourrait les ramener a leur premier etat. *C. R. Acad. Sci., Paris, 70*, 1323–1325.

Moftakhar, A., Buczynski, A., & Glinka, G. (1995). Calculation of elasto-plastic strains and stress in notches under multiaxial loading. *International Journal of Fracture, 70*, 357–373.

Molsky, K., & Glinka, G. (1981). A method of elastic-plastic stress and strain calculation at a notch root. *Material Science and Engineering, 50*, 93–100.

Mroz, Z. (1967). On the description of anisotropic workhardening. *Journal of the Mechanics and Physics of Solids, 15*, 163–175.

Neuber, H. (1961). Theory of stress concentration for shear-strained prismatic bodies with arbitrary nonlinear stress-strain law. *Journal of Applied Mechanics, Transactions of ASSM, Section E, 28*, 544–550.

Ohno, N., & Wang, J. D. (1993a). Kinematic hardening rules with critical state of dynamic recovery, part I—Formulation and basic features for ratcheting behavior. *International Journal of Plasticity, 9*, 375–390.

Ohno, N., & Wang, J. D. (1993b). Kinematic hardening rules with critical state of dynamic recovery, part II—Application to experiments of ratcheting behavior. *International Journal of Plasticity, 9*, 391–403.

Phillips, A., & Lee, C. (1979). Yield surfaces and loading surfaces, experiments and recommendations. *International Journal of Solids and Structures, 15*, 715–729.

Prager, W. (1955). The theory of plasticity: A survey of recent achievement. *Proceedings of the Institution of Mechanical Engineers, 169*, 41–57.

Prandtl, L. (1925). Spannungsverteilung in plastischen koerpern. *Proceedings of the First International Congress of Applied Mechanics, Delft*, 43–54.

Reuss, E. (1930). Beruecksichtigung der elastischen formaenderungen in der plastizitaetstheorie. *Zeitschriftfür Angewandte Mathematik und Mechanik, Journal of Applied Mathematics and Mechanics, 10*, 266–274.

Rolovic, R., & Tipton, S. M. (2000a). Multiaxial cyclic ratcheting in coiled tubing—Part I: Theoretical modeling. *ASME Journal of Engineering Materials and Technology, 122*, 157–161.

Rolovic, R., & Tipton, S. M. (2000b). Multiaxial cyclic ratcheting in coiled tubing—Part II: Experimental program and model evaluation. *ASME Journal of Engineering Materials and Technology, 122,* 162–167.

Socie, D. F., & Marquis, G. B. (2000). *Multiaxial fatigue.* Warrendale, PA: SAE International.

Tanaka, A., Murakami, S., & Ooka, M. (1985). Effects of strain path shapes on nonproportional cyclic plasticity. *Journal of the Mechanics and Physics of Solids, 33*(6), 559–575.

Tanaka, A. (1994). A nonproportional parameter and a cyclic viscoplastic constitutive model taking into account amplitude dependences and memory effects of isotropic hardening. *European Journal of Mechanics, A/Solids, 13*(2), 155–173.

von Mises, R. (1913). Mechanik der festen körperim plastisch deformablen zustand. *Nachrichten der Gesellschaft der. Wissenschaften in Göttingen, Mathematisch Physikalische Klasse,* 582–592.

von Mises, R. (1928). Mechanik der plastischen formänderung von kristallen. *Zeitschriftfür Angewandte Mathematik und Mechanik (Applied Mathematics and Mechanics), 8,* 161–185.

Wang, C. H., & Brown, M. W. (1993). Inelastic deformation and fatigue complex loading. In K. Kussmaul (Ed.), *SMiRT-12* (pp. 159–170). New York: Elsevier Science.

Ziegler, H. (1959). A modification of prager's hardening rule. *Quarterly of Applied Mathematics, 17,* 55–65.

Zhang, J., & Jiang, Y. (2008). Constitutive modeling of cyclic plasticity deformation of a pure polycrystalline copper. *International Journal of Plasticity, 24,* 1890–1915.

CHAPTER 8

Strain-Based Multiaxial Fatigue Analysis

Mark E. Barkey
The University of Alabama

Yung-Li Lee
Chrysler Group LLC

Chapter Outline

Introduction

This chapter's emphasis is on critical plane methods and appropriate damage parameter selection, since the more general critical plane methods simplify to equivalent stress–strain approaches for proportional multiaxial loading. Damage parameters and calculations based on energy density terms that contain combinations of stress and strain components are also discussed.

In addition to performing fatigue calculations during the design phase, the CAE analyst may also be asked to examine test data collected using strain gages for life prediction or for correlation to CAE models. Therefore, the basic data reduction approaches for strain-gage rosettes are also discussed. Additionally, an example of multiaxial strain-based fatigue analysis is presented using strain gage rosette data, and additional examples using multiaxial fatigue criteria and the critical plane approach are presented.

Fatigue Damage Models

Strain-based uniaxial approaches were discussed in Chapter 6. Approaches for multiaxial low-cycle fatigue can be broadly categorized as equivalent stress–strain approaches, energy approaches, and damage parameter, critical plane approaches. A review of those approaches proposed from 1980 to 1996 is presented by You and Lee (1996), and more recently by Wang and Yao (2004). These various approaches have been extended and modified since, and are still current areas of research and discussion.

All of these approaches rely on an accurate characterization of the time-varying stress and strain state at the critical location where the fatigue damage is to be assessed. Material constitutive models were discussed in Chapter 7, initially for uniaxial loading conditions and then extended to proportional cyclic loading and for general multiaxial stress states. In a similar manner, the type of strain-based low-cycle fatigue approach will depend on the manner in which the critical location is stressed.

Equivalent Strain Approaches

Under laboratory conditions, multiaxial loading is often accomplished by subjecting a cylindrical solid or hollow test specimen to tension–torsion loading. In some cases, a cruciform type specimen is subjected to in-plane biaxial tension/compression loading. The loading conditions are usually represented by a phase plot of the loads, stresses, or strains in the primary loading directions.

In the case of tension and torsion, the phase plots of stress are usually shown in equivalent stress axes; for the von Mises yield criterion:

$$\sigma_o = \frac{1}{\sqrt{2}}\left[(\sigma_x - \sigma_y)^2 + (\sigma_y - \sigma_z)^2 + (\sigma_z - \sigma_x)^2 + 6\tau_{xy}^2 + 6\tau_{xz}^2 + 6\tau_{yz}^2\right]^{1/2}. \quad (8.1)$$

If the specimen is subjected to only an axial stress σ_x, the equivalent stress $\sigma_o = \sigma_x$ and for only a torsional stress, $\sigma_o = \sqrt{3}\tau_{xy}$. If combined axial and torsional loads are applied, then the expression for the von Mises equation results in

$$\sigma_o = \sqrt{(\sigma_x^2 + 3\tau_{xy}^2)}.$$

A 45° path on this phase plot represents a proportional tension–torsion loading where the equivalent stress contribution from the tension and torsion terms are equal. A circular path on these axes represents 90° out-of-phase loading. Examples of stress phase plots are shown in Figure 8.1. Note that these plots are not Mohr's circle of stress plots.

Proportional and nonproportional loading paths are easily distinguished by using these phase plots—proportional loading paths are straight lines and non-proportional paths are any other paths. Under laboratory conditions, a complete loading cycle can be defined as one circuit around the path.

An equivalent plastic strain equation can be defined in a manner consistent with the definition of the von Mises equation (Mendelson, 1983). The equivalent plastic strain, ε_o^p, is given by:

$$\varepsilon_o^p = \frac{\sqrt{2}}{3}\left[(\varepsilon_x^p - \varepsilon_y^p)^2 + (\varepsilon_y^p - \varepsilon_z^p)^2 + (\varepsilon_z^p - \varepsilon_x^p)^2 + 6(\varepsilon_{xy}^p)^2 + 6(\varepsilon_{xz}^p)^2 + 6(\varepsilon_{yz}^p)^2\right]^{1/2}$$

(8.2)

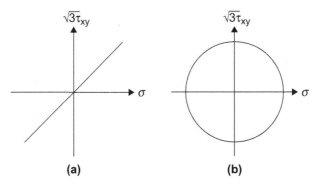

Figure 8.1
Proportional (a, in-phase) and nonproportional (b, 90° out-of-phase) loading plots in equivalent stress phase space (90° out-of-phase).

where

ε_x^p, ε_y^p, ε_z^p, ε_{xy}^p, ε_{xz}^p, and ε_{yz}^p = the plastic strain tensor components

When a uniaxial stress is applied in the x direction, strains occur in the axial direction and both transverse directions.

For an isotropic material that follows the conservation of volume during plastic deformation, the transverse plastic strains are $\varepsilon_y^p = \varepsilon_z^p = -\frac{1}{2}\varepsilon_x^p$. Thus for a uniaxial stress state,

$$\varepsilon_o^p = \frac{\sqrt{2}}{3}\left[\left(\varepsilon_x^p + \frac{1}{2}\varepsilon_x^p\right)^2 + \left(-\frac{1}{2}\varepsilon_x^p - \varepsilon_x^p\right)^2\right]^{1/2} = \frac{\sqrt{2}}{3}\left[\frac{18}{4}(\varepsilon_x^p)^2\right]^{1/2} = \varepsilon_x^p. \tag{8.3}$$

For a state of pure shear loading,

$$\varepsilon_o^p = \frac{\sqrt{2}}{3}\left[6(\varepsilon_{xy}^p)^2\right]^{1/2} = \frac{\sqrt{2}}{3}\sqrt{6}\,\varepsilon_{xy}^p = \frac{\sqrt{2}}{3}\sqrt{6}\,\frac{\gamma_{xy}^p}{2} = \frac{1}{\sqrt{3}}\gamma_{xy}^p$$

where the term γ_{xy}^p is the engineering plastic shear strain.

For a combined state of uniaxial tension and torsion loading,

$$\varepsilon_o^p = \left[(\varepsilon_x^p)^2 + \frac{1}{3}(\gamma_{xy}^p)^2\right]^{1/2}. \tag{8.4}$$

For low-cycle fatigue applications, the elastic portion of the strains is often neglected, and the equivalent strain is computed based on total plastic strain quantities.

In other cases, the elastic strains are incorporated into the equivalent strain through the use of a modified Poisson's ratio term, as shown in Equation (8.5) (Shamsaei & Fatemi, 2010):

$$\varepsilon_o = \frac{1}{\sqrt{2}(1+\bar{\nu})}\left[2(\varepsilon_x)^2(1+\bar{\nu})^2 + \frac{3}{2}(\gamma_{xy})^2\right]^{1/2} \tag{8.5}$$

where

$$\bar{\nu} = \frac{\nu_e\bar{\varepsilon}_e + \nu_p\bar{\varepsilon}_p}{\bar{\varepsilon}}$$

In this expression, the modified Poisson's ratio depends on the elastic Poisson's ratio (the usual material property), the plastic Poisson's ratio (a result of volume conserving plastic flow of metals), and equivalent elastic, plastic, and total

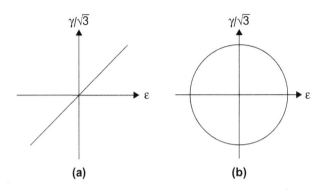

Figure 8.2
Proportional (a, in-phase) and nonproportional (b, 90° out-of-phase) loading plots in equivalent strain phase space.

strains. In Equation (8.5), the equivalent plastic strain, $\bar{\varepsilon}_p = \varepsilon_o^p$, is determined from Equation (8.2), the equivalent elastic strain is determined by $\bar{\varepsilon}_e = \sigma_0/E$, and the total equivalent strain is given by $\bar{\varepsilon} = \bar{\varepsilon}_e + \bar{\varepsilon}_p$.

The typical presentation of equivalent strain phase plots takes the form shown in Figure 8.2, where plots are presented on axes of total axial and total engineering shear strain over the square root of three. Note that these plots are not Mohr's circle of strain plots.

Equivalent Strain Approaches for Proportional Loading

The goal of the equivalent strain approach is to use the equivalent strain amplitude, $\bar{\varepsilon}_a$, in a damage parameter equation to determine fatigue life. For example, modifying the uniaxial strain–life equation to write it in terms of equivalent strain amplitude results in

$$\bar{\varepsilon}_a = \frac{\sigma_f'}{E}(2N)^b + \varepsilon_f'(2N)^c \qquad (8.6)$$

where

$$2N = \text{the number of reversals to crack initiation}$$
$$\sigma_f' = \text{the fatigue strength coefficient}$$
$$\varepsilon_f' = \text{the fatigue ductility coefficient}$$
b, c, and E = the fatigue strength exponent, fatigue ductility exponent, and modulus of elasticity, respectively

This method reduces to the uniaxial strain–life equation in the absence of shearing strains. Additionally, if only shearing stress is applied, it provides a way to relate shear strain–life material properties to uniaxial strain–life material properties. As described in Bannantine et al. (1990), if the slopes of the uniaxial and shear strain–life curves are the same (the material constants b and c), then an equivalence can be seen between the shear strain–life equation:

$$\gamma_a = \frac{\tau_f'}{G}(2N)^b + \gamma_f'(2N)^c \qquad (8.7)$$

and the equivalent strain–life equation for torsional-only loading:

$$\bar{\varepsilon}_a = \frac{\gamma_a}{\sqrt{3}} = \frac{\sigma_f'}{E}(2N)^b + \varepsilon_f'(2N)^c,$$

resulting in

$$\gamma_f' = \sqrt{3}\varepsilon_f', \quad \text{and} \quad \tau_f' = \sqrt{3}\sigma_f'\frac{G}{E} = \frac{\sqrt{3}\sigma_f'}{2(1+\nu)} \approx \frac{\sigma_f'}{\sqrt{3}}$$

when plastic strains dominate.

In these equations, τ_f' is the torsional fatigue strength coefficient, γ_f' is the torsional fatigue ductility coefficient, and b, c, ν, and G are the fatigue strength exponent, fatigue ductility exponent, Poisson's ratio, and shear modulus of elasticity, respectively.

It should be noted however, that although the equations developed with this approach predict that equivalent strain should result in the same fatigue life for both the axial-only and torsional-only loading, experimental discrepancies are to be expected and have been noted (Krempl, 1974). Torsional-only loading can often have a factor of two on fatigue lives as compared with axial-only loading at the same equivalent strain range.

In the low-cycle fatigue regime, this discrepancy is often attributed to the fatigue damage mechanism being sensitive to the hydrostatic stress (Mowbray, 1980). Mowbray proposed a modification of the strain–life equation as:

$$\frac{\Delta\varepsilon_1}{2} = \frac{\sigma_f'}{E}f(\lambda,\nu)(2N)^b + \left(\frac{3}{3-m}\right)\varepsilon_f'g(\lambda,m)(2N)^c \qquad (8.8)$$

where

$$f(\lambda, \nu) = \frac{(1 - \nu\lambda)}{(1 - \lambda + \lambda^2)^{1/2}} \quad \text{and}$$

$$g(\lambda, m) = \frac{(2 - \lambda)\left[3(1 - \lambda + \lambda^2)^{1/2} - m(1 + \lambda)\right]}{6(1 - \lambda + \lambda^2)}$$

in which ν is Poisson's ratio, m is a material constant with an upper bound of 1.5 that incorporates the effect of hydrostatic stress, $\lambda = \Delta\sigma_2/\Delta\sigma_1$ is the biaxiality ratio computed from magnitude ordered principal stress ranges, and the remaining terms in Equation (8.8) are from the strain–life equation.

Mowbray's equivalent strain parameter, shown in Equation (8.8), was developed using the deformation theory of plasticity, which restricts the use of this equation to proportional loading paths. Although this approach has not been widely adopted by the fatigue community, it does demonstrate that damage parameters based on equivalent strain terms alone are unable to accommodate experimental observations and that stress terms must be included if a multiaxial damage parameter is to be suitable for both low-cycle and mid-cycle fatigue regimes.

Although the most common equivalent strain approaches are based on the von Mises stress and strain, equivalent approaches based on the maximum principal strain and maximum shear strain (Tresca) criteria have also been developed. The advantages of equivalent strain approaches for multiaxial fatigue is that the methods are easy to apply for proportional strain loading histories and usually require only commonly available strain–life material properties.

Equivalent Strain Approaches for Nonproportional Loading

Equivalent strain approaches have also been developed for nonproportional loading. Most of them incorporate a parameter that describes the degree of out-of-phase loading within the loading path. Applications of these approaches are limited, and are usually used for load paths that can be readily applied in a laboratory setting. Even so, these approaches can be useful for describing and quantifying material behavior differences between proportional and nonproportional loading paths.

For tension–torsion multiaxial loading, the largest difference in fatigue behavior is often between proportional and 90° out-of-phase loading. Equivalent strain

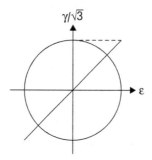

Figure 8.3
In-phase and out-of-phase loading paths based on equal maximum
equivalent strains for an entire cycle.

paths that exhibit the same maximum equivalent strain ranges can be constructed
for in-phase and out-of-phase loading, and are illustrated in Figure 8.3. Using
equivalent strain approaches for nonproportional loading can generate additional
difficulties.

As discussed previously in Chapter 5, some materials can experience additional
hardening for out-of-phase or nonproportional loading—meaning that under
strain controlled conditions, stresses in the material are higher for out-of-phase
loading than for in-phase loading. These additional stresses can cause more
damage for out-of-phase loading as compared to in-phase loading.

The degree of nonproportional hardening is dependent on the microstructure of
the material. It can be characterized by the ratio of equivalent stress developed
from out-of-phase loading and in-phase loading as follows:

$$\alpha = \frac{\overline{\sigma}_{OOP}}{\overline{\sigma}_{IP}} - 1 \qquad (8.9)$$

where

α = the nonproportional hardening coefficient
$\overline{\sigma}_{OOP}$ = the equivalent out-of-phase stress
$\overline{\sigma}_{IP}$ = the equivalent in-phase stress

Testing to determine the nonproportional hardening coefficient is more complicated than is standard strain controlled fatigue testing. Shamsaei and Fatemi
(2010) have observed that materials that tend to cyclically harden during

uniaxial testing exhibit a higher degree of nonproportional hardening than materials that do not cyclically harden.

Shamsaei and Fatemi further proposed the following relation to determine the nonproportional hardening coefficient:

$$\alpha = 1.6 \left(\frac{K}{K'}\right)^2 \left(\frac{\Delta\varepsilon}{2}\right)^{2(n-n')} - 3.8 \left(\frac{K}{K'}\right) \left(\frac{\Delta\varepsilon}{2}\right)^{(n-n')} + 2.2 \tag{8.10}$$

where

$\quad$ K and K' = the Ramberg–Osgood coefficients for the monotonic and cyclic
$\qquad\qquad\quad$ stress–strain curves
$\quad$ n and n' = the monotonic and cyclic exponents
$\qquad \Delta\varepsilon/2$ = the strain amplitude.

The Ramberg–Osgood properties for materials are usually available or can themselves be estimated from fatigue properties of materials. If a high degree of nonproportional low-cycle plasticity is expected for an analysis, the out-of-phase hardening parameters can be accounted for and the equivalent stress amplitude-plastic strain amplitude curve can be adjusted by

$$\frac{\Delta\sigma}{2} = K'(1 + \alpha F) \left(\frac{\Delta\bar{\varepsilon}_p}{2}\right)^{n'} \tag{8.11}$$

where F represents the factor of nonproportionality due to a loading path effect. Therefore, the equivalent strain approaches can be used for more general nonproportional loading if enough data is available to assess the nonproportional hardening affects.

Another equivalent strain approach has been developed by Itoh et al. (1999). In their study, Itoh and coworkers conducted nonproportional tension–torsion fatigue testing on materials subjected to as many as 14 different loading paths based on equivalent stress and strain ranges, such as earlier in Figure 8.3, and other paths including box-paths, stair-step paths, and X-paths. One cycle was considered one circuit around the loading path.

At any instant in time, the maximum absolute value of principle strain can be calculated by:

$$\varepsilon_I(t) = \begin{cases} |\varepsilon_1(t)| & \text{for} \quad |\varepsilon_1(t)| \geq |\varepsilon_3(t)| \\ |\varepsilon_3(t)| & \text{for} \quad |\varepsilon_1(t)| < |\varepsilon_1(t)| \end{cases} \tag{8.12}$$

and the maximum value of $\varepsilon_I(t)$ is defined as

$$\varepsilon_{Imax} = Max[\varepsilon_I(t)]. \tag{8.13}$$

During the cycle, the principal directions of strain will change, in general. The angle between the current direction of $\varepsilon_I(t)$ to the direction of the maximum value of $\varepsilon_I(t)$ over the cycle (ε_{Imax}) is defined as $\xi(t)$. The maximum principal strain range over the cycle can then be defined by

$$\Delta\varepsilon_I = Max[\varepsilon_{Imax} - \cos(\xi(t))\,\varepsilon_I(t)]. \tag{8.14}$$

Itoh and coworkers' (1999) tests on 304 steel and 6061 aluminum alloy both exhibited additional hardening for nonproportional loading. Since the directions of principal strains are changing, the different planes in the material exhibit varying amounts of shear stress and strain range during the cycle. Depending on the material microstructure, the slip systems within the material may interact.

As explained by Itoh, for the 304 steel, the interaction was such that large additional hardening occurred during nonproportional cycling. For the 6061 aluminum alloy, much less interaction occurred since this aluminum alloy has a higher stacking fault energy and exhibits wavy dislocation slip, resulting in an easier change of slip glide planes.

Itoh then proposed a nonproportional strain range as

$$\Delta\varepsilon_{NP} = (1 + \alpha f_{NP})\,\Delta\varepsilon_I \tag{8.15}$$

where

α = a material constant related to the additional hardening (defined as in Equation 8.9)

f_{NP} = the nonproportionality factor that expresses the severity of nonproportional straining

The latter depends on the strain path that is taken. Itoh and coworkers defined this term as

$$f_{NP} = \frac{k}{T\varepsilon_{Imax}} \int_0^T \left|\sin(\xi(t))\right| [\varepsilon_I(t)]\,dt \tag{8.16}$$

where

k = chosen to make f_{NP} equal to one for 90° out-of-phase loading

T = the time period of the cycle, and the result normalized by ε_{Imax}

f_{NP} = zero for proportional loading

Itoh reported that the fatigue test data resulted in less scatter when using this equation, for both materials.

In general, equivalent strain approaches do not consider combinations of stress and strain acting together on critical planes within the material. These approaches do not consider complicated material models; they attempt to reduce the strain tensor components to a single value, and incorporate observed effects that do not match with the basic theory with adjustable parameters. However, equivalent strain approaches can be used effectively when the loading behavior and material behavior are well characterized.

Energy Approaches

The goal of an energy-based fatigue damage parameter is to use the product of stress and strain quantities to determine fatigue damage at a location in a material. The success of the von Mises yield criterion—also known as the maximum distortion energy criterion—in reducing a complex stress state to an equivalent stress that can be compared to the uniaxial yield stress of a metal gives some rationale for this approach. These approaches can be used for high-cycle and low-cycle fatigue.

The Smith–Watson–Topper damage parameter is often cited as an energy-based damage parameter:

$$\sigma_{max}\varepsilon_a = \frac{(\sigma_f')^2}{E}(2N)^{2b} + \sigma_f'\varepsilon_f'(2N)^{b+c} \quad \sigma_{max} > 0 \quad (8.17)$$

where

ε_a = the strain amplitude on the tensile crack plane

σ_{max} = the maximum normal stress on this plane

The right side of the equation is a mean stress influenced representation of the strain–life equation.

The product of the strain amplitude and the maximum stress on the left side of the equation is a strain energy density term. Knowing the material properties on the right side from strain–life fatigue tests, and the left side from the loading history and constitutive equation, the fatigue life/damage can be calculated. In practice, the equation would only be evaluated when the left side is positive, meaning that the maximum stress term has at least some value of tension. Physically, this scenario would represent a material with perhaps even a very small crack being held open while the strain is cycling.

In uniaxial fatigue conditions, the Smith–Watson–Topper damage parameter has been used with a great degree of success. For multiaxial fatigue conditions, the use of a single parameter has met with some challenges in the implementation, particularly for general nonproportional loading.

One approach is to compute the multiaxial strain energy density, as done by Garud (1981). In this approach, the plastic work is computed for a cycle:

$$\Delta W_p = \int_{\text{cycle}} \sigma_{ij} d\varepsilon_{ij}^p \qquad (8.18)$$

where

σ_{ij} = the stress tensor components
$d\varepsilon_{ij}^p$ = the plastic strain tensor components
ΔW_p = the plastic work done in a cycle

It is noted that the implied summation is used for tensor components over i and j.

Fatigue life is related to the plastic work by a function calibrated from uniaxial test data, such as in the power law form $N = A \Delta W_p^{\alpha^*}$, where A and α^* are the power law fitting constants. The approach is similar in concept to that of Ellyin (1974) and variants that use total strain energy:

$$\Delta W_{total} = \Delta W_p + \Delta W_e = \int_{\text{cycle}} \sigma_{ij} d\varepsilon_{ij}. \qquad (8.19)$$

The advantage of these approaches is that the energy approaches are derived from general principals and can be applied to multiaxial nonproportional loading. Such approaches, however, rely on the accurate characterization of the constitutive model of the material under multiaxial loading conditions. General, multiaxial constitutive models are discussed in Chapter 7. The applications of these approaches to materials that exhibit failure modes dependent on stress level and loading mode is a current area of research.

Critical Plane Approaches

In the previous section we have seen that low-cycle fatigue damage parameters are based on both strains and stresses. As in the case of stress-based multiaxial fatigue, a constitutive model is needed to calculate the stress tensor time history or the strain tensor time history from either known strains or stresses, respectively. Constitutive modeling for relatively high strains can be challenging to do. Additionally, some materials can experience additional hardening for out-of-phase or nonproportional loading—meaning that under strain-controlled conditions, stresses in the material are higher for out-of-phase loading than for in-phase loading.

These additional stresses can cause an accelerated amount of fatigue damage on critical planes as compared to in-phase loading. For fatigue calculations, constitutive models that do not account for nonproportional hardening effects may result in nonconservative fatigue life calculations.

An early strain-based critical plane approach was proposed by Kandil, Brown, and Miller (1982), who proposed that a combination of shear strain and normal strain acting on a plane was responsible for crack initiation and growth:

$$\frac{\Delta\gamma_{max}}{2} + S\varepsilon_n = C \tag{8.20}$$

where

$\Delta\gamma_{max}/2 =$ the strain amplitude on the maximum shear strain plane
$\varepsilon_n =$ the normal strain on this plane
$S =$ a material fitting constant

This parameter provided a physical basis for crack growth for shear cracks opened by normal strains perpendicular to the crack surface. However, it has been shown by Socie and coworkers (Socie & Marquis, 2000; Fatemi & Socie, 1988) that strain parameters alone cannot correlate fatigue behavior for a range of materials subjected to both in-phase and out-of-phase loading.

Calculations using the critical plane approach have been discussed in Chapter 5 and can be made in a similar manner for strain-based or low-cycle fatigue calculations. Damage parameters and cracking mechanisms are subsequently reviewed, with an emphasis on low-cycle fatigue applications.

Damage parameters are functions that relate physically meaningful variables, such as stress or strain, to fatigue damage. The fatigue damage of metals is understood to involve the movement of dislocations, the formation of slip bands, and the development of a small crack in the material due to alternating loads.

Therefore, terms that can affect the dislocations, slip bands, and small crack formation in materials subjected to alternating loads would be justifiable to include in a fatigue damage parameter. Indeed, as expected, many damage parameters include shear stress amplitude, mean stress, and crack opening strains fitted for a particular material by the use of material constants.

In addition to these basic terms, it also understood that load level and type of loading can influence the damage evolution in a material. While the use of fitted material constants to a proposed damage parameter is sometimes the only recourse an analyst may have, some rationale for choosing appropriate parameters can be made based on the physical development of small cracks in a particular material.

Socie (1993), through the use of damage mechanism maps, explained a rationale for using different terms in damage parameters based on the respective damage mechanisms that are activated. For example, in his paper, AISI 304 stainless steel was tested in torsion and in tension and the cracking behavior was described:

> *Cracking behavior could be categorized into two regions: Region A and Region B. Region A behavior was observed at short lives. Microcracks initiated on shear planes. Once initiated, the cracks became more distinct but showed no significant increase in length. At failure, a large density of small coarse cracks dominated the surface of the specimen. A small amount of branching onto tensile planes (Stage II planes) was observed. Failure cracks grew on either shear planes (Stage I planes) or tensile planes (Stage II planes) by a slow linking of previously initiated shear cracks. Region B is characterized by shear crack nucleation followed by crack growth on planes of maximum principal strain amplitude (Stage II planes). Shear crack growth consumes a small fraction of the fatigue life. Region C behavior was observed at the longest lives in torsion. The fraction of life spent growing the crack on shear planes was reduced, as was the crack density. A small*

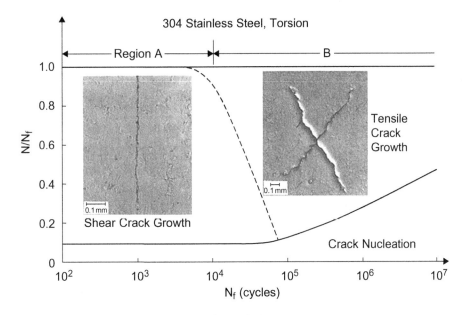

Figure 8.4

Damage mechanism map for 304 stainless steel tested in torsion (Socie, 1993).

number of cracks initiated on shear planes but quickly branched to Stage II planes. Growth on these planes occurred by the propagation of the mail crack rather than by a linking process.[1]

Figure 8.4 shows the life fraction on the vertical axis, and the cycles to failure on the horizontal axis. For the AISI 304 stainless steel loaded in torsion, a portion of the life was spent nucleating small cracks. As the fatigue lives became longer, the portion of fatigue nucleation life increased as a percentage of total life.

For low-cycle torsional fatigue, the nucleated cracks grew in a vertical orientation and horizontal directions on the specimen, as defined by the loading axis of the specimen. These orientations have the highest shear stress and shear plastic strain amplitudes. Socie denoted this region as region A.

---

[1] This paragraph, Figure 8.4, and Figure 8.5 are reprinted, with permission, from *STP 1191 Advances in Multiaxial Fatigue*; copyright ASTM International, 100 Barr Harbor Drive, West Conshohocken, PA 19428.

At longer lives, in the mid-cycle to high-cycle fatigue regime, crack growth was dominated by cracks growing at approximately a 45° angle to the axis of the specimen. For torsional loading, these planes experienced reversed tension and compression loading. Socie denoted this region as region B. In the high-cycle regime, the nucleated cracks transitioned immediately to a tensile growth mode. In the mid-cycle fatigue life regime, some shear and tensile crack growth may be expected.

Figure 8.5 shows the life fraction on the vertical axis, and the cycles to failure on the horizontal axis, this time for AISI 304 stainless steel loaded in tension. A large portion of the life was spent nucleating small cracks. The low-cycle fatigue regime indicated tensile crack growth from the initiated cracks. These planes would be aligned perpendicular to the loading axis of the specimen.

Socie also presented damage mechanism maps for Inconel-718 and 1045 steel. The different materials exhibited different mechanism maps due to their different

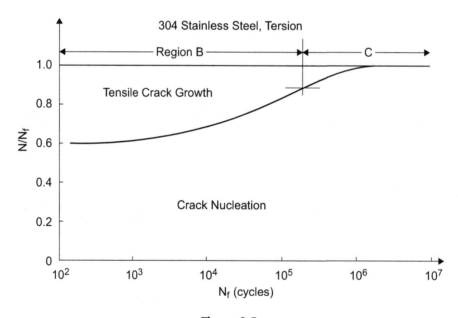

Figure 8.5
Damage mechanism map for 304 stainless steel
tested in tension (Socie, 1993).

microstructures and abilities to impede dislocations and small cracks. Damage parameters were proposed for each region (Socic, 1993) as follows:

Region A: low-cycle, shear-crack growth, plastic shear-strain–dominated and affected by crack opening normal stress

$$\gamma_a \left(1 + k \frac{\sigma_{n,max}}{\sigma_y} \right) = \frac{\tau_f'}{G}(2N)^b + \gamma_f'(2N)^c \tag{8.21}$$

where γ_a is the shear-strain amplitude on the cracking plane, k is a fitting parameter dependent on the material of interest, $\sigma_{n,max}/\sigma_y$ is the maximum normal stress opening the crack faces normalized by the material yield stress, and the right side of the equation represents a shear strain–life fit of experimental test data (Fatemi & Socie, 1988).

Region B: low- or mid-cycle, normal-stress–driven crack growth, described by Smith–Watson–Topper or similar damage parameter

$$\sigma_{max}\varepsilon_a = \frac{(\sigma_f')^2}{E}(2N)^{2b} + \sigma_f'\varepsilon_f'(2N)^{b+c} \quad \sigma_{max} > 0 \tag{8.22}$$

where ε_a is the strain amplitude on the tensile crack plane, σ_{max} is the maximum normal stress on this plane, and the right side of the equation represents a mean-stress–influenced representation of the strain–life equation.

Region C: high-cycle fatigue, dominated by crack nucleation on shear planes due to elastic-shear strains or elastic stress, influenced by tensile stress

$$\tau_a + k_\tau \sigma_{n,max} = \tau_f'(2N)^b \tag{8.23}$$

where τ_a is the shear-stress amplitude, k_τ is a fitting constant, $\sigma_{n,max}$ is the maximum normal stress opening the crack faces, and the right side represents the elastic-dominated shear-stress–fatigue-life curve.

Different materials can be modeled with one or more of these damage parameters. These damage parameters can be used in conjunction with a critical plane sweep method to accumulate calculated fatigue damage on candidate planes. The plane experiencing the most calculated damage is determined to be the cracking plane. In instances where the transition from one damage parameter to another is not known in advance (i.e., the material lacks a damage mechanism map), each damage parameter can be evaluated for the entire loading history and the most damaging parameter can be used to give a conservative estimate on calculated fatigue life.

The advantages of the critical plane approach for strain-based fatigue analysis are the same as in the stress-based approach. The approach is general enough to apply to both proportional and nonproportional multiaxial loading conditions. The physical nature of the development of small cracks can be captured by the use of appropriate damage parameters.

The expected crack orientation in the material can be calculated. However, an accurate constitutive model is essential in obtaining the best results, as is knowledge about the appropriate damage parameter to use for a given material at a particular load level.

Strain Gage Rosette Analysis

An electrical resistance foil-strain gage is a strain-sensing element made of multiple metal foil loops that is calibrated to measure strain in the axis of the strain gage. For uniaxial stress states, the gage factor of the strain gage can be used to determine the strain in the axis of the strain gage when the gage axis is aligned with the stress. The gage factor is determined by placing the strain gage in a uniaxial stress field on a reference material with a Poisson's ratio of 0.285.

In many practical situations, however, the stress state may be multiaxial with fixed directions of principal stress—proportional stressing or loading—or with changing directions of principal stresses—nonproportional stressing or loading. Additionally, even if the stress state is known to be proportional, the principal directions of stress may be unknown when the strain gage is installed. In these cases, it is essential to use a strain-gage rosette.

A strain-gage rosette is a pattern of multiple single-axis strain gages. The most common type of strain-gage rosette is the three-element 0°–45°–90° rosette. In this type of rosette, two linear strain gages are aligned perpendicular to each other, and another is located at a 45° angle in-between the other two. For multiaxial states of stress, the corresponding strain field components perpendicular to the gage axis may affect the reading of the strain gage. The degree to which the strain gage is affected by a strain perpendicular to its axis is called the *transverse sensitivity* of the strain gage. A detailed discussion of transverse sensitivity of strain gages is presented in Vishay Micro-Measurements (2007).

For the $0°$–$45°$–$90°$ rosette, with gages a, b, and c associated with each angular direction, respectively, the corrected gage readings are given by the following set of equations (Vishay Micro-Measurements, 2007):

$$\varepsilon_a = \frac{\varepsilon_a^*(1 - \nu_o K_{ta}) - K_{ta}\varepsilon_c^*(1 - K_{tc})}{1 - K_{ta}K_{tc}} \tag{8.24}$$

$$\varepsilon_b = \frac{\varepsilon_b^*(1 - \nu_o K_{tb})}{1 - K_{tb}} - \frac{K_{tb}\left[\varepsilon_a^*(1 - \nu_o K_{ta})(1 - K_{tc}) + \varepsilon_c^*(1 - \nu_o K_{tc})(1 - K_{ta})\right]}{(1 - K_{ta}K_{tc})(1 - K_{tb})} \tag{8.25}$$

$$\varepsilon_c = \frac{\varepsilon_c^*(1 - \nu_o K_{tc}) - K_{tc}\varepsilon_a^*(1 - K_{ta})}{1 - K_{ta}K_{tc}} \tag{8.26}$$

where the uncorrected measured strains are $\varepsilon_a^*, \varepsilon_b^*$, and ε_c^*, their respective transverse sensitivities are given by K_{ta}, K_{tb}, and K_{tc}, and ν_o is Poisson's ratio of the material on which the strain gage factors and sensitivities were determined, and is usually 0.285 for Vishay Micro-Measurements strain gages. Transverse sensitivity correction equations for other types of strain-gage rosettes are presented in Vishay Micro-Measurements (2007).

It should be noted that the transverse sensitivities are given in the strain gage data sheet for the gage as percentages, and must be converted into numbers before the application of these equations. The rolling process of the strain gage metal and manufacture of the strain gages typically results in equal transverse sensitivity factors for the a and c gages.

Although the correction for transverse sensitivity is usually small in comparison to the strain readings, it can be large depending on the stress field and how the strain gages are aligned within the stress field. For this reason, it is recommended that transverse sensitivity correction be made to gage readings as part of the data-reduction process.

After the strains have been corrected, the local in-plane strain state can be determined by the use of strain transformation equations. In the local axis of the strain-gage rosette where the x-axis is aligned along the grid axis of gage a, and the y-axis is aligned along the grid axis of gage c, the local coordinate strain components are given by:

$$\varepsilon_x = \varepsilon_a$$
$$\varepsilon_y = \varepsilon_c \tag{8.27}$$
$$\gamma_{xy} = 2\varepsilon_b - \varepsilon_a - \varepsilon_c.$$

The local in-plane strains can be used in a constitutive model to calculate stresses. Since the strain-gage rosette is on a traction-free surface of a material, the plasticity model that is used will assume that the stress state is in-plane stress. In general, the constitutive model may include the effects of both elastic and plastic material behavior as described in Chapter 7.

Strain Data Acquisitions

A strain-gage rosette was placed on a steel rail, and data acquisition channels 1, 2, and 3 were collected from the strain gages that are composed of the strain-gage rosette gages *a*, *b*, and *c*. A photo of the strain gage installation is shown in Figure 8.6. Data was collected from the strain gages as a train passed over the section of rail.

Figure 8.7 shows a time history of a portion of the data as the wheel sets passed over the strain gage location. In this application, the data acquisition rate was 100 points per second, and the horizontal axis is the time in seconds. The units

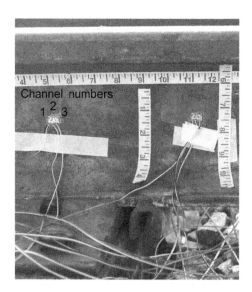

Figure 8.6
Strain-gage rosettes installed on a section of rail.
Source: *Photo of rail provided by J. McDougall of ESI.*

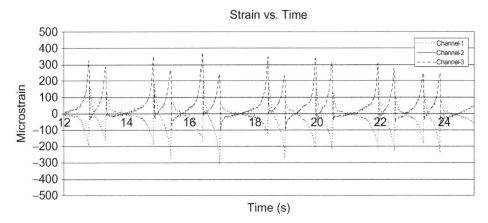

Figure 8.7
Gage time history.

on the vertical axis of the plot are microstrain. This data was collected with a 10-kHz low-pass filter, but otherwise was not filtered or edited.

The strain gage rosette data was corrected for transverse sensitivity by the use of Equations (8.24) through (8.26) and the information on the strain gage data sheet shown in Figure 8.8, transformed into the local strain gage coordinate system strains by the use of Equation (8.27), and then transformed into a coordinate system aligned with the vertical and horizontal directions of the rail by the use of standard strain transformation equations. These coordinate systems are indicated in Figure 8.9, which shows a strain-gage rosette.

The data must be examined to determine if a proportional or nonproportional analysis should be conducted. A phase plot of the strains is shown in Figure 8.10. There is a large cluster of data near zero in the plot. These points are typical of low-amplitude noise in the data. Of more interest is the data near an equivalent strain magnitude of 200 microstrain or greater.

Due to the alternating sign of the shear strain as the wheel sets pass over the strain gage location as shown in Figure 8.11, the plot basically takes the idealized shape as shown on the phase plot in Figure 8.12. This type of path is a nonproportional loading path, indicating that a critical plane approach must be used to analyze the fatigue life.

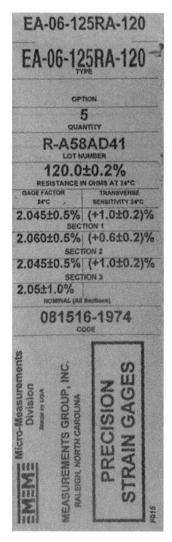

Figure 8.8
Strain-gage rosette data sheet.

The next step in the analysis is to calculate the stress-tensor history. The material is assumed to be in a state of plane stress, since the rosette is on a traction-free surface. Because of the nature of this application and the magnitude of the strains, the stress calculations can be done based on a linear elastic stress model; that is, 3-D Hooke's law. However, if the strains were larger, resulting

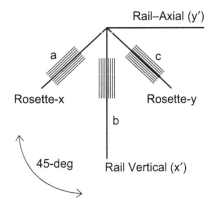

Figure 8.9
Strain gage coordinate system and coordinate system
aligned with the rail.

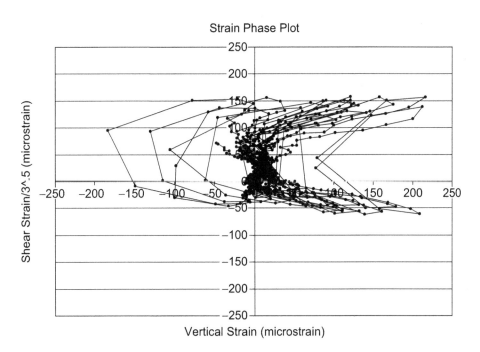

Figure 8.10
Phase plot of vertical and shear strain on equivalent strain axes.

Figure 8.11
Train wheel sets passing over gage location.

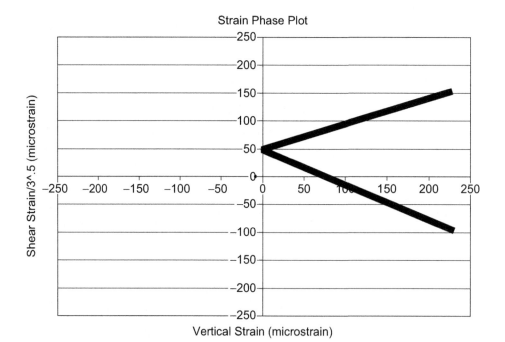

Figure 8.12
Idealized phase plot.

in stress above the yield stress of the material, the material models discussed in Chapter 7 would be used to calculate the stress tensor history.

The centerline of the rail where the strain-gage rosette is located is expected to carry mainly transverse shearing stress in addition to an axial component of stress from the wheel set approaching the point and finally passing over the point. This would be expected to be a shear-dominated, high-cycle fatigue application, and therefore Region C behavior is expected.

A damage parameter in the form of Equation (8.21) is most appropriate. The critical plane approach described in Chapter 5 would be used to calculate the fatigue damage and expected cracking plane. In this example, the strains at this location in the rail are such that an infinite life is calculated using material properties of the rail steel.

Fatigue Analysis with Plasticity

The preceding example measured strains from an instrumented section of rail. The analysis resulted in an infinite fatigue life as was expected for that location on the rail. To illustrate the use of the damage parameters for finite life fatigue analysis using the critical plane approach, the same basic time histories will be used but this time will be scaled to represent the stresses at a location on the surface of a component, which will be assumed to be in a state of plane stress. The state of stress to be examined will be as indicated in Figure 8.13. The time histories of the stress components are shown in Figure 8.14.

Material properties listed in Table 8.1 representing a carbon steel were used in a multiaxial cyclic plasticity material model to obtain the corresponding in-plane strains ε_x, ε_y, and γ_{xy}, and the out-of-plane normal strain, ε_z. The material constitutive model was a strain controlled Mroz model for isotropic materials.

Plasticity was obtained during this analysis, as shown in the hysteresis loop plots of the shear stress and engineering shear strain components that is illustrated in Figure 8.15. Strain–life fatigue properties shown in Table 8.2 were used in the fatigue analysis.

As mentioned previously, the selection of damage parameter to be used in an analysis depends on the behavior of the material and the manner in which it is sensitive to the formation of small cracks. The cracking behavior of the material

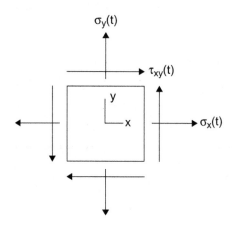

Figure 8.13
Plane stress state.

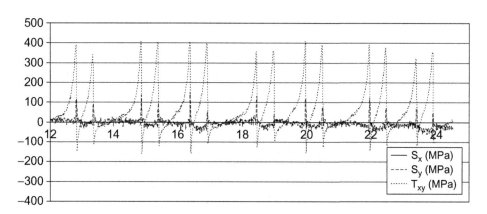

Figure 8.14
Stress-time history.

**Table 8.1: Material Properties Used
in Plasticity Analysis**

$E = 210,000$ MPa
$\nu = 0.3$
$\sigma_0 = 100$ MPa
$n' = 0.239$
$K' = 1289$ MPa

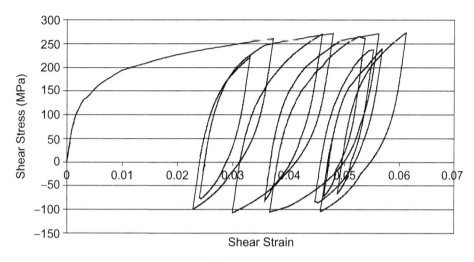

Figure 8.15
Hysteresis plot of shear stress and engineering shear strain.

Table 8.2: Material Properties Used
in Fatigue Analysis

$E = 210,000$ MPa
$\sigma'_f = 1555$ MPa
$\varepsilon'_f = 0.811$
$b = -0.076$
$c = -0.732$

can depend on the level of load that is applied and the amount of mean stress and hydrostatic stress within the material. Material constants for some damage parameters can be difficult to obtain.

In these cases, the fatigue analyst may choose to evaluate the fatigue damage for a loading history using more than one damage parameter to examine the calculated results for general trends. Prior experience of the analyst and the examination of fatigue test results can eventually lead to the determination of an appropriate damage parameter for a material.

In this example, the critical plane approach will be used for in-plane normal strains using the Region B Smith–Watson–Topper equation (Equation 8.22) and for in-plane and out-of-plane shear stresses using the region A shear strain–life equation (Equation 8.21) and shear fatigue properties derived from the axial strain–life properties. In this case, in-plane shearing stresses refer to the shearing stress in the plane of the rail surface, and out-of-plane shearing stresses refer to the shearing stresses that are at a 45° angle between the surface of the rail and the direction normal to the rail surface.

For the region B Smith–Watson–Topper equation, the strain tensor time history determined by the use of the strain controlled multiaxial material model was resolved to candidate-critical planes by the use strain transformation equations. In this approach, only the normal strain on the plane was considered to influence the fatigue life. The resulting normal strain time history was picked for peaks, rearranged to start with the highest peak, and then the rainflow cycle counted to determine closed hysteresis loops as well as their corresponding maximum normal stresses.

Reversals to failure were determined from Equation (8.16), and damage per hysteresis loop was assigned based on the cycles to failure. A low amplitude strain cycle cutoff threshold was determined based on 10^7 reversals to failure. For cycles with less than or equal to this strain range, no damage was assigned.

Figure 8.16 shows the resulting damage per candidate-critical plane. In this analysis, candidate-critical planes were analyzed at both 10° and 2° increments. Based on the critical planes every 2° apart, the maximum damage that was calculated was 6.33e-3 at 56°, resulting in a fatigue life of 158 repeated blocks of the stress history. If only the 10° planes were considered, the maximum damage would have been determined to be at the 50° plane, resulting in a fatigue life of 159 repeated blocks of the stress history.

The analysis was repeating using the region A shear strain–life equation, considering in-plane and out-of-plane shearing strains. The results are plotted in Figure 8.17. The results indicate that the in-plane shear strain is more damaging than the normal strain or out-of-plane shear strain. The calculated critical plane is at 0°, resulting in a life of 79 repeats of the stress.

To further examine this stress history, load scaling factors were applied to the stress history, making new stress histories that were 1.10, 1.25, and 1.5 times the original stress history. Figures 8.18 through 8.20 show the effect on damage

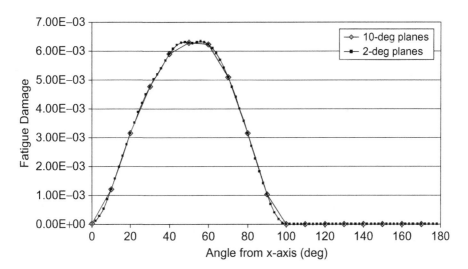

Figure 8.16
Damage per candidate-critical plane.

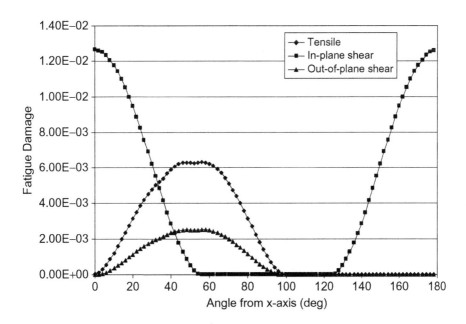

Figure 8.17
Damage per candidate-critical plane determined by normal strain, in-plane shear strain, and out-of-plane shear strain parameters.

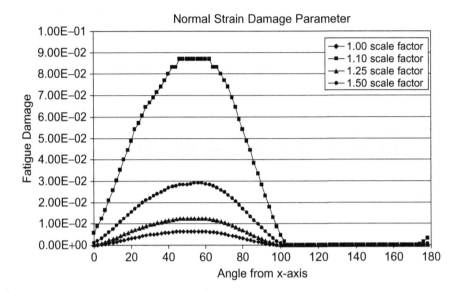

Figure 8.18
Damage per candidate-critical plane determined by normal strain
for various scale factors on load.

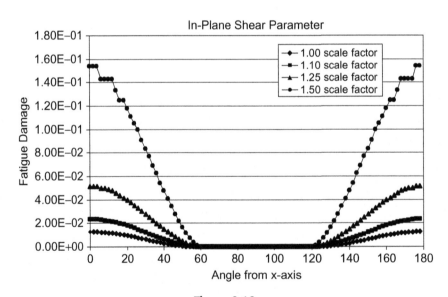

Figure 8.19
Damage per candidate-critical plane determined by in-plane shear strain
for various scale factors on load.

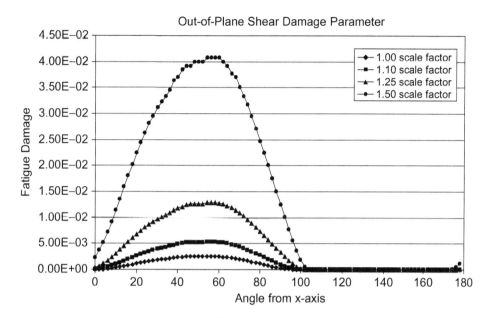

Figure 8.20
Damage per candidate-critical plane determined by out-of-plane shear strain
for various scale factors on load.

calculation for each candidate-critical plane of scaling the stress history for the
normal strain, in-plane shear strain, and out-of-plane shear strain damage parameters. The location of the candidate-critical plane remains the same, although
the damage increases disproportionately with the scale factor.

Summary

Strain tensor components can be used as damage parameters for multiaxial fatigue analysis. To capture physical features of the cracking behavior of the materials, the strain terms are often used in conjunction with stress terms to account
for mean stresses or hydrostatic stresses. Strain-based damage parameters can be
used for high-, mid-, and low-cycle fatigue life regimes with the proper selection
of damage parameter.

Fatigue life estimates for proportional multiaxial loading can be obtained
with equivalent strain equations based on a yield criterion. Special loading conditions, such as 90° out-of-phase loading can also be accounted for by using an

out-of-phase hardening parameter and a basic material constitutive equation. However, for general nonproportional loading, a critical plane approach matched with an appropriate damage parameter is necessary to determine fatigue life and the expected cracking plane.

For damage parameters that include stress terms, an accurate calculation of the stress tensor history from the strain tensor history requires sophisticated multiaxial material models and material constants. These material constants and damage parameter material constants can be challenging to obtain for an absolute fatigue life prediction, although comparative analysis can be conducted to examine general trends in fatigue life using multiple damage parameters.

References

Bannantine, J., Comer, J., & Handrock, J. (1990). *Fundamentals of Metal Fatigue Analysis.* Englewood Cliffs, NJ: Prentice-Hall.

Ellyin, F. (1974). A criterion for fatigue under multiaxial states of stress. *Mechanics Research Communications, 1,* 219–224.

Fatemi, A., & Socie, D. F. (1988). A critical plane to multiaxial fatigue damage including out-of-phase loading. *Fatigue and Fracture of Engineering Materials and Structures, 11*(3), 149–165.

Garud, Y. S. (1981). A new approach to the evaluation of fatigue under multiaxial loadings. *ASME Journal of Engineering Materials and Technology, 103,* 118–125.

Itoh, T., Nakata, T., Sakane, M., & Ohnami, M. (1999). Nonproportional low cycle fatigue of 6061 aluminum alloy under 14 strain paths. In E. Macha, W. Bedkowski, & T. Lagoda (Eds.), *Multiaxial fatigue and fracture, fifth international conference on biaxial/multiaxial fatigue and fracture: Vol. 25* (pp. 41–54). European Structural Integrity Society, Elsevier.

Kandil, F. A., Brown, M. W., & Miller, K. J. (1982). Biaxial low cycle fatigue fracture of 316 stainless steel at elevated temperatures. *Book 280, The Metals Society* (pp. 203–210). London.

Krempl, E. (1974). *The influence of state of stress on low cycle fatigue of structural materials: A literature survey and interpretative report.* ASTM STP 549. West Conshohocken, PA: ASTM.

Mendelson, A. (1983). *Plasticity: Theory and application.* Malabar, FL: Robert E Krieger Publishing Company, Inc.

Mowbray, D. F. (1980). A hydrostatic stress-sensitive relationship for fatigue under biaxial stress conditions. *ASTM Journal of Testing and Evaluation, 8*(1), 3–8.

Shamsaei, N., & Fatemi, A. (2010). Effect of microstructure and hardness on non-proportional cyclic hardening coefficient and predictions. *Materials Science and Engineering, A, 527,* 3015–3024.

Socie, D. (1993). Critical plane approaches for multiaxial fatigue damage assessment. In D. L. McDowell, & E. Ellis (Eds.), *Advances in multiaxial fatigue, ASTM STP 1191* (pp. 7–36). Philadelphia: America Society for Testing and Materials.

Socie, D. F., & Marquis, G. B. (2000). *Multiaxial fatigue.* Warrendale, PA: SAE International.

Vishay Micro-Measurements. (2007). *Errors due to transverse sensitivity in strain gages, Tech Note TN-509*, revised August 15.

Wang, Y., & Yao, W. (2004). Evaluation and comparison of several multiaxial fatigue criteria. *International Journal of Fatigue, 26,* 17–25.

You, B., & Lee, S. (1996). A critical review on multiaxial fatigue assessments of metals. *International Journal of Fatigue, 18*(4), 235–244.

CHAPTER 9

Vibration Fatigue Testing and Analysis

Yung-Li Lee
Chrysler Group LLC

Hong-Tae Kang
The University of Michigan–Dearborn

Chapter Outline

Introduction

Traditionally vibration test specifications, such as IEC 60068-2-6 (2007) and ISO-16750-3 (2003), were generated by using an envelope of generic customer usage vibration profiles. In general, these generic test standards are extremely severe, which may lead to different failure modes than the ones found in the field, and occasionally can be poorly adapted to the present needs. These test standards also specify the testing environmental values (accelerations, temperature, etc.) for a product according to the purpose of its usage. Most often, the two vibration test methods are allowed to apply to the same description of the product of interest.

However, there are several fundamental differences between the two vibration tests. Sinusoidal vibration has a bathtub-shaped histogram for its probability density function (PDF) and random vibration has a bell-shaped histogram. Also a single sine tone frequency is excited by sinusoidal vibration, whereas a broad spectrum of frequency components is presented simultaneously in random vibration. It will be costly to conduct both test methods to validate a product for durability and life requirements. Thus, it becomes a challenging task for product manufacturers to decide which test standard to follow in order to save cost and test time. There is a need to develop an analytical solution to evaluate the fatigue damage severity of the vibration tests.

Another challenging task to assess the product durability is to develop an accelerated vibration test specification. MIL-STD-810F (2000) and GAM EG-13 (1986) have addressed the concept of "test tailoring," which tailors a product's environmental design and test limits to the conditions that it will experience throughout its service life and develops an accelerated test method that replicates the effects of environments on the product rather than imitating the environments themselves.

On the other hand, the test-tailoring method is the method to develop a vibration specification based on the customer usage conditions and the use of the material. It is a two-stage process that consists of mission profiling and test synthesis, as described by Halfpenny and Kihm (2006).

In the mission profiling stage, the measured customer usage (CU) events and data are required to be identified first. For each CU event, two damage criteria such as the shock response spectrum (SRS) and the fatigue damage spectrum (FDS) are used to represent the customer usage profile in terms of damage severity. The shock response is referred to the largest displacement or acceleration response of a system subjected to a time-based excitation. For a linear SDOF system with a given natural frequency and damping ratio, the shock response (SR) and fatigue damage (FD) can be easily calculated.

Both SRS and FDS in each CU event can be generated by varying the natural frequency one at a time in a frequency range of interest. The SRS is sometimes termed the extreme response spectrum (ERS) if the system is subjected to a frequency-based excitation. To represent the worst local response likely to be seen by the linear SDOF system during the entire life of the product in all the CU events, the lifetime SRS can be obtained by using an envelope of all the measured SRSs. Moreover, the lifetime FDS is determined by the sum of all the FDSs, based on the Palmgren–Miner linear damage rule (Palmgren, 1924; Miner, 1945).

In the test synthesis stage, a close-form solution to calculate ERS and FDS due to a vibration test profile such as the power spectral density (PSD) has been successfully derived by Lalanne (2002) based on test duration. Therefore, the test PSD is obtained from the lifetime FDS. The accelerated test time can be determined based on the principle that the ERS of the test PSD should be compared with the lifetime SRS by adjusting the test duration, and be less than the lifetime SRS to minimize the risk of shock failure during testing.

The objective of this chapter is to present an analytical solution that can be used to assess fatigue damage severity for various vibration test specifications. The close form calculation (Lalanne, 2002; Halfpenny & Kihm, 2006) for the FDS calculated directly from the base acceleration PSD is a noble approach to the test-tailoring method. However, other than the use of the linear SDOF system, the formula was derived based on the following assumptions:

- Using Miles' equation for calculation of the root mean square acceleration in Gs

- A linear stiffness constant to relate the relative displacement to the local stress

- The narrow-band random stress process

- The Rayleigh PDF for the stress amplitude

Therefore, there is a need to evaluate this FDS calculation process by developing an analytical solution to include the state-of-art frequency-based fatigue damage theories such as Wirsching–Light's method (1980), Ortiz–Chen's method (1987), and Dirlik's method (1985). These frequency-based fatigue theories are addressed in this chapter. This chapter also presents the fundamentals of sinusoidal and random vibration test methods and their fatigue damage spectrum calculations.

Swept Sinusodial or Single-Frequency Sweep Test

In the sweep test, a controller inputs a pure sine tone to a shaker where the tone may sweep in frequency and vary in amplitude. The sine tone is obtained from performing a peak hold Fast Fourier Transform (FFT) of the raw periodic data and the excitation frequency is continuously varying between the minimum and maximum frequencies.

The changing rate of the excitation frequency and the method of varying this rate as a function of test frequency have a significant effect on the response of test parts. The sweep method controls the amount of time and the number of cycles accumulated in any frequency range and the sweep rate affects the amplitude of resonant response.

There are two standard sweep methods, namely the logarithmic sweep and the linear sweep. In the logarithmic sweep, the excitation frequency f varies at a rate $\dot{f}$ proportional to itself. Hence,

$$\dot{f} = \frac{df}{dt} = A \cdot f \qquad (9.1)$$

where

   A = a proportional constant that can be determined by a given sweep time
   T in seconds from the minimum excitation frequency f_{min} to the
   maximum frequency f_{max}

For example, A is obtained by

$$A = \frac{\ln(f_{max}) - \ln(f_{min})}{T}. \qquad (9.2)$$

Equation (9.1) can be rewritten in the following incremental form,

$$\Delta t = \frac{\Delta f}{A \cdot f}. \qquad (9.3)$$

So the number of cycles accumulated for each frequency interval is

$$n_i = f \cdot \Delta t = \frac{\Delta f}{A} = \frac{T \cdot \Delta f}{\ln(f_{max}) - \ln(f_{min})}. \qquad (9.4)$$

In the linear sweep, the excitation frequency rate equals a constant A′. As a result,

$$\dot{f} = \frac{df}{dt} = A' \tag{9.5}$$

where

$$A' = \frac{f_{max} - f_{min}}{T}. \tag{9.6}$$

With the incremental form, Equation (9.5) becomes

$$\Delta t = \frac{\Delta f}{A'}. \tag{9.7}$$

Also the number of cycles experienced for each frequency interval is

$$n_i = f \cdot \Delta t = \frac{\Delta f \cdot f}{A'} = \frac{T \cdot \Delta f \cdot f}{f_{max} - f_{min}}. \tag{9.8}$$

During sinusoidal sweep testing it is necessary to control the equal number of cycles or the time at each resonance. According to Equation (9.4), the logarithmic frequency sweep gives the same number of cycles for a given frequency interval and is independent of the frequency level, whereas the linear sweep produces a large number of cycles of high frequency as seen in Equation (9.8). Thus, a logarithmic frequency sweep is the preferable one for use in sinusoidal sweep testing because of its easy way to dictate an equal number of cycles at each resonance.

Response to a Linear Single-Degree-of-Freedom System Subjected to Sinusoidal-Based Excitation

Figure 9.1(a) shows a mathematical model for a linear single-degree-of-freedom (SDOF) system where m_o, c_o, and k_o represent the mass, viscous damping coefficient, and stiffness, individually. The displacement of the mass equals to $x_o(t)$ and the base input displacement equals to $y_o(t)$.

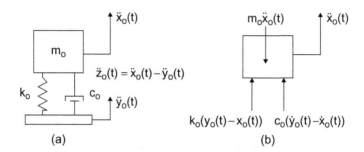

Figure 9.1
(a) Mathematical model of an SDOF system
and (b) a free body diagram.

From the free body diagram of Figure 9.1(b), summing the forces in the vertical direction leads to the following differential equation of motion:

$$m_o \ddot{x}_o(t) = c_o(\dot{y}_o(t) - \dot{x}_o(t)) + k_o(y_o(t) - x_o(t)) \tag{9.9}$$

where

$\ddot{x}_o(t), \dot{x}_o(t) =$ the acceleration and velocity of $x_o(t)$
$\ddot{y}_o(t), \dot{y}_o(t) =$ the acceleration and velocity of $y_o(t)$

By introducing a relative displacement, $z_o(t) = x_o(t) - y_o(t)$, Equation (9.9) can be written as

$$m_o \ddot{z}_o(t) + c_o \dot{z}_o(t) + k_o z_o(t) = -m_o \ddot{y}_o(t). \tag{9.10}$$

Dividing Equation (9.10) by m yields

$$\ddot{z}_o(t) + (c_o/m_o)\dot{z}_o(t) + (k_o/m_o)z_o(t) = -\ddot{y}_o(t). \tag{9.11}$$

By definition,

$$(c_o/m_o) = 2\xi \cdot \omega_n \tag{9.12}$$

$$(k_o/m_o) = \omega_n^2 \tag{9.13}$$

where

$\omega_n =$ the natural frequency in (radians/second)
$\xi =$ the damping ratio

An assumed value of 0.05 is commonly used (Lalanne, 2002) for the damping ratio.

Substituting Equations (9.12) and (9.13) into Equation (9.10) has

$$\ddot{z}_o(t) + 2\xi \cdot \omega_n \dot{z}_o(t) + \omega_n^2 z_o(t) = -\ddot{y}_o(t). \tag{9.14}$$

By replacing ω_n by $2\pi \cdot f_n$ and then solving Equation (9.14), the steady-state relative response $|\ddot{z}_o(f_i, \xi, f_n)|$ to an SDOF system with a resonant frequency f_n in Hz subjected to a base input sine vibration with an excitation frequency of f_i $|\ddot{y}_o(f_i)|$ can be obtained by

$$|\ddot{z}_o(f_i, \xi, f_n)| = |H_1(r_i)||\ddot{y}_o(f_i)|; \quad r_i = \frac{f_i}{f_n} \tag{9.15}$$

where

$\quad |H_1(r_i)| = $ the gain function or the modulus of the transfer function

This is defined as

$$|H_1(r_i)| = \frac{r_i^2}{\sqrt{(1 - r_i^2)^2 + (2\xi \cdot r_i)^2}}. \tag{9.16}$$

For a steady state sinusoidal forcing, the maximum relative response will occur at the excitation frequency approximately equal to the natural frequency, $r_i = f_i/f_n = 1$. In this case, Equation (9.16) reduces to the following gain function:

$$Q = \frac{|\ddot{z}_o(f_i, \xi, f_n)|}{|\ddot{y}_o(f_i)|} = \frac{1}{2\xi} \tag{9.17}$$

where

$\quad Q = $ the dynamic amplification factor

If the damping ratio is assumed to be 0.05, then $Q = 10$. Also the maximum relative displacement can be determined by dividing the maximum relative acceleration by $(2\pi \cdot f_i)^2$ as follows:

$$|z_o(f_i, \xi, f_n)| = \frac{|\ddot{z}_o(f_i, \xi, f_n)|}{(2\pi \cdot f_i)^2}. \tag{9.18}$$

Therefore, Equation (9.18) can be expressed as

$$|z_o(f_i, \xi, f_n)| = |H_2(r_i)||\ddot{y}_o(f_i)| \tag{9.19}$$

where $|H_2(r_i)|$ is obtained as follows:

$$|H_2(r_i)| = \frac{1}{(2\pi \cdot f_n)^2 \sqrt{(1-r_i^2)^2 + (2\xi \cdot r_i)^2}}. \tag{9.20}$$

When a system is excited by a constant excitation at a resonant frequency, the amplitude of the response will gradually build up to a level proportional to the amplification of the resonance, also termed the maximum steady state response. The number of constant excitation cycles to obtain the maximum steady state response is proportional to the resonance amplification. It is preferable that the sweep rate is slow enough to allow a sufficient number of cycles to occur in the resonance bandwidth.

To answer the question of how slow of a sweep rate is considered to be too slow, a sweep parameter, η, was developed by Cronin (1968) to relate the response level to the properties of the SDOF system. This parameter is defined as follows:

$$\eta = \frac{Q}{n_{fn}} \tag{9.21}$$

where

n_{fn} = the number of cycles of excitation between the half-power bandwidth B

As illustrated in Figure 9.2, the half-power bandwidth B is defined as the frequency bandwidth at $\sqrt{1/2}$ of the peak response amplitude, and can be related to the resonant frequency f_n and the dynamic amplification factor Q as

$$B = \frac{f_n}{Q}. \tag{9.22}$$

In a logarithmic sweep test, the number of cycles for a given frequency interval B is then obtained from Equation (9.4) as follows:

$$n_{fn} = \frac{T \cdot B}{\ln(f_{max}) - \ln(f_{min})}. \tag{9.23}$$

Substituting Equations (9.22) and (9.23) into Equation (9.21) results in

$$\eta = \frac{Q^2 [\ln(f_{max}) - \ln(f_{min})]}{T \cdot f_n}. \tag{9.24}$$

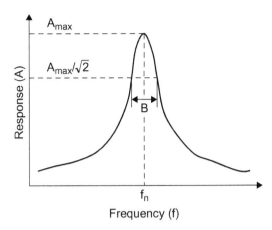

Figure 9.2

Illustration of the half-power bandwidth in a response spectrum
with a resonant frequency f_n.

Based on Cronin's study (1968), it was estimated that the fraction of the maximum steady state response (G) of a system subjected to a sinusoidal excitation passing through the resonant frequency can be approximated by

$$G = 1 - e^{(-2.86\eta^{-0.445})}.\tag{9.25}$$

This equation states that in theory, a minimum sweep parameter exists in order to achieve a response level at a fraction of the maximum steady state response (G) for a system subjected to a sinusoidal excitation at the resonant frequency. With Equation (9.25), the minimum logarithmic sweep time T from f_{min} to f_{max} can be determined by substituting Equation (9.24) into (9.25), required to obtain a fraction of maximum steady state response (G) for the half-power bandwidth of an SDOF system under a resonant excitation.

As a result, the minimum sweep time T in seconds is written as follows:

$$T = \frac{Q^2[\ln(f_{max}) - \ln(f_{min})]}{f_n} \left(\frac{\ln(1-G)}{-2.86}\right)^{2.247}.\tag{9.26}$$

Fatigue Damage Calculation

A linear SDOF system with a resonant frequency f_n is subjected to a base sinusoidal forcing input with an excitation frequency f_i. The maximum relative displacement, in terms of the sinusoidal acceleration input, can be obtained from Equation (9.19) as follows:

$$|z_o(f_i, \xi, f_n)| = \frac{|\ddot{y}_o(f_i)|}{(2\pi \cdot f_n)^2 \sqrt{(1 - r_i^2)^2 + (2\xi \cdot r_i)^2}}. \tag{9.27}$$

It is assumed that the relative displacement can be related to the stress amplitude $S_a(f_i, \xi, f_n)$ by a constant K. Thus, it is written

$$S_a(f_i, \xi, f_n) = K \cdot |z_o(f_i, \xi, f_n)|. \tag{9.28}$$

If an S-N curve exists and has the following Basquin expression (1910),

$$N_{f,i} \cdot S_a^m(f_i, \xi, f_n) = C \tag{9.29}$$

then the fatigue life $N_{f,i}$ at the stress amplitude level $S_a(f_i, \xi, f_n)$ would be

$$N_{f,i} = \frac{C}{S_a^m(f_i, \xi, f_n)} = \frac{C}{K^m \cdot |z_o(f_i, \xi, f_n)|^m} \tag{9.30}$$

where

    m = the slope factor of the S-N curve
    C = the material constant

For the logarithmic frequency sweep from the minimum excitation frequency f_{min} to the maximum frequency f_{max}, the number of cycles n_i is given as

$$n_i = \frac{T \cdot \Delta f}{\ln(f_{max}) - \ln(f_{min})}. \tag{9.31}$$

According to the Palmgren–Miner linear damage rule (Palmgren, 1924; Miner, 1945), the fatigue damage $d_i(f_n)$ to a system with a resonant frequency f_n, subjected to a sinusoidal input with an excitation frequency f_i, can be calculated as follows:

$$d_i(f_n) = \frac{n_i}{N_{fi}}. \tag{9.32}$$

If the sinusoidal forcing input sweeps from a minimum frequency to a maximum frequency, the linear damage rule would yield the following fatigue damage spectrum, $FDS(f_n)$, as

$$FDS(f_n) = \sum_{f_{min}}^{f_{max}} d_i(f_n) = \sum_{f_{min}}^{f_{max}} \frac{n_i}{N_{f,i}}. \tag{9.33}$$

By substituting Equations (9.30) and (9.31) into Equation (9.33), the FDS becomes

$$FDS(f_n) = \frac{T \cdot K^m}{C \cdot (4\pi^2 f_n^2)^m \left[\ln(f_{max}) - \ln(f_{min})\right]} \sum_{f_{min}}^{f_{max}} \frac{|\ddot{y}(f_i)|^m \Delta f}{\left[(1 - r_i^2)^2 + (2\xi \cdot r_i)^2\right]^{\frac{m}{2}}}. \tag{9.34}$$

Example 9.1

IEC 68-2-6, Classification III specifies the following swept sinusoidal test for a mechanical system. It follows a logarithmic sweep with a 20-minute sweep from 5 Hz to 200 Hz for 6 hours (total of 18 sweeps). Its vibration profile is tabulated in Table 9.1 and shown in Figure 9.3.

Determine the fatigue damage spectrum based on the IEC specification.

Solution
The first step is to check if the logarithmic sweep time of 20 minutes exceeds the minimum sweep time required to reach 99% of the maximum steady state response, say $G = 0.99$, at a resonant frequency. For the sweep frequency varying from 5 Hz to 200 Hz, the minimum sweep time

Table 9.1: Tabulated Vibration Profile Based on IEC 68-2-6, Classification III Specification

Frequency (Hz)	Peak Acceleration (G)
5	0.5
18.6	7
50	7
50	4.5
100	4.5
100	3
200	3

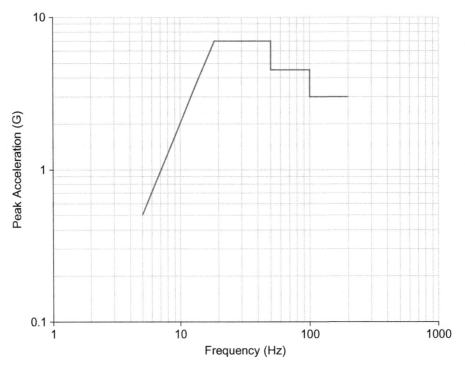

Figure 9.3
Vibration profile based on IEC 68-2-6, Classification III Specification.

can be calculated as 215 seconds or 3.6 minutes from Equation (9.26) based on the assumption of Q = 10 and f_n = 5 Hz. So a 20-minute sweep time is reasonable for this test.

The second step is to determine Δf for the integration of Equation (9.34). The half-power bandwidth B is naturally the choice that is related to the resonant frequency and the dynamic amplification factor as expressed in Equation (9.20). The minimum Δf = 0.5 Hz is obtained in the frequency sweep range by using Q = 10 and f_n = 5 Hz.

The last step is to perform the numerical integration to calculate the relative fatigue damage at each resonant frequency by arbitrarily assuming K = 10^6, m = 4, and C = 1. Please note that for relative comparison, the value of m for steel is used and is approximated as m = 4. The final FDS is illustrated in Figure 9.4.

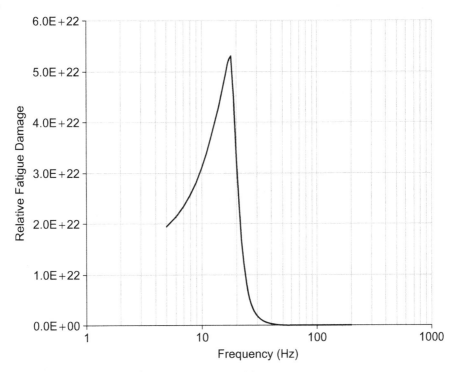

Figure 9.4
Relative fatigue damage spectrum plot based on IEC 68-2-6,
Classification III Specification.

Random Vibration Test

Random vibration testing is used to simulate a stationary random field excitation to a system or equipment. Stationary random data from a field is random in nature and has a consistent energy or root mean square (RMS) value of the data over time. But nonstationary random data is random in terms of the energy or RMS level. It is assumed that the random data used for creating a random test specification are stationary.

Characteristics of Random Vibration

The following subsections describe sample time history, ensembles, correlations, and Fourier transforms, which are characteristics of random vibration.

Sample time history

A system produces a certain response under excitation. If the excitation or the response motion $X(t)$ is unpredictable, the system is in random vibration because the exact value of $X(t)$ cannot be precisely predicted in advance. It can only be described probabilistically. The probability density function of a time history $X(t)$ can be obtained by calculating its statistical properties by the following means.

First, for an example of a time history for a random process $X(t)$ during a time interval T, if $X(t)$ exists between the values of x and $x + dx$ for a total time of $(dt_1 + \cdots + dt_k)$, then the probability that $x \leq X(t) \leq x + dx$ is therefore given by

$$P[x \leq X(t) \leq x + dx] = \frac{dt_1 + \cdots + dt_k}{T}. \tag{9.35}$$

If the duration T is long enough, the probability density function $f_X(x)$ is given by

$$f_X(x) = P[x \leq X(t) \leq x + dx] = \frac{\sum_{i=1}^{k} dt_i}{T}. \tag{9.36}$$

Alternatively, the probability density function can be determined by the fraction of the total number of samples in the band between x and $x + dx$. This can be done by digitizing the time history at a certain sampling rate in the time interval T. For example, if the total number of the sample points between x and $x + dx$ is $\sum_{i=1}^{k} no_i$ and the total sample points in T is NO, then $f_X(x)$ is given by

$$f_X(x) = P[x \leq X(t) \leq x + dx] = \frac{\sum_{i=1}^{k} no_i}{NO}. \tag{9.37}$$

Equations (9.36) and (9.37) are correct if the time duration T goes to infinity, which implies the sample time history continues forever. But measurement of the time intervals $\sum_{i=1}^{k} dt_i$ or the sample points $\sum_{i=1}^{k} no_i$ for the probability density function $f_X(x)$ is very cumbersome.

The statistical properties of $X(t)$ in describing a probability density function $f_X(x)$ are addressed here. The mean value μ_X or expected value $E(X(t))$ describes the central tendency of the random process, defined as

$$\mu_X = E[X(t)] = \frac{1}{T} \int_0^T X(t)dt = \int_0^T X(t)f_X(x)dx. \tag{9.38}$$

The mean-square value $E[X^2(t)]$ is the average value of $X^2(t)$

$$E[X^2(t)] = \frac{1}{T} \int_0^T X(t)^2 dt = \int_0^T X(t)^2 f_X(x)dx. \tag{9.39}$$

The variance σ_X^2 of the process is the dispersion of the data measured from the mean value, given as

$$\sigma_X^2 = \frac{1}{T} \int_0^T \left[X(t) - \mu_X\right]^2 dt = \int_{-\infty}^{+\infty} \left[X(t) - \mu_X\right]^2 f_X(x)dx = E\left[(X(t) - E(X(t))^2\right] \tag{9.40}$$

where

$\sigma_X =$ the standard deviation of $X(t)$

Equation (9.40) can be further reduced to

$$\sigma_X^2 = E[X^2(t)] - (E[X(t)])^2. \tag{9.41}$$

Quite often the mean value of a random process is zero, and the variance equals the mean square value. The root mean square RMS_X of the random process is defined as

$$RMS_X = \sqrt{E[X^2(t)]} = \sigma_X. \tag{9.42}$$

A random process $X(t)$ is called the Gaussian random process if its probability density function $f_X(x)$ follows the normal distribution. Thus, the probability density function is given by

$$f_X(x) = \frac{1}{\sqrt{2\pi}\sigma_X} \exp\left[-\frac{1}{2}\left(\frac{x - \mu_X}{\sigma_X}\right)^2\right] \quad -\infty < x < +\infty. \tag{9.43}$$

When a normally distributed random variable is normalized in terms of a new variable $z = \frac{x - \mu_X}{\sigma_X}$, the probability density function is known as the standard normal distribution and can be described in the following form:

$$f_Z(z) = \frac{1}{\sqrt{2\pi}} \exp\left[-\frac{1}{2}(z)^2\right] \quad -\infty < z < +\infty. \tag{9.44}$$

Figure 9.5 shows the standard normal distribution for a Gaussian random process with a mean value of μ_X. Since the secondary vibration environment of concern is sinusoidal vibration, the probability density function $f_Z(z)$ of a sinusoidal wave is shown in Figure 9.5 and is defined by

$$f_Z(z) = \frac{1}{\pi\sqrt{2 - z^2}} \quad -\infty < z < +\infty. \tag{9.45}$$

This figure shows that the probability density functions for sinusoidal and random vibrations differ significantly. Both probability density functions of Equations (9.44) and (9.45) are symmetrical about zero. The maximum z value is known as a crest factor. A crest factor for a sine wave is approximate to $\sqrt{2} = 1.414$, while the crest factor for a random signal is usually chosen as 3.0 (a 3 sigma design).

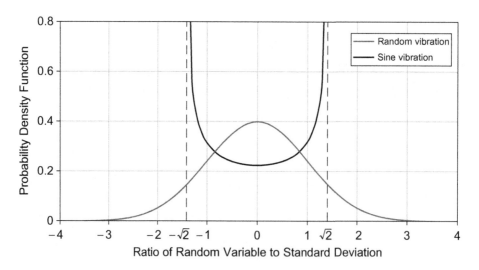

Figure 9.5
Normalized probability density functions for random vibration and sinusoidal vibration.

Ensemble

A collection of an infinite number of sample time histories such as $X_1(t)$, $X_2(t), \ldots, X_k(t)$, and so on makes up the ensemble $X(t)$ as shown in Figure 9.6. The statistical properties of an ensemble can be easily computed at any time instant. A random process is said to be stationary if the probability distributions for the ensemble remain the same (stationary) for each time instant. This implies that the ensemble mean, standard deviation, variance, and mean square are all time invariant.

A stationary process is called "ergodic" if the statistical properties along any single sample time history are the same as the properties taken across the ensemble. That means each sample time history completely represents the ensemble. Note that if a random process is ergodic, it must be stationary. However, the converse is not true; a stationary process is not necessarily ergodic. It is assumed here that all random processes are stationary and ergodic.

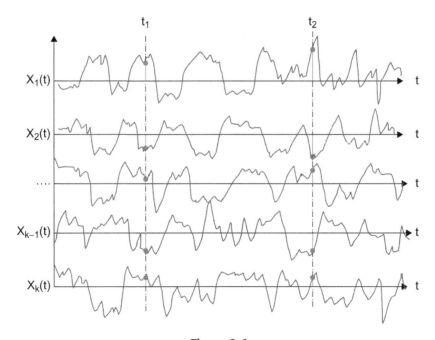

Figure 9.6
Random process ensemble of random sample time histories.

Correlation

Correlation is a measure of the dependence between two random processes. It is known that larger correlation exists for two similar random processes and smaller correlation for two dissimilar processes. If two random processes are stationary but differ by a time lag τ, the correlation between $X(t)$ and $X(t+\tau)$ is termed the autocorrelation function $R(\tau)$ of a random process, expressed as

$$R_X(\tau) = \lim_{T \to \infty} \frac{1}{T} \int_0^T X(t)X(t+\tau)dt = E[X(t)X(t+\tau)]. \qquad (9.46)$$

It is evident from Equation (9.46) that $R(\tau)$ is an even function $(R_X(\tau) = R_X(-\tau))$ and the autocorrelation becomes the mean square value when $\tau = 0 (\text{i.e., } R_X(0) = E[X^2(t)].)$. Figure 9.7 schematically illustrates the autocorrelation function in the positive time lag axis.

Fourier transforms

Many times a transformation is performed to provide a better or clear understanding of phenomena. The time representation of a sine wave may be difficult

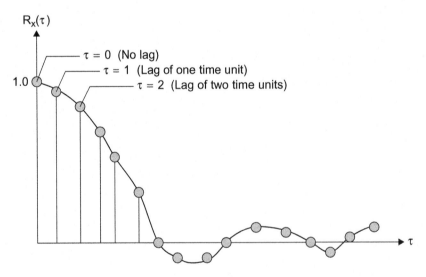

Figure 9.7
Autocorrelation function $R(\tau)$ of a stationary random process.

to interpret. By using a Fourier series representation, the original periodic signals in time can be easily transformed and much better understood.

Transformation is also performed to represent the same data with significantly less information. In general, random vibrations are not periodic, and the frequency analysis requires the extension of a Fourier series to the Fourier integrals for nonperiodic functions. A Fourier transform as a result of Fourier integrals are used extensively for solving random vibration problems.

The Fourier series can be used to represent any periodic time history by the summation of a series of sinusoidal waves of various amplitude, frequency, and phase. If $X(t)$ is a periodic function of time with a period T, $X(t)$ can be expressed by an infinite trigonometric series of the following form:

$$X(t) = A_0 + \sum_{n=1}^{\infty} \left[A_n \cdot \cos\left(\frac{2\pi}{T} nt\right) + B_n \cdot \sin\left(\frac{2\pi}{T} nt\right) \right] \qquad (9.47)$$

where

$$A_0 = \frac{1}{T} \int_{-T/2}^{T/2} X(t) dt = \frac{1}{T} \int_0^T X(t) dt$$

$$A_n = \frac{2}{T} \int_{-T/2}^{T/2} X(t) \cos\left(\frac{2\pi}{T} nt\right) dt = \frac{2}{T} \int_0^T X(t) \cos\left(\frac{2\pi}{T} nt\right) dt$$

$$B_n = \frac{2}{T} \int_{-T/2}^{T/2} X(t) \sin\left(\frac{2\pi}{T} nt\right) dt = \frac{2}{T} \int_0^T X(t) \sin\left(\frac{2\pi}{T} nt\right) dt.$$

The Fourier series can be expressed in exponential form by introducing $X_n = A_n - i \cdot B_n$ and $e^{\pm i\theta} = \cos(\theta) \mp i \cdot \sin(\theta)$,

$$X(t) = \sum_{n=-\infty}^{\infty} X_n \cdot e^{i\left(\frac{2\pi}{T} nt\right)} \qquad (9.48)$$

where the complex coefficients X_n is the nth coefficient and relates to a sinusoidal wave of frequency n/T Hz, given by

$$X_n = \frac{1}{T} \int_0^T X(t) \cdot e^{-i\left(\frac{2\pi}{T} nt\right)} dt = \frac{1}{T} \int_{-T/2}^{T/2} X(t) \cdot e^{-i\left(\frac{2\pi}{T} nt\right)} dt. \qquad (9.49)$$

When the perioric time history $X(t)$ is digitized by N equally spaced time intervals $t_0, t_1, t_2, \ldots, t_{N-1}$, where $t_j = j \cdot \Delta t$ and $T = N \cdot \Delta t$, the complex coefficients X_n of the discrete Fourier transform of the time series $X(t_j)$ is obtained as follows:

$$X_n = \frac{1}{N} \sum_{n=0}^{N-1} X(t_j) \cdot e^{-i\left(2\pi \frac{nj}{N}\right)} \tag{9.50}$$

and its inverse discrete Fourier transform is

$$X(t_j) = \sum_{n=0}^{N-1} X_n \cdot e^{i\left(2\pi \frac{nj}{N}\right)}. \tag{9.51}$$

The Fourier integral can be viewed as a limiting case of the Fourier series as the period T approaches infinity. This can be illustrated as follows by rewriting Equation (9.48) with infinite T:

$$X(t) = \lim_{T \to \infty} \sum_{n=-\infty}^{\infty} \left(\frac{1}{T} \int_{-T/2}^{T/2} X(t) e^{-i\left(\frac{2\pi}{T} nt\right)} dt \right) e^{i\left(\frac{2\pi}{T} nt\right)}. \tag{9.52}$$

If the frequency of the k-th harmonic ω_k in radians per second is

$$\omega_k = \frac{2\pi}{T} k \tag{9.53}$$

and the spacing between adjacent periodic functions $\Delta\omega$ is

$$\Delta\omega = 2\pi. \tag{9.54}$$

Equation (9.52) becomes

$$X(t) = \lim_{T \to \infty} \sum_{n=-\infty}^{\infty} \left(\frac{\Delta\omega}{2\pi} \int_{-T/2}^{T/2} X(t) e^{-i(n\Delta\omega \cdot t)} dt \right) e^{i(n\Delta\omega \cdot t)}. \tag{9.55}$$

As T goes to infinity, the frequency spacing, $\Delta\omega$, becomes infinitesimally small, denoted by $d\omega$, and the sum becomes an integral. As a result, Equation (9.55) can be expressed by the well-known Fourier transform pair $X(t)$ and $X(\omega)$:

$$X(\omega) = \frac{1}{2\pi} \int_{-\infty}^{\infty} X(t) e^{-i(\omega t)} dt \tag{9.56}$$

$$X(t) = \int\limits_{-\infty}^{\infty} X(\omega)e^{i(\omega t)}d\omega. \tag{9.57}$$

The function $X(\omega)$ is *the forward Fourier transform* of $X(t)$ and $X(t)$ is *the inverse Fourier transform* of $X(\omega)$. Similarily, if the frequency f is used, $d\omega = 2\pi \cdot \Delta f$. Equations (9.56) and (9.57) can be written by the following Fourier transform pair $X(t)$ and $X(f)$

$$X(f) = \int\limits_{-\infty}^{\infty} X(t)e^{-i(2\pi \cdot ft)}dt \tag{9.58}$$

$$X(t) = \int\limits_{0}^{\infty} X(f)e^{+i(2\pi \cdot f \cdot t)}df. \tag{9.59}$$

The Fourier transform exists if the following conditions are met:

1. The integral of the absolute function exists; that is, $\int\limits_{-\infty}^{\infty} |X(t)|dt < \infty$.

2. Any discontinuities are finite.

The Fourier transform of a stationary random process $X(t)$ usually does not exist because the condition $\int\limits_{-\infty}^{\infty} |X(t)|dt < \infty$ is not met. However, the Fourier transform of the autocorrelation function $R_X(\tau)$ for a stationary random process $X(t)$ with $\mu_X = 0$ always exists. In this case, the forward and inverse Fourier transforms of $R_X(\tau)$ are given by

$$S_X(\omega) = \frac{1}{2\pi} \int\limits_{-\infty}^{\infty} R_X(\tau)e^{-i\omega \cdot \tau}d\tau \tag{9.60}$$

$$R_X(\tau) = \int\limits_{-\infty}^{\infty} S_X(\omega)e^{i\omega \cdot \tau}d\omega \tag{9.61}$$

where

$S_X(\omega) = $ the spectral density of a stationary random process $X(t)$ with $\mu_X = 0$

If $\tau = 0$, Equation (9.61) reduces to

$$R_X(0) = \int\limits_{-\infty}^{\infty} S_X(\omega)d\omega = E[X^2] = \sigma_X^2. \tag{9.62}$$

This means that the square root of the area under a spectral density plot $S_X(\omega)$ is the root mean square (RMS) of a normalized stationary random process. $S_X(\omega)$ is also called *mean square spectral density* and is illustrated in Figure 9.8.

The idea of a negative frequency has been introduced for mathematical completeness. However, it has no physical meaning. It is common in practice to consider the frequency from zero to infinity and to have the frequency f expressed in Hz (cycles/second), rather than ω in radians/second. Therefore, the two-sided spectral density $S_X(\omega)$ can be transformed into an equivalent one-sided spectral density $W_X(f)$ as follows:

$$E[X^2] = \sigma_X^2 = \int\limits_{0}^{\infty} W_X(f)df \tag{9.63}$$

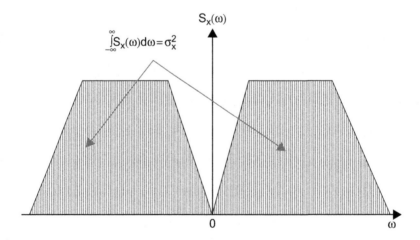

Figure 9.8
Relationship between the spectral density and RMS of a normalized
stationary random process.

where $W_X(f)$ is termed as the power spectral density (PSD), defined as

$$W_X(f) = 4\pi \cdot S_X(\omega). \tag{9.64}$$

For a discrete random time series $X(t_j)$, which has the discrete Fourier transforms as Equations (9.50) and (9.51), its mean square value $E[X^2(t)]$ can be obtained as follows:

$$E[X^2(t)] = \frac{1}{T} \sum_{j=0}^{N-1} X^2(t_j) \Delta t. \tag{9.65}$$

Equation (9.65) is further reduced by substituting Equations (9.50) and (9.51) as

$$
\begin{aligned}
E[X^2(t)] &= \frac{1}{N} \sum_{j=0}^{N-1} X(t_j) X(t_j) = \frac{1}{N} \sum_{j=0}^{N-1} X(t_j) \sum_{n=0}^{N-1} X_n \cdot e^{i\left(2\pi \frac{nj}{N}\right)} \\
&= \sum_{j=0}^{N-1} X_n \left[\frac{1}{N} \sum_{n=0}^{N-1} X(t_j) e^{i\left(2\pi \frac{nj}{N}\right)} \right] = \sum_{j=0}^{N-1} X_n X_n^*
\end{aligned}
\tag{9.66}
$$

where

$X_n^* =$ the complex conjugate of X_n

The PSD function is usually presented on a log-log scale. An octave (oct) is a doubling of frequency. The increase in octaves from f_1 to f_2 is

$$\text{Octaves} = \frac{\ln(f_{max}) - \ln(f_{min})}{\ln(2)}. \tag{9.67}$$

A bel is the common logarithm of the ratio of two measurements of power. A decibel (dB) is one-tenth of a bel and is defined by

$$dB = 10 \log\left(\frac{W_2}{W_1}\right). \tag{9.68}$$

A doubling of power spectral density corresponds to an increase of approximately 3 dB. If the power spectral density doubles for each doubling of frequency, the spectrum increases at 3 dB per octave or the spectrum has a positive roll-off rate of 3 dB per octave.

Responses to a Linear Single-Degree-of-Freedom System Subjected to Base Random Vibration

The following material describes the calculation of responses to a linear single-degree-of-freedom system to base random accelerations. The convolution theory is used to define a relationship between input and output signals for linear time-invariant (LTI) systems described by an impulse response function.

For a linear, discrete time, time-invariant system, an arbitrary input $X[n]$ can be expressed as a weighted sum of time-shift impulses as

$$X[n] = \sum_{k=-\infty}^{\infty} X[k]\delta[n-k] \tag{9.69}$$

where

$\delta[n-k] =$ the Dirac delta function

This is defined as

$$\delta[n-k] = \begin{cases} 1 & \text{for } n=k \\ 0 & \text{for } n \neq k. \end{cases}$$

If H is the sum operator on the input, Equation (9.69) is written as

$$H\{X[n]\} = H\left\{\sum_{k=-\infty}^{\infty} X[k]\delta[n-k]\right\} = \sum_{k=-\infty}^{\infty} X[k]H\{\delta[n-k]\}. \tag{9.70}$$

For a linear operation, Equation (9.70) becomes

$$Y[n] = \sum_{k=-\infty}^{\infty} X[k]h[n-k] \tag{9.71}$$

where

$h[n] =$ a response to the linear time invariant system due to an impulse $\delta[n]$
$Y[n] =$ the output with a weighted sum of time-shift impulse responses

Similarly, for a linear, continuous time, time-invariant system, an arbitrary input can be expressed as

$$X(t) = \int_{-\infty}^{\infty} X(\tau)\delta(t-\tau)d\tau \tag{9.72}$$

where

$$\delta(t-\tau) = \text{the Dirac delta function}$$

This is defined as

$$\delta(t-\tau) = \begin{cases} 1 & \text{for } t=\tau \\ 0 & \text{for } t \neq \tau. \end{cases}$$

Then, the output $Y(t)$ of a linear time invariant system described by an impulse response $h(t)$ is obtained as

$$Y(t) = \int_{-\infty}^{\infty} X(\tau)h(t-\tau)d\tau. \tag{9.73}$$

Equation (9.73) is called the convolution integral. Another derivation of the convolution integral is given by introducing $\theta = t - \tau$,

$$Y(t) = \int_{-\infty}^{\infty} X(t-\theta)h(\theta)d\theta. \tag{9.74}$$

To determine the frequency content $H(\omega)$ of the impulse response $h(t)$, let $X(t) = e^{i\omega \cdot t}$. Then Equation (9.74) becomes

$$Y(t) = \int_{-\infty}^{\infty} e^{i\omega(t-\theta)}h(\theta)d\theta = e^{i\omega \cdot t}\int_{-\infty}^{\infty} e^{i\omega(\theta)}h(\theta)d\theta = H(\omega)e^{i\omega \cdot t}. \tag{9.75}$$

The relationship between the Fourier transforms of $X(t)$ and $Y(t)$ is used to derive responses of an SDOF system to random vibration input. Take the Fourier transform of both sides of Equation (9.75),

$$Y(\omega) = \frac{1}{2\pi}\int_{-\infty}^{\infty} \left[\int_{-\infty}^{\infty} X(t-\theta)h(\theta)d\theta\right] e^{-i\omega \cdot t}dt. \tag{9.76}$$

Introducing $\tau = t - \theta$ and $dt = d\tau$,

$$Y(\omega) = \frac{1}{2\pi}\int_{-\infty}^{\infty} \left[\int_{-\infty}^{\infty} X(\tau)h(\theta)d\theta\right] e^{-i\omega \cdot (\tau+\theta)}d\tau. \tag{9.77}$$

Rearranging,

$$Y(\omega) = \left(\int_{-\infty}^{\infty} h(\theta) e^{-i\omega\cdot\theta} d\theta \right) \cdot \left(\frac{1}{2\pi} \int_{-\infty}^{\infty} X(\tau) e^{-i\omega\cdot\tau} d\tau \right). \tag{9.78}$$

Because $h(t)$ and $H(\omega)$ are the Fourier transform pairs,

$$H(\omega) = \int_{-\infty}^{\infty} h(\theta) e^{-i\omega\cdot\theta} d\theta \tag{9.79}$$

and

$$h(t) = \frac{1}{2\pi} \int_{-\infty}^{\infty} H(\omega) e^{-i\omega\cdot t} d\omega. \tag{9.80}$$

Thus, Equation (9.77) follows that

$$Y(\omega) = H(\omega)X(\omega) \tag{9.81}$$

where

$H(\omega) =$ the transfer function or the frequency response function

The spectral density of the output equals the spectral density of the input multiplying the squares of the gain function, which is expressed in terms of the frequency ω in radians/second as

$$S_Y(\omega) = |H(\omega)|^2 S_X(\omega) \tag{9.82}$$

where

$|H(\omega)| =$ the gain function (the modulus of the transfer function)

This is defined as

$$|H(\omega)| = \sqrt{H(\omega)H^*(\omega)} \tag{9.83}$$

where

$H^*(\omega) =$ the complex conjugate of $H(\omega)$

In terms of the frequency f in cycles/second, the power spectral density of the response is

$$W_Y(f) = |H(f)|^2 W_X(f). \tag{9.84}$$

The variance of this response σ_Y^2 can be calculated as the area under the response spectral density function as

$$\sigma_Y^2 = \int_{-\infty}^{\infty} S_Y(\omega)d\omega = \int_{-\infty}^{\infty} |H(\omega)|^2 S_X(\omega)d\omega \tag{9.85}$$

or

$$\sigma_Y^2 = \int_0^{\infty} S_Y(f)df = \int_0^{\infty} |H(f)|^2 S_X(f)df. \tag{9.86}$$

As discussed in previous sections, the steady-state relative acceleration $\ddot{z}(f_i, \xi, f_n)$ to an SDOF system with a resonant frequency f_n subjected to a base zero-mean stationary acceleration $\ddot{y}(f_i)$ with an excitation frequency of f_i can be obtained by

$$|\ddot{z}(f_i, \xi, f_n)| = |H_1(r_i)||\ddot{y}(f_i)|; \quad r_i = \frac{f_i}{f_n} \tag{9.87}$$

where $|H_1(r_i)|$ is defined as

$$|H_1(r_i)| = \frac{r_i^2}{\sqrt{(1 - r_i^2)^2 + (2\xi \cdot r_i)^2}} \tag{9.88}$$

where ξ is the damping ratio, and the maximum relative displacement $|z(f_i, \xi, f_n)|$ can be determined by

$$|z(f_i, \xi, f_n)| = |H_2(r_i)||\ddot{y}(f_i)| \tag{9.89}$$

where $|H_2(r_i)|$ is obtained as follows:

$$|H_2(r_i)| = \frac{1}{(2\pi \cdot f_n)^2 \sqrt{(1 - r_i^2)^2 + (2\xi \cdot r_i)^2}}. \tag{9.90}$$

Therefore, the power spectral density of the relative acceleration $W_{\ddot{z}}(f_i, \xi, f_n)$, to a linear single-degree-of-freedom system on which to base random accelerations $W_{\ddot{y}}(f_i)$, is obtained:

$$W_{\ddot{z}}(f_i, \xi, f_n) = |H_1(r_i)|^2 W_{\ddot{y}}(f_i) \tag{9.91}$$

and the power spectral density of the relative displacement $W_z(f_i, \xi, f_n)$ is obtained:

$$W_z(f_i, \xi, f_n) = |H_2(r_i)|^2 W_{\ddot{y}}(f_i). \tag{9.92}$$

It is assumed that the relationship between the stress amplitude S_a and relative displacement $|z|$ follows:

$$S_a = K \cdot |z| \tag{9.93}$$

where

$K =$ the coefficient relating a relative displacement to a stress amplitude

In this case, the power spectral density of the response stress amplitude $W_{S_a}(f_i, \xi, f_n)$ is obtained:

$$W_{S_a}(f_i, \xi, f_n) = K^2 \cdot W_z(f_i, \xi, f_n). \tag{9.94}$$

With Equations (9.92) and (9.94), the root mean square of this stress response $S_{a,RMS}$ can be calculated by

$$S_{a,RMS} = \sqrt{\int_0^\infty K^2 \cdot W_z(f_i, \xi, f_n) df} = \sqrt{\int_0^\infty K^2 \cdot |H_2(r_i)|^2 W_{\ddot{y}}(f_i) df}. \tag{9.95}$$

The following approach developed by Miles (1954) is an approximation to a root mean square stress response. The Miles equation was derived for the absolute response of an SDOF system excited by a "white noise" base acceleration of a constant level. If the base acceleration excitation is white noise, $S_{\ddot{y}}(\omega) = S_o$, and the spectral density function of the absolute acceleration of the mass $S_{\ddot{x}}(\omega)$ can be derived as

$$S_{\ddot{x}}(\omega) = |H(\omega)|^2 S_{\ddot{y}}(\omega) = \left| \frac{k + ic\omega}{(k - m\omega^2) + ic\omega} \right|^2 S_o = \frac{[k^2 + (c\omega)^2] S_o}{(k - m\omega^2)^2 + (c\omega)^2}. \tag{9.96}$$

The variance of the absolute acceleration is

$$\sigma_{\ddot{x}}^2 = \int_{-\infty}^\infty S_{\ddot{x}}(\omega) d\omega = S_o \int_{-\infty}^\infty \left| \frac{k + ic\omega}{(k - m\omega^2) + ic\omega} \right|^2 d\omega = S_o \left[\frac{\pi(kc^2 + mk^2)}{kcm} \right]. \tag{9.97}$$

Finally,

$$\sigma_{\ddot{x}}^2 = S_o \left[\frac{\pi(c^2 + mk)}{cm} \right]. \tag{9.98}$$

In terms of the power spectral density function, $W_o = 4\pi \cdot S_o$, the variance can be written as

$$\sigma_{\ddot{x}}^2 = \frac{W_o}{4\pi}\left[\frac{\pi(c^2 + mk)}{cm}\right] = \frac{\pi}{4}\frac{f_n W_o(1 + 4\xi^2)}{\xi} = \frac{\pi}{2}f_n QW_o(1 + 4\xi^2). \qquad (9.99)$$

For small damping ratio, $\xi \ll 1$, it is shown that

$$\sigma_{\ddot{x}}^2 = \frac{\pi}{2}f_n QW_o. \qquad (9.100)$$

Then the root mean square of the absolute acceleration response $\ddot{x}_{RMS}$ is

$$\ddot{x}_{RMS} = \sqrt{\frac{\pi}{2}f_n QW_o}. \qquad (9.101)$$

Also the root mean square of the relative displacement response is approximated as

$$z_{RMS} = \frac{\ddot{x}_{RMS}}{(2\pi \cdot f_n)^2}. \qquad (9.102)$$

With the assumption of Equation (9.93), the root mean square of this stress amplitude response $S_{a,RMS}$ can be approximated by

$$S_{a,RMS} = K \cdot z_{RMS} = \frac{K \cdot \ddot{x}_{RMS}}{(2\pi \cdot f_n)^2} = K\sqrt{\frac{QW_o}{4 \cdot (2\pi \cdot f_n)^3}}. \qquad (9.103)$$

It should be noted that the Miles equation should be used only if the power spectral density amplitude is flat within one octave on either side of the natural frequency.

Fatigue Damage Models under Random Stress Process

Fatigue damage models under narrow- and wide-band random stress processes will be addressed in the following sections with the emphasis on the frequency-based cycle counting techniques.

Level crossing rate of narrow-band random processes

For a continuous and differentiable stationary process $X(t)$, the expected number of positively sloped crossing (up-crossing) in an infinitesimal interval is only dependent on dt. We have

$$E[N_{a^+}(dt)] = v_{a^+}\,dt \qquad (9.104)$$

where

v_{a+} = the expected rate of up-crossing per time unit

If A denotes the event that any random sample from $X(t)$ has an up-crossing $x = a$ in an infinitesimal time interval dt, the propability of such an event A is

$$P(A) = v_{a+} dt. \qquad (9.105)$$

Equation (9.105) allows us to express v_{a+} in terms of $P(A)$. In order for the event A to exist, we must have

$$a - \dot{X}(t) < X(t) < a \text{ and } \dot{X}(t) > 0. \qquad (9.106)$$

Combining these two conditions, $P(A)$ can be written as

$$P(A) = P\left(a - \dot{X}(t) < X(t) < a \cap \dot{X}(t) > 0\right). \qquad (9.107)$$

These conditons define a triangle area in the $X(t) - \dot{X}(t)$ plane, as shown in Figure 9.9.

The probability of event A is calculated by integrating the joint probability density function of $X(t)$ and $\dot{X}(t)$ over this region; that is,

$$P(A) = \int\limits_{0}^{\infty} \int\limits_{a-vdt}^{a} f_{X\dot{X}}(u,v) du dv. \qquad (9.108)$$

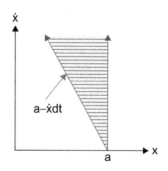

Figure 9.9
The region where event A occurs.

Substitution of Equation (9.108) into Equation (9.105) leads to the following expression of the level up-crossing rate for a stationary random process:

$$v_{a^+} = \int_0^\infty v f_{X\dot{X}}(a, v) dv. \qquad (9.109)$$

If $X(t)$ is Gaussian, the expected up-crossing rate of $x = a$ is

$$v_{a^+} = \frac{1}{2\pi} \frac{\sigma_{\dot{X}}}{\sigma_X} e^{\frac{-a^2}{2\sigma_X^2}}. \qquad (9.110)$$

The expected rate of zero up-crossings $E[0^+]$ is found by letting $a = 0$ in Equation (9.110):

$$E[0^+] = \frac{1}{2\pi} \frac{\sigma_{\dot{X}}}{\sigma_X}. \qquad (9.111)$$

The mean square displacement and velocity can be related to the moment of the spectral sensity function as

$$\sigma_X^2 = \int_{-\infty}^\infty S_X(\omega) d\omega = \int_0^\infty W_X(f) df \qquad (9.112)$$

$$\sigma_{\dot{X}}^2 = \int_{-\infty}^\infty \omega^2 S_X(\omega) d\omega = (2\pi)^2 \int_0^\infty f^2 W_X(f) df. \qquad (9.113)$$

Using Equations (9.112) and (9.113), the expected rate of zero up-crossing is then obtained as

$$E[0^+] = \sqrt{\frac{\int_0^\infty f^2 W_X(f) df}{\int_0^\infty W_X(f) df}}. \qquad (9.114)$$

The expected rate of peak crossing $E[P]$ is found from a similar analysis of the velocity process $\dot{X}(t)$. The rate of zero down-crossing of the velocity process corresponds to the occurrence of a peak in $\dot{X}(t)$. The result for a Gaussian process is

$$E[P] = \frac{1}{2\pi} \frac{\sigma_{\ddot{X}}}{\sigma_{\dot{X}}}.$$

(9.115)

In terms of the moment of the spectral density function, we have

$$E[P] = \sqrt{\frac{\int_0^\infty f^4 W_X(f) df}{\int_0^\infty f^2 W_X(f) df}}.$$

(9.116)

A narrow-band process is smooth and harmonic. For every peak there is a corresponding zero up-crossing, meaning $E[0^+]$ is equal to $E[P]$. However, the wide-band process is more irregular. A measure of this irregularity is the ratio of the zero up-crossing rate to the peak-crossing rate. The ratio is known as the irregularity factor γ expressed as

$$\gamma = \frac{E[0^+]}{E[P]}.$$

(9.117)

Alternatively, a narrow- or wide-band process can be judged by the width of its spectrum. For this reason, the spectral width parameter λ is introduced as

$$\lambda = \sqrt{1 - \gamma^2}.$$

(9.118)

Note that $\lambda \to 0$ represents a narrow-band random process.

If M_j is the j-th moment of a one-sided power spectral density function for random vibration stress amplitude (see Figure 9.10) defined as

$$M_j = \int_0^\infty f^j W_{S_a}(f) df$$

(9.119)

then the rate of zero crossings $E[0^+]$ and the rate of peaks $E[P]$ are given by

$$E[0^+] = \sqrt{\frac{M_2}{M_0}}$$

(9.120)

$$E[P] = \sqrt{\frac{M_4}{M_2}}$$

(9.121)

and the the irregularity factor γ and the spectral width parameter λ are rewritten as

$$\gamma = \sqrt{\frac{M_2^2}{M_0 M_4}} \qquad (9.122)$$

$$\lambda = \sqrt{1 - \frac{M_2^2}{M_0 M_4}}. \qquad (9.123)$$

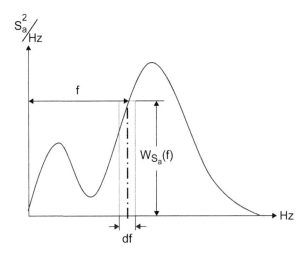

Figure 9.10
Moments from a one-sided power spectral density.

Fatigue damage under narrow-band random stresses

The Palmgren–Miner linear damage rule (Palmgren, 1924; Miner, 1945) and an S-N curve following the Basquin expression (1910) are employed for fatigue damage calculation under narrow-band random stresses. Variable amplitude loading is simulated by a sequence of blocks of constant amplitudes. The linear damage D is defined as

$$D = \sum_{i=1}^{kn} \frac{n_i}{N_{f,i}} \qquad (9.124)$$

where

$n_i =$ the total number of cycles in the i-th block of constant stress amplitude $S_{a,i}$
$kn =$ the total number of the stress blocks
$N_{f,i} =$ the fatigue life as the number of cycles to failure under $S_{a,i}$

According to the Palmgren–Miner linear damage rule, failure occurs when $D \geq 1$. The S-N curve follows the following Basquin expression:

$$N_{f,i} \cdot S_{a,i}^m = C. \tag{9.125}$$

Alternatively, the fatigue life $N_{f,i}$ at the stress amplitude level $S_{a,i}$ would be

$$N_{f,i} = \frac{C}{S_{a,i}^m} \tag{9.126}$$

where

  m = the slope factor of the S-N curve
  C = the material constant

The cycle-counting histogram for a narrow-band stress process $S(t)$ can be established by either performing the rainflow cycle counting technique or by counting the number of peaks n_i in the window Δs_i around a stress level. Suppose that the total number of peaks counted in the stress process is denoted by

$$\sum_{j=1}^{kn} n_j.$$

The probability pdf (f_i) that the stress amplitude $S_a = s_{a,i}$ may occur is

$$f_i = \frac{n_i}{\sum_{j=1}^{kn} n_j}. \tag{9.127}$$

Thus, Equation (9.127) is the probability density function of the random variable S_a. In this case, the total fatigue damage can be written as

$$D = \sum_{i=1}^{kn} \frac{n_i}{N_{f,i}} = \sum_{i=1}^{kn} \frac{f_i \sum_{j=1}^{kn} n_j}{N_{f,i}}. \tag{9.128}$$

Using the linear S-N model as in Equation (9.125), the expression for fatigue damage is

$$D = \frac{\sum_{j=1}^{kn} n_j}{C} \sum_{i=1}^{kn} f_i S_{a,i}^m. \tag{9.129}$$

Also, the expected value of S_a^m is

$$E(S_a^m) = \sum_{i=1}^{kn} f_i S_i^m. \tag{9.130}$$

For narrow-band random stresses, the total count of cycles $\sum_{j=1}^{k} n_j$ is equal to the rate of zero up-crossing multiplying the total time period T. Thus, the fatigue damage can be expressed as

$$D = \frac{\sum_{j=1}^{kn} n_j}{C} E(S_a^m) = \frac{E[0^+] \cdot T}{C} E(S_a^m). \tag{9.131}$$

Assume that the probability density function of stress amplitude S_a can be treated as a continuous random variable, as illustrated in Figure 9.11. The expected value of S_a^m is

$$E(S_a^m) = \int_0^\infty s_a^m f_{S_a}(s_a) ds_a. \tag{9.132}$$

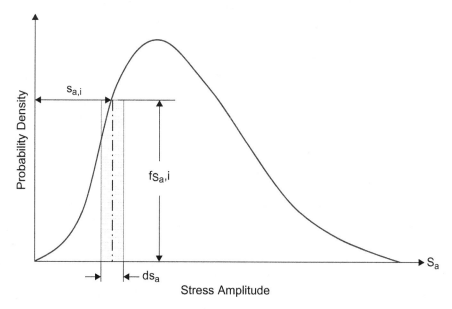

Figure 9.11
Continuous probability density function of stress amplitude.

Even though any statistical model of S_a can be employed, it is common to use the Weibull distribution with the following cumulative distribution function:

$$F_{Sa}(s_a) = 1 - \exp\left[-\left(\frac{s_a}{\alpha}\right)^\beta\right] \tag{9.133}$$

where

α and β = the scale parameter (characteristic life) and the shape parameter (Weibull slope), respectively

For the Weibull distribution,

$$E(S_a^m) = \alpha^m \Gamma\left(\frac{m}{\beta} + 1\right) \tag{9.134}$$

where

$\Gamma()$ = the gamma function

In the special case where $\beta = 2$, the Weibull distribution reduces to the Rayleigh distribution. This is an important case because Rayleigh is the distribution of peaks or ranges or amplitude in a stationary narrow-band Gaussian process that has RMS value of σ_{S_a}. Also it can be shown that

$$\alpha = \sqrt{2}\sigma_{S_a}. \tag{9.135}$$

Therefore, if S(t) is a zero-mean stationary narrow-band Gaussian and the stress amplitudes follow the Rayleigh distribution, the expected value of S_a^m becomes

$$E(S_a^m) = (\sqrt{2}\sigma_{S_a})^m \Gamma\left(\frac{m}{2} + 1\right) \tag{9.136}$$

where

$$\sigma_{S_a} = \sqrt{M_0}. \tag{9.137}$$

Finally, the fatigue damage D_{NB} of a stress process over a time interval T can be written as

$$D_{NB} = \frac{E[0^+] \cdot T}{C} (\sqrt{2M_0})^m \Gamma\left(\frac{m}{2} + 1\right). \tag{9.138}$$

Fatigue damage under wide-band random stresses

Based on the rainflow counting method, a model for predicting fatigue damage under a stationary wide-band Gaussian stress process has been proposed by Wirsching and Light (1980). Using the narrow-band approach as a starting point, the general expression for the damage $D_{WB,Wirsching}$ over a time interval T is

$$D_{WB,Wirsching} = \zeta_W D_{NB} \qquad (9.139)$$

where

$\quad D_{NB}$ = the fatigue damage under a narrow-band random process
$\quad \zeta_W$ = the rainflow correction factor

ζ_W is an empirical factor derived from extensive Monte Carlo simulations that include a variety of spectral density functions. It is expressed as follows:

$$\zeta_W = a_W + [1 - a_W](1 - \lambda)^{b_W} \qquad (9.140)$$

where

$\quad a_W = 0.926 - 0.033\,m$
$\quad b_W = 1.587\,m - 2.323$

Note that m is the slope of the S-N curve, and λ is the spectral width parameter.

Ortiz and Chen (1987) also derived another similar expression for fatigue damage $D_{WB,Oritz}$ under wide-band stresses as

$$D_{WB,Oritz} = \zeta_O D_{NB} \qquad (9.141)$$

where

$$\zeta_O = \frac{1}{\gamma}\sqrt{\frac{M_2 M_k}{M_0 M_{k+2}}} \quad \text{and} \quad k = \frac{2.0}{m}.$$

The irregularity factor γ is defined in Equation (9.117) or (9.122).

Instead of using the damage correction factor from the narrow-band random stresses to the wide-band random stresses, Dirlik (1985) has developed an empirical closed-form expression for the probability density function of stress amplitude $f_{S_a}(s_a)$ based on the rainflow cycle counting results from extensive Monte Carlo simulations of random stress time histories.

Dirlik's solutions were sucessfully verified by Bishop in theory (Bishop, 1988). Dirlik's damage model for a time period of T is presented here:

$$D_{WB,Dirlik} = \frac{E[P]T}{C} \int_0^\infty s_a^m f_{S_a}(s_a) ds_a \tag{9.142}$$

$$f_{S_a}(s_a) = \frac{1}{\sqrt{M_0}} \left[\frac{D_1}{Q} e^{\frac{-Z}{Q}} + \frac{D_2 Z}{R^2} e^{\frac{-Z^2}{2R^2}} + D_3 Z e^{\frac{-Z^2}{2}} \right] \tag{9.143}$$

where

$$Z = \frac{s_a}{\sqrt{M_0}}$$

is the nomalized stress amplitude with respect to the RMS of random stress amplitude, and

$$\gamma = \frac{M_2}{\sqrt{M_0 M_4}}$$

$$X_m = \frac{M_1}{M_0} \sqrt{\frac{M_2}{M_4}}$$

$$D_1 = \frac{2(X_m - \gamma^2)}{1 + \gamma^2}$$

$$R = \frac{\gamma - X_m - D_1^2}{1 - \gamma - D_1 + D_1^2}$$

$$D_2 = \frac{1 - \gamma - D_1 + D_1^2}{1 - R}$$

$$D_3 = 1 - D_1 - D_2$$

$$Q = \frac{1.25(\gamma - D_3 - D_2 R)}{D_1}.$$

To perform the numerical integration analysis, the Dirlik equation needs to be expressed in the following discrete format:

$$D_{WB,Dirlik} = \frac{E[P]T}{C} \sum_0^\infty s_a^m f_{s_a}(s_a) \Delta s_a = \sum_0^\infty \frac{(E[P]T)(f_{s_a}(s_a)\Delta s_a)}{\dfrac{C}{s_a^m}} = \sum_0^\infty \frac{n_i}{N_{f,i}}. \qquad (9.144)$$

Equation (9.144) states that the Dirlik damage calculation is the sum of the incremental damage value in each Δs_a.

If the stress range, instead of stress amplitude, is the preferable variable to be used in the S-N expression and the damage calculation, then the PDF of stress range $f_{s_r}(s_r)$ would follow this expression:

$$f_{s_r}(s_r) = \frac{1}{2\sqrt{M_0}} \left[\frac{D_1}{Q} e^{\frac{-Z}{Q}} + \frac{D_2 Z}{R^2} e^{\frac{-Z^2}{2R^2}} + D_3 Z e^{\frac{-Z^2}{2}} \right] \qquad (9.145)$$

where

$$Z = \frac{s_r}{2\sqrt{M_0}}.$$

Please note that the difference in Equations (9.143) and (9.145) by a factor of 2 is due to the simple variable transformation from Δs_a to $\Delta s_r/2$.

Bishop (1989, 1994) concluded that the Dirlik formula is far superior to other existing methods for estimating rainflow fatigue damage. The Dirlik method is preferable for fatigue damage calculations based on the PSD and has been widely adopted by many commercial fatigue software packages.

However, the Dirlik method has some drawbacks. First of all, it is an empirical approach that is not supported by any kind of theoretical framework. Second, the proposed rainflow distribution does not account for the mean stress effects, making it impossible for further extension to cover non-Gaussian problems.

Example 9.2

A hot-rolled component made of SAE 1008 steel is subjected to random loading process. The stress response at a critical location is calculated in terms of the power spectrum density in Figure 9.12. The PSD has two frequencies of 1 Hz and 10 Hz, corresponding to 10,000 MPa^2/Hz and 2500 MPa^2/Hz, respectively.

The material S-N curve is given as follows:

$$S_{a,i} = S'_f(2N_{f,i})^b$$

where

S'_f = the fatigue strength coefficient of 1297 MPa
b = the fatigue strength exponent of -0.18

Please determine the fatigue damage of this component, using the preceding equations for wide-band stresses. Note that a sine wave has a crest factor of $\sqrt{2} = 1.414$.

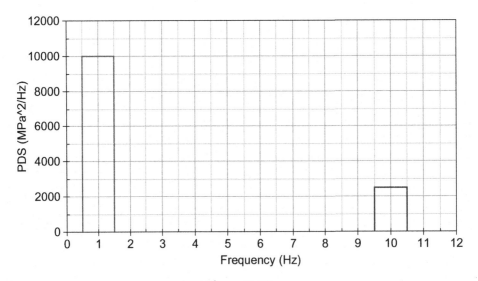

Figure 9.12
Power spectral density of the stress response of a component
made of SAE 1008 steel.

Solution

We will calculate the fatigue damage of the component in time domain first and use it as baseline information to compare with the predicted damage based on the PSD.

The original stress time history can be obtained by adding two sine waves, one for each block in the PSD. For a sine wave, the amplitude of each is calculated from 1.414 times the RMS value, that is, the area of each PSD block.

The stress amplitude $S_{a@1HZ}$ of the sine wave at 1 Hz is

$$S_{a@1HZ} = \sqrt{10,000 \times 1} \times 1.414 = 141.4 \, \text{MPa}.$$

The stress amplitude $S_{a@10HZ}$ of the second sine wave at 10 Hz is

$$S_{a@10HZ} = \sqrt{2500 \times 1} \times 1.414 = 70.7 \, \text{MPa}.$$

The application of the rainflow cycle-counting technique to the superimposed sine waves would result in the stress amplitude of 212.1 MPa ($= 141.4 + 70.7$ MPa) with 1 Hz frequency and the stress amplitude of 70.7 MPa with 10 Hz frequency, excluding the mean stress effect.

The material S-N curve for vibration fatigue usually follows this S-N expression:

$$N_{f,i} S_{a,i}{}^m = C.$$

Thus,

$$m = -1/b = -1/(-0.18) = 5.56$$
$$C = 0.5 \times (S_f')^m = 0.5 \times 1297^{5.56} = 1.02 \times 10^{17} \, \text{MPa}.$$

With the preceding S-N equation, we determine the fatigue life for each sine wave as follows:

$$N_{f,1} = C \cdot S_{a,1}^{-m} = 1.02 \times 10^{17} \times 212.1^{-5.56} = 1.18 \times 10^4 \, \text{cycles}$$
$$N_{f,2} = C \cdot S_{a,2}^{-m} = 1.02 \times 10^{17} \times 70.7^{-5.56} = 5.32 \times 10^6 \, \text{cycles}.$$

In a one-second-time interval, the sine waves at 1 Hz and 10 Hz represent 1 cycle ($n_1 = 1$) and 10 cycles ($n_2 = 10$) of reversed loading, respectively. The linear damage calculation for this time interval gives

$$D_{t=1} = \frac{n_1}{N_{f,1}} + \frac{n_2}{N_{f,2}} = \frac{1}{1.18 \times 10^4} + \frac{10}{5.32 \times 10^6} = 8.66 \times 10^{-5}.$$

This corresponds to a fatigue life of 11,500 seconds ($= 1/8.66 \times 10^{-5}$).

We then proceed to calculate the fatigue damage to the component based on the given PSD. The aforementioned frequency domain methods will be used for the damage estimation. The j-th moment of the PSD can be easily calculated as

$$M_j = \int_0^\infty f^j W_{S_a}(f)df = 1^j \times 10{,}000 \times 1 + 10^j \times 2500 \times 1$$

$$M_0 = 1^0 \times 10{,}000 \times 1 + 10^0 \times 2500 \times 1 = 12{,}500$$

$$M_2 = 1^2 \times 10{,}000 \times 1 + 10^2 \times 2500 \times 1 = 260{,}000$$

$$M_4 = 1^4 \times 10{,}000 \times 1 + 10^4 \times 2500 \times 1 = 25{,}010{,}000$$

from which we can compute

$$E[0^+] = \sqrt{\frac{M_2}{M_0}} = \sqrt{\frac{260{,}000}{12{,}500}} = 4.56$$

$$E[P] = \sqrt{\frac{M_4}{M_2}} = \sqrt{\frac{25{,}010{,}000}{260{,}000}} = 9.81 \text{ per second}$$

$$\gamma = \frac{E[0^+]}{E[P]} = \frac{4.56}{9.81} = 0.465$$

$$\lambda = \sqrt{1 - \gamma^2} = \sqrt{1 - 0.465^2} = 0.885.$$

The Wirsching and Light Method

Fatigue damage D_{NB} of a zero-mean stationary narrow-band Gaussian stress process over a time interval $T = 1$ second can be written as

$$D_{NB} = \frac{E[0^+] \times T}{A} \left(\sqrt{2M_0}\right)^m \Gamma\left(\frac{m}{2} + 1\right)$$

$$= \frac{4.56 \times 1}{1.02 \times 10^{17}} \left(\sqrt{2 \times 12{,}500}\right)^{5.56} \Gamma\left(\frac{5.56}{2} + 1\right)$$

$$D_{NB} = 0.000345.$$

The rainflow correction factor is calculated as

$$a_W = 0.926 - 0.033\,m = 0.926 - 0.033 \times 5.56 = 0.743$$
$$b_W = 1.587\,m - 2.323 = 1.587 \times 5.56 - 2.323 = 6.501$$
$$\zeta_W = a_W + [1 - a_W](1 - \lambda)^{b_W} = 0.743 + (1 - 0.743) \times (1 - 0.885)^{6.501} = 0.743.$$

Finally, the fatigue damage $D_{WB,Wirsching}$ is computed as

$$D_{WB,Wirsching} = \zeta_W D_{NB} = 0.743 \times 0.000345 = 0.000256 \text{ per second.}$$

This corresponds to a fatigue life of 3900 seconds ($= 1/0.000256$), which is very conservative as compared to baseline fatigue life (11,500 seconds).

The Ortiz and Chen Method

$D_{NB} = 0.000345$ is the same as the one calculated previously. Calculation of the rainflow correction factor ζ_O is required. Given the slope of the S-N curve, $m = 5.56$,

$$k = \frac{2.0}{m} = \frac{2.0}{5.56} = 0.3597$$

$$M_k = 10^{0.3597} \times 10,000 \times 1 + 10^{0.3597} \times 2500 \times 1 = 15,723$$

$$M_{k+2} = 10^{0.3597+2} \times 10,000 \times 1 + 10^{0.3597+2} \times 2500 \times 1 = 582,338$$

$$\zeta_O = \frac{1}{\gamma} \sqrt{\frac{M_2 M_k}{M_0 M_{k+2}}} = \frac{1}{0.465} \sqrt{\frac{260,000 \times 15,723}{12,500 \times 582,338}} = 1.612.$$

The fatigue damage $D_{WB,Oritz}$ is computed as

$$D_{WB,Oritz} = \zeta_O D_{NB} = 1.612 \times 0.000345 = 0.000556 \text{ per second.}$$

This corresponds to a fatigue life of 1800 seconds ($= 1/0.000556$), which is very conservative as compared to the baseline fatigue life (11,500 seconds).

The Dirlik Method

It is necessary to determine the following parameters for the probability density function of stress amplitudes that have been rainflow cycle counted.

$$X_m = \frac{M_1}{M_0}\sqrt{\frac{M_2}{M_4}} = \frac{35,000}{12,500}\sqrt{\frac{260,000}{25,010,000}} = 0.2859$$

$$D_1 = \frac{2(X_m - \gamma^2)}{1+\gamma^2} = \frac{2(0.2859 - 0.465^2)}{1+0.465^2} = 0.1146$$

$$R = \frac{\gamma - X_m - D_1^2}{1-\gamma - D_1 + D_1^2} = \frac{0.465 - 0.2859 - 0.1146^2}{1-0.465 - 0.1146 + 0.1146^2} = 0.3828$$

$$D_2 = \frac{1-\gamma - D_1 + D_1^2}{1-R} = \frac{1-0.465 - 0.1164 + 0.1164^2}{1-0.3828} = 0.7023$$

$$D_3 = 1 - D_1 - D_2 = 1 - 0.1146 - 0.7023 = 0.1831$$

$$Q = \frac{1.25(\gamma - D_3 - D_2 R)}{D_1} = \frac{1.25(0.465 - 0.1831 - 0.7023 \times 0.3828)}{0.1146}$$

$$= 0.1425.$$

Substituting the preceding values into Equation (9.143) provides the Dirlik's probability density function of stress amplitudes as follows:

$$f_{S_a}(s_a) = 0.0071907 e^{-7.01523Z} + 0.042867 e^{-3.41213Z^2} + 0.0016377 e^{-0.5Z^2}.$$

The numerical integration technique for the Dirlik formula, Equation (9.144), is illustrated in Table 9.2. For a given time exposure of 1 second, the calculation leads to $D_{WB,Dirlik} = 1.38 \times 10^{-4}$, the sum of all the damage values in the last column of the table. This corresponds to a fatigue life of 7250 seconds, which correlates better to the baseline fatigue life (11,500 seconds).

Table 9.2: Calculation Procedures for $D_{WB,Dirlik}$ ($\Delta s_a = 10$ MPa; $T = 1$ seconds)

$s_{a,i}$ MPa	$Z = s_{a,i}/\sqrt{M_0}$	$f_i = f_{S_a}(s_{a,i}) \cdot \Delta s_a$	$n_i = (E[P]T) \cdot f_i$	$N_{f,i} = C/S_a^m$	$d_i = n_i/N_{f,i}$
10	0.0894	0.077162	0.757	2.8×10^{11}	2.7×10^{-12}
20	0.1789	0.092135	0.904	6.0×10^{11}	1.5×10^{-10}
30	0.2683	0.105155	1.032	6.2×10^{11}	1.7×10^{-9}
⋮	⋮	⋮	⋮	⋮	⋮
980	8.7654				1.2×10^{-17}
990	8.8548				5.9×10^{-18}
1000	8.9443				2.9×10^{-18}
				$\sum d_i = 1.38 \times 10^{-4}$	

Example 9.3

An electronic system mounted directly on a vehicle body (a sprung mass) is subjected to road-load driving, inducing random vibration to the system. It is recommended that the vehicle manufacturer and supplier perform the random vibration test based on the ISO (International Organization for Standardization) 16750-3 standard.

The test duration should be 8 hours for each principal axis of the system. The power spectral density function of the base random excitation to the system is shown in Figure 9.13 and tabulated in Table 9.3.

It is assumed that the S-N curve follows $N_{f,i} \cdot S_a^m = C$ where $m = 4$ and $C = 1$ and that the linear relationship between the stress amplitude and the relative displacement exists as $S_a = K \cdot z$ where $K = 10^6$. The $Q = 10$ for

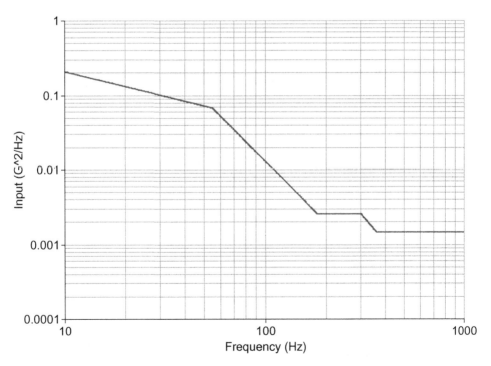

Figure 9.13
PSD plot based on the ISO 16750-3 random vibration profile.

**Table 9.3: Tabulated PSD Data for the ISO 16750-3
Random Vibration Profile**

f (Hz)	G²/Hz
10	0.208
55	0.0677
180	0.0026
300	0.0026
360	0.00146
1000	0.00146

dynamic amplification factor is assumed, equivalent to the damping ratio of $\xi = 5\%$.

Calculate the fatigue damage of the electronic system based on the Wirshing–Light method, the Ortiz–Chen method, and the Dirlik method.

Solution
The electronic system of interest consists of a number of subsystems and components whose natural frequencies are not easily obtained experimentally or analytically. In order to assess the fatigue damage of the system, it is very common to construct a fatigue damage spectrum by estimating the damage for an individual subsystem or component that can be modeled as a single-degree-of-freedom system subjected to base random excitation, whose natural frequency is allowed to vary as an independent variable.

The analytical solution to assess the fatigue damage severity of a linear SDOF system subjected to base excitations is derived and discussed. The general solution for relative displacement PSD for each natural frequency is given as

$$W_z(f_i, \xi, f_n) = W_{\ddot{y}}(f_i)|H_2(r_i)|^2.$$

With the assumption of $S_a = K \cdot |z|$, the stress amplitude PSD for each natural frequency can be determined by

$$W_{S_a}(f_i, \xi, f_n) = K^2 \cdot W_z(f_i, \xi, f_n) = K^2 \cdot |H_2(r_i)|^2 \cdot W_{\ddot{y}}(f_i).$$

According to the S-N curve expressed as $N_{f,i} \cdot S_a^m(\xi, f_n) = C$, the fatigue damage value based on the test duration T and the stress amplitude

PSD at *each natural frequency*, f_n can be calculated by the following theories:

1. Calculate the narrow-band damage

$$D_{NB} = \frac{E[0^+]T}{C} E[S_a^m] = \frac{E[0^+]T}{C} \left(\sqrt{2M_0}\right)^m \Gamma\left(\frac{m}{2}+1\right).$$

2. Calculate the narrow-band damage on Miles' equation

$$D_{NB,Miles} = f_n \cdot T \cdot \frac{K^m}{C} \cdot \left(\sqrt{\frac{Q \cdot W_{\ddot{y}}(f_n)}{2 \cdot (2\pi \cdot f_n)^3}}\right)^m \Gamma\left(1+\frac{m}{2}\right).$$

3. Calculate the wide band damage based on the Wirsching–Light method

$$D_{WB,Wirsching} = \zeta_W D_{NB} \quad \text{and} \quad \zeta_W = a_W + [1 - a_W](1 - \lambda)^{b_W}.$$

4. Calculate the wide-band damage based on the Ortiz–Chen method

$$D_{WB,Oritz} = \zeta_O D_{NB} \quad \text{and} \quad \zeta_O = \frac{1}{\gamma}\sqrt{\frac{M_2 M_k}{M_0 M_{k+2}}}$$

where

$$k = \frac{2.0}{m}.$$

5. Calculate the wide-band damage based on the Dirlik method

$$D_{WB,Dirlik} = \frac{E[P]T}{C} \int_0^\infty s_a^m f_{S_a}(s_a) ds_a$$

$$f_{S_a}(s_a) = \frac{1}{\sqrt{M_0}} \left[\frac{D_1}{Q} e^{\frac{-Z}{Q}} + \frac{D_2 Z}{R^2} e^{\frac{-Z^2}{2R^2}} + D_3 Z e^{\frac{-Z^2}{2}}\right]$$

where

$$Z = \frac{s_a}{\sqrt{M_0}}$$

$$\gamma = \frac{M_2}{\sqrt{M_0 M_4}}$$

$$X_m = \frac{M_1}{M_0}\sqrt{\frac{M_2}{M_4}}$$

$$D_1 = \frac{2(X_m - \gamma^2)}{1 + \gamma^2}$$

$$R = \frac{\gamma - X_m - D_1^2}{1 - \gamma - D_1 + D_1^2}$$

$$D_2 = \frac{1 - \gamma - D_1 + D_1^2}{1 - R}$$

$$D_3 = 1 - D_1 - D_2$$

$$Q = \frac{1.25(\gamma - D_3 - D_2 R)}{D_1}.$$

The FDS can be constructed by varying the natural frequency changes from 10 Hz to 1000 Hz according to the ISO specification. According to the aforementioned frequency-based fatigue theories, the normalized fatigue damage spectrum plots with respect to the Dirlik method are presented in Figure 9.14. This figure shows

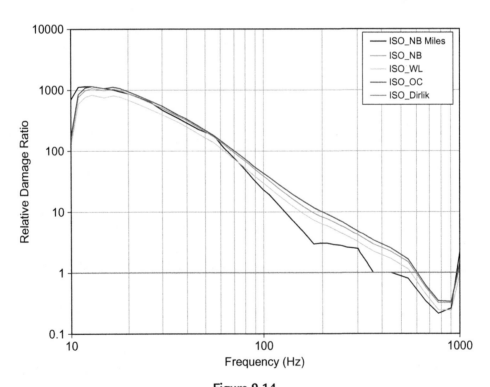

Figure 9.14
Normalized fatigue damage spectrum plots based on various frequency-based fatigue theories, with respect to Dirlik's method.

that the narrow-band Miles' equation is conservative below 350 Hz and becomes less conservative above 350 Hz as compared to Dirlik's method.

Also of the wide-band damage methods, Ortiz–Chen's method is the most conservative while Dirlik's has been found by Bishop (Bishop & Sherratt, 1989; Bishop, 1994) to be the most accurate in practice. Furthermore, Figure 9.14 shows relative FDS plots based on IEC 60068-2-6, Classification III, and ISO 16750-3. This figure indicates the IEC test specification induces the most severe FDS to a linear single-degree-of-freedom system than those based on the ISO test standard.

Summary

This chapter has presented a standard FDS calculation procedure that can be employed for assessing durability of automotive parts subjected to vibrational loading conditions. It also explained fatigue damage calculation methods for sinusoidal and random vibration tests in detail.

As discussed in previous sections, vibration fatigue test methods should be chosen based on the characteristics of dominant forcing inputs. For example, a logarithmic frequency sweep is the preferable one for use in sinusoidal sweep testing because of its easy way to dictate an equal number of cycles at each resonance.

Finally, this chapter has compared the fatigue damage calculation results obtained from various fatigue damage calculation methods for random vibration excitations. It shows that the narrow-band Miles equation is conservative below 350 Hz and becomes less conservative above 350 Hz, as compared to Dirlik's method. Also for the wide-band damage methods, Ortiz–Chen's method is the most conservative while Dirlik's has been found to be the most accurate in practice. It is also seen that IEC 60068-2-6 is more severe than ISO 16750-3 standards, in terms of fatigue damage severity comparison.

References

Basquin, O. H. (1910). The exponential law of endurance tests. *Proceedings of American Society for Testing and Materials, 19*, 625–630.

Bishop, N. W. M. (1988). The use of frequency domain parameters to predict structural fatigue. Ph.D. thesis. West Midlands, UK: Warwick University.

Bishop, N. W. M. (1994). Spectral methods for estimating the integrity of structural components subjected to random loading. In A. Carpinteri (Ed.), *Handbook of fatigue crack propagation in metallic structures* (Vol. 2, pp. 1685–1720). New York: Elsevier Science.

Bishop, N. W. M., & Sherratt, F. (1989). Fatigue life prediction from power spectral density data. Part 2: Recent development. *Environmental Engineering, 2*(1 and 2), 5–10.

Cronin, D. L. (1968). Response spectra for sweeping sinusoidal excitations. *Shock and Vibration Bulletin, 38*(1), 133–139.

Curtis, A. J., Tinling, N. G., & Abstein, H. T. (1971). *Selection and performance of vibration tests* (Contract Number: N00173-69-C-0371). Technical Information Division, Naval Research Laboratory.

Dirlik, T. (1985). *Application of computers in fatigue analysis.* Ph.D. thesis. West Midlands, UK: Warwick University.

GAM EG-13. (1986). Essais generaux en environement des materials (general tests of materials in environment). *Delegation Generale pour l'Armement,* Ministere de la Defense, France.

Halfpenny, A., & Kihm, F. (2006). Mission profiling and test synthesis based on fatigue damage spectrum. In *9th international fatigue congress* (p. 342), Atlanta.

IEC-60068-2-6. (2007). Environmental testing. Part 2: Tests Fc: Vibration (sinusoidal).

ISO 16750-3. (2003). Road vehicles. Environmental conditions and testing for electrical and electronic equipment. Part 3: Mechanical loads. Geneva: International Organization for Standardization.

Lalanne, C. (2002). *Mechanical vibration and shock: Fatigue damage* (Vol. IV). London: Hermes Penton Ltd.

Miles, J. W. (1954). On structural fatigue under random loading. *Journal of the Aeronautical Sciences,* November, 573.

MIL-STD-810F. (2000). Department of Defense Test Method Standard for Environmental Engineering Considerations and Laboratory. U.S. Army, Developmental Test Command.

Miner, M. A. (1945). Cumulative damage in fatigue. *Journal of Applied Mechanics, 67,* A159–A164.

Ortiz, K., & Chen, N. K. (1987). Fatigue damage prediction for stationary wideand stresses. Presented at the 5th International Conference on the Applications of Statistics and Probability in Civil Engineering, Vancouver, BC, Canada.

Palmgren, A. (1924). Die lebensdauer von kugellagern (The service life of ball bearings). *Zeitschrift des Vereinesdeutscher Ingenierure (VDI Journal), 68*(14), 339–341.

Wirsching, P. H., & Light, M. C. (1980). Fatigue under wide band random stresses. *ASCE Journal of the Structural Division, 106,* 1593–1607.

CHAPTER 10

Fatigue Life Prediction Methods of Seam-Welded Joints

Hong-Tae Kang
The University of Michigan–Dearborn

Yung-Li Lee
Chrysler Group LLC

Chapter Outline

Introduction

All the seam-welding techniques require high thermal energy input to weld work pieces together. Contrary to a spot weld, a seam weld is a continuous weld in various welding geometries such as fillet and butt welds. In this chapter, only seam-welded joints are considered, whereas the analysis and behavior of spot-welded joints is a specific area that has been treated differently and will be discussed in Chapter 11.

During the service life of welded structures exposed to various service loading conditions, welded joints are usually the potential fatigue failure sites due to the highest stress concentration areas and altered material properties. Thus, engineers and scientists are always interested in understanding fatigue characteristics of the welded joints, and are trying to develop analytical tools to estimate fatigue lives of welded joints. However, prediction of fatigue life of welded joints is frequently complicated and inaccurate because many parameters affect the fatigue life of welded joints.

Welding strongly affects the materials by the process of heating and subsequent cooling as well as by the fusion process with additional filler material, resulting in inhomogeneous and different materials. Furthermore, a weld is usually far from being perfect, containing inclusions, pores, cavities, undercuts, and so on. The shape of the weld profile and nonwelded root gaps creates high stress concentrations with varying geometry parameters. Moreover, residual stresses and distortions due to the welding process affect the fatigue behavior.

In view of the complexity of this subject on fatigue life prediction models of seam-welded joints and the wide area of applications, it is not surprising that several analytical approaches exist and none of them could account for the aforementioned process variables. Therefore, it has been an ongoing research area of interest to all the engineering disciplines to improve the life predictive capability for seam-welded joints.

Due to the vast amount of relevant literature, this chapter will present only Dong's and Fermer's structural stress approaches (Dong, 2001a,b; Fermer et al., 1998) and the notch pseudo stress approach because the three approaches have been coded in some commercial fatigue analysis modules as one of the Computer Aided Engineering (CAE) tools used in automotive engineering. Refer to the book by Radaj et al. (2006) for a detailed review of all other methods. However, this chapter will start with an introduction of the parameters affecting the fatigue life of welded joints to help you understand the possible sources of the variability of fatigue data.

Parameters Affecting Fatigue Lives of Seam-Welded Joints

Generally fatigue test results of seam-welded joints contain various levels of scatter. Much of this scatter is caused by geometric and processing variations such as part fit-up, weld gap, variation in feed rates, travel rates, weld angles, and so on. This scatter confuses the interpretation of test results, and it is often nearly impossible to discern the effects of the material and other factors.

Numerous researchers have indicated that one of the most critical factors affecting the fatigue life of a welded joint is the consistency of the cross-sectional weld

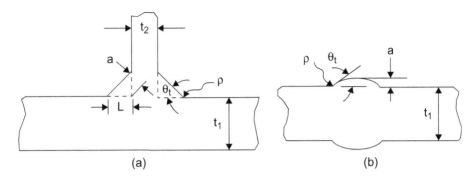

Figure 10.1
Weld geometric parameters for fillet (a) and butt (b) joints.
Source: Adapted from Berge (1985).

geometry (Jiang, 2006; Ninh Nguyen & Wahab, 1995; Ferreira & Branco, 1989; Maddox, 1991; Caccese et al., 2006; Berge, 1985; Branco et al., 1985; Ferreira & Branco, 1991; Seto et al., 2004; Kang et al., 2011a). Weld geometric parameters including plate thickness (t_1, t_2), effective weld throat thickness (a), weld leg length (L), weld throat angle (θ_t), and weld toe radius (ρ) are shown in Figure 10.1.

Ninh Nguyen and Wahab (1995) investigated the effect of a toe radius on fatigue lives of butt-weld joints while the toe radius was increased from 0.2 mm to 2.5 mm, but other weld geometric parameters (weld bead flank angle, plate edge preparation angle, plate thickness, and tip radius of undercut at weld toe) were not changed. The fatigue life increased whereas the toe radius increased. The same trends were reported by Lawrence (1973), Lawrence and Munse (1973), Gurney (1979), and Maddox (1991).

Ninh Nguyen and Wahab (1995) also reported the effect of a flank angle on fatigue lives of butt-weld joints. The variations of the flank angles were from 0° to 60° but other weld geometric parameters were kept constant. While the flank angle decreased, the fatigue life of the butt-weld joints increased. However, the increment of the fatigue life was not significant when the flank angle decreased between 60° and 20°. Similar observations were also reported by Lawrence (1973), Lawrence and Munse (1973), Gurney (1979), and Maddox (1991).

The plate thickness effect on fatigue lives of butt-weld joints was reported by several researchers (Gurney, 1979; Parker, 1981; Newman & Raju, 1981; Nelson, 1982; Berge, 1985; Foth et al., 1986; Yee & Burns, 1988; Niu & Glinka, 1989;

Ohta et al., 1990; Maddox, 1991; Ninh Nguyen & Wahab, 1995). They found that the fatigue life of welded joints decreased while the plate thickness increased.

Ninh Nguyen and Wahab (1995) observed this phenomenon for their specimens varied from 9 to 32 mm. The decrement of the fatigue life was noticeable when the plate thickness increased from 9 to 32 mm. However, the fatigue life was not affected by the plate thickness changes from 20 mm to 9 mm. On the contrary, thin steels used in the automotive industry showed that fatigue life increased as sheet thickness increased from 1.6 to 3.4 mm in single lap-shear specimens (Bonnen et al., 2009; Kang et al., 2011a,b).

Many researchers (Gurney, 1979; Berge, 1985; Branco et al., 1985; Ferreira & Branco, 1989, 1991) have found that the fatigue life of welded joints also depends on the attachment thickness, the main plate thickness, and the weld toe radius of curvature. Branco et al. (1985) investigated the effect of the weld geometry parameters on the stress intensity factor that is directly related to the fatigue life of a welded joint. They found that the stress intensity factor increased as the weld angle increased.

The same trends were observed for the attachment thickness and main plate thickness in T-joint specimens. Ferreira and Branco (1989) showed that the fatigue life increased as the toe radius decreased. However, this effect was not noticeable when the main plate thickness was less than 6 mm.

The effect of the edge preparation angle on the fatigue life of butt-weld joints was investigated by Ninh Nguyen and Wahab (1995). The angle was reduced from 90° to 45° with keeping other geometry parameters constant. It showed that when the edge preparation angle was smaller, the fatigue life of the weld increased. However, the variation was insignificant and ignorable. They also reported that the fatigue life of butt joints increased as the radius of the undercut at weld toe decreased. The order of weld geometry parameters influencing fatigue life of butt-weld joints was the flank angle, weld toe radius, plate thickness, tip radius of undercut, and edge preparation angle.

Caccese et al. (2006) reported that the effect of weld profile on fatigue performance of cruciform specimens fabricated with laser welding. The specimens with the concave round fillet produced better fatigue characteristics than those with the straight fillet.

On the other hand, many researchers (McGlone & Chadwick, 1978; Doherty et al., 1978; Yang et al., 1993; Huissoon et al., 1994; Maul et al., 1996; Chandel et al., 1997; Kim et al., 2003) worked on the welding process control to obtain better and consistent welding geometry that could result in better fatigue performance of the welded joint. The weld geometry is directly related to bead height, width, and penetration. Thus numerous researchers focused on the relationships with bead dimensions and welding process control variables.

Kim et al. (2003) conducted a study on the relationship between welding process variables and bead penetration for robotic Gas Metal Arc Welding (GMAW). They found that the bead penetration increased as the welding current, welding voltage, and welding angle increased. However, the bead penetration decreased as welding speed increased.

Huissoon et al. (1994) also studied welding process variables for robotic GMAW. Voltage, wire feed rate, travel speed of the torch, and the contact tip to workpiece distance were controlled to obtained the optimum weld width and the throat thickness.

Chandel et al. (1997) investigated the effect of increasing deposition rate on the bead geometry of SAW. The deposition rate increased with the electrode negative but bead penetration and bead width decreased, which may result in lack of fusion at the welded joint. In the same way with polarity, smaller electrode diameter increased the deposition rate but produced unfavorable bead geometry.

In addition to the previously mentioned geometric welding parameters, the complications of fatigue damage assessment of the welded joints include (1) the inhomogeneous material due to the added filler to the base material; (2) the welding defects and imperfections such as cracks, pores, cavities, undercut, inadequate penetration, and such; and (3) welding residual stresses induced from the rapid cooling process and distortions.

Fatigue Life Prediction Methods

Fatigue life of seam-welded joints is generally influenced by weld geometry, service loading history, and material properties. Thus, fatigue life calculation methods should be developed to account for those influencing factors. This section introduces various prediction methods for the fatigue life of seam-welded joints, including nominal stress approaches, structural stress approaches, and notch pseudo stress approaches.

Nominal Stress Approaches

Traditionally, the fatigue life of welded joints was assessed with the nominal stress-based S-N curves generated from fatigue tests of welded specimens for different weld notch classifications. The S-N curves are obtained from extensive experimental data of welded steel bridges, with life varying from 105 to 107 cycles. The complex weld geometries are placed in groups having similar fatigue strengths and identified by weld class depending on the weld types and loading conditions. It is found that fatigue strength is not sensitive to mean stress or ultimate strength.

Tensile residual stresses from the welding process are already at the yield strength of the material and will be larger than any applied mean stresses. The slope factors for normal stress and shear stress are found to be 3.0 and 5.0, respectively, indicating fatigue of welded joints is governed by crack propagation. It is also assumed that the fatigue strength could be modified due to the influence of the sheet thickness.

Further information on nominal stress approaches can be found from numerous papers (Maddox, 2003), books (Gurney, 1979; Radaj, 1990; Maddox, 1991; Radaj et al., 2006), and design guidelines and codes (ASME boiler and pressure vessel code, 1989; British standards, 1980, 1988, 1991, 1993, 2004; European recommendations and standards, 1985, 2005; German standard (Deutsches Institut Fur Normung), 1984; IIW recommendations, 1982, 1990; Japanese standard, 1995).

These approaches are relatively simple but are limited to disclosing stresses and strains at the critical regions of the welded joint. The manufacturing effects are directly included in the large empirical database for structural steels, but the residual stress effect due to different manufacturing processes is not taken into account. It also appears difficult to determine weld class for complex weld shapes and loadings.

For variable amplitude multiaxial fatigue damage assessment, IIW and Eurocode 3 (IIW recommendations, 1990; Eurocode 3, 2005) recommend the damage values (D_σ and D_τ) due to normal stress and shear stress be calculated separately, using the Palmgren–Miner rule (Palmgren, 1924; Miner, 1945), and then combined using the following interaction equations:

- Proportional loading:

$$D_\sigma + D_\tau \leq 1.0 \qquad\qquad (10.1)$$

- Nonproportional loading:

$$D_\sigma + D_\tau \leq 0.5 \qquad\qquad (10.2)$$

Structural Stress Approaches

In structural stress approaches, the structural stress is determined based on macro-behavior of a structure at the location where the fatigue crack is most likely to initiate and propagate. The structural stress is defined as the nominal stress at the weld toe or root cross-section, which excludes the local geometric (weld toe or root radius) effect in the stress calculation. Therefore, a structural stress is not a true local stress.

In the present structural stress approaches (Femer et al., 1998; Dong, 2001, 2005; Dong & Hong, 2002; Potukutchi et al., 2004; Poutiainen & Marquis, 2006), the structural stress of a welded joint is calculated based on nodal forces and moments obtained from a linear elastic finite element analysis (FEA).

In addition, an S-N curve is generated by fatigue testing fabricated welded laboratory specimens to failure at various structural stress levels. Thus, the fatigue life of a real welded structure can then be calculated by this structural stress at the critical welded joint and the S-N curve from laboratory testing.

Dong's and Femer's structural stress approaches (Dong, 2001a,b; Femer et al., 1998) are popular among all the structural stress approaches because they define a systematic way to calculate structural stresses based on nodal forces and moments extracted from any linear elastic FEA. The two commonly used approaches will be described in the following sections.

Dong's Approach

Dong and his coworkers (Dong, 2001a,b; Dong & Hong, 2002; Dong et al., 2003) proposed a mesh-insensitive structural stress parameter based on the stress intensity factor concept derived on structural stresses along a weld line. This approach has been validated and documented elsewhere (Dong, 2001a,b; Dong & Hong, 2002; Dong et al., 2003).

More recently, this method has been included in the American Society of Mechanical Engineers' update to the Boiler and Pressure Vessel Codes, and is available in Section VIII Division 2 (ASME Boiler and Pressure Vessel Code, 2007). Thus, the mesh-insensitive approach is introduced here, particularly for estimating structural stresses and stress intensities at notches and for generating master S-N curves for fatigue analyses of welded joints.

Figure 10.2 shows the physical model of a welded joint whose characteristics can be described by the partial penetration depth (d), weld roots (locations 1 and 2), weld toe (location 3), and the effective throat (a). It is assumed in the figure that the crack will initiate at the weld root radius (location 1) and propagate along the member thickness (t) direction. So the stress distribution normal to the crack surface is responsible for the crack opening and propagation.

The typical stress distribution in a thickness direction at a critical location under arbitrary loading can be simplified as shown in Figure 10.2. Here the total stress distribution can be decomposed into two components: one for the structural stress distribution without a weld root radius (Figure 10.2(a)) and

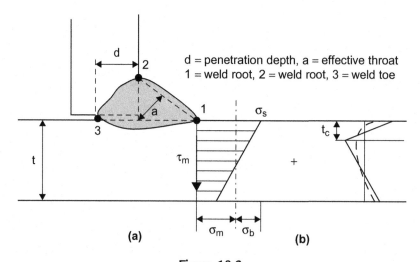

Figure 10.2
Structural stress and notch stress definitions for a fatigue crack in thickness direction at the edge of a weld.

the other for the local notch stress distribution as a result of the root radius effect (Figure 10.2(b)).

As illustrated in Figure 10.2(a), the structural stress (σ_s) normal to the crack surface can be expressed in terms of membrane (σ_m) and bending (σ_b) stresses. In mathematical expression, the structural stress is presented as

$$\sigma_s = \sigma_m + \sigma_b. \tag{10.3}$$

The structural stress definition that follows the elementary structural mechanics theory becomes the far-field stress definition in the linear fracture mechanics context. As shown in Figure 10.2(b), the true notch stress distribution as a result of the consideration of a weld root radius is represented by the dotted line, but, based on the study by Dong et al. (2003), this distribution is assumed to be approximated by a bilinear distribution with a characteristic depth t_c.

The static equilibrium condition holds for the structural stress distribution. Thus, the structural stress distribution along a weld line can be easily calculated by the balanced nodal forces and moments in a linear elastic FEA, using the elementary structural mechanics theory. A weld line is defined as the boundary between the weld and base metal, and generally located along the weld toe or root.

To accurately capture these forces and moments, the weld can be modeled by shell/plate elements, in which the thickness of each element should be selected to reflect the local equivalent stiffness of the joint, as shown in Figure 10.3. In the structural stress calculation procedure, the balanced nodal forces and moments originally solved in a global coordinate system should be converted to a local coordinate system such that the resulting membrane stress and bending stress components will act normal to the weld line.

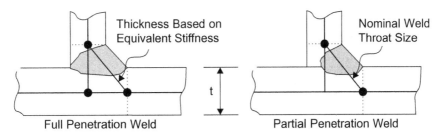

Figure 10.3
Weld representations with plate/shell elements.

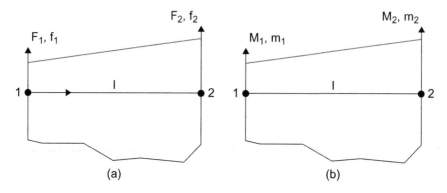

Figure 10.4
Equilibrium-based transformation of nodal forces (a) and moments
to unit line forces and moments (b).

Now the local balanced nodal forces and moments are converted to the line forces and moments using the static equilibrium concept. For example, Figure 10.4 shows the nodal forces/moments and unit weld line force/moment distribution for a first-order plate or shell element with a length of l.

Here F_1 and F_2 are the beginning and the end nodal forces; M_1 and M_2 are the beginning and the end moments; f_1 and f_2 are the unit nodal weld forces at the beginning and end nodes, following a linear weld force distribution function; and m_1 and m_2 are the unit nodal weld moments at the beginning and end nodes, having a linear weld moment distribution relation.

Let $f(x')$ be the linear weld force distribution force as a function of a distance x' from node 1; the following static equilibrium conditions holds:

$$\sum_{i=1}^{2} F_i = \int_0^l f(x')dx' \tag{10.4}$$

$$\sum_{i=1}^{2} F_i x_i' = \int_0^l x'f(x')dx'. \tag{10.5}$$

Solving Equations (10.4) and (10.5) leads to

$$f_1 = \frac{2}{l}(2F_1 - F_2) \tag{10.6}$$

$$f_2 = \frac{2}{l}(2F_2 - F_1).$$
(10.7)

With Equations (10.6) and (10.7), the weld line force distribution can be easily obtained. In similar fashion, m_1 and m_2 can be derived in the following relations:

$$m_1 = \frac{2}{l}(2M_1 - M_2)$$
(10.8)

$$m_2 = \frac{2}{l}(2M_2 - M_1).$$
(10.9)

If a weld is modeled by two linear plate or shell elements with three nodes, as shown in Figure 10.5, it is assumed that the weld line force distribution follows the shape function of each nodal displacement. The static equilibrium condition leads to the following relations between nodal balanced forces and unit nodal weld line forces:

$$F_1 = \frac{f_1 l_1}{3} + \frac{f_2 l_1}{6}$$
(10.10)

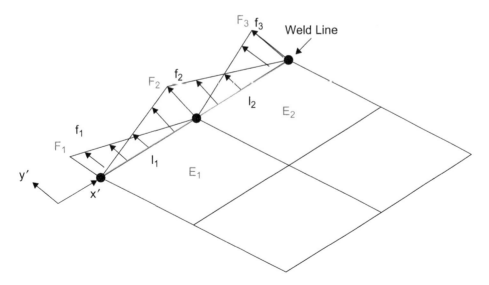

Figure 10.5
Equilibrium-based transformation of nodal forces to line forces.

$$F_2 = \frac{f_1 l_1}{6} + \frac{f_2 l_1}{3} + \frac{f_2 l_2}{3} + \frac{f_3 l_2}{6} \tag{10.11}$$

$$F_3 = \frac{f_3 l_2}{3} + \frac{f_2 l_2}{6}. \tag{10.12}$$

Similarly, if the weld modeled by $(n-1)$ plate or shell elements with n nodes has an open weld end, the relation between nodal forces and unit nodal weld line forces can be expressed as follows:

$$\begin{Bmatrix} F_1 \\ F_2 \\ F_3 \\ \cdot \\ \cdot \\ F_n \end{Bmatrix} = \begin{bmatrix} \frac{l_1}{3} & \frac{l_1}{6} & 0 & 0 & .. & 0 \\ \frac{l_1}{6} & \frac{(l_1+l_2)}{3} & \frac{l_2}{6} & 0 & .. & .. \\ 0 & \frac{l_2}{6} & \frac{(l_2+l_3)}{3} & \frac{l_3}{6} & .. & .. \\ .. & .. & .. & .. & .. & .. \\ .. & .. & .. & .. & .. & .. \\ 0 & 0 & 0 & 0 & \frac{l_{n-1}}{6} & \frac{l_{n-1}}{3} \end{bmatrix} \begin{Bmatrix} f_1 \\ f_2 \\ f_3 \\ \cdot \\ \cdot \\ f_n \end{Bmatrix} \tag{10.13}$$

where

$l_1, l_2, l_3...l_n$ = the element lengths
$f_1, f_2, f_3...f_n$ = the weld line forces
$F_1, F_2, F_3...,F_n$ = nodal balanced forces in local coordinate systems at the nodal points

Or, if the weld has a close weld end, the relation becomes

$$\begin{Bmatrix} F_1 \\ F_2 \\ F_3 \\ \cdot \\ \cdot \\ F_n \end{Bmatrix} = \begin{bmatrix} \frac{l_1}{3} & \frac{l_1}{6} & 0 & 0 & .. & \frac{l_n}{6} \\ \frac{l_1}{6} & \frac{(l_1+l_2)}{3} & \frac{l_2}{6} & 0 & .. & .. \\ 0 & \frac{l_2}{6} & \frac{(l_2+l_3)}{3} & \frac{l_3}{6} & .. & .. \\ .. & .. & .. & .. & .. & .. \\ .. & .. & .. & .. & .. & .. \\ \frac{l_n}{6} & 0 & 0 & 0 & \frac{l_{n-1}}{6} & \frac{(l_{n-1}+l_n)}{3} \end{bmatrix} \begin{Bmatrix} f_1 \\ f_2 \\ f_3 \\ \cdot \\ \cdot \\ f_n \end{Bmatrix}. \tag{10.14}$$

Unit weld line forces can be obtained by inverting the preceding matrix form of the simultaneous equations as Equations (10.13) or (10.14), depending on the weld end description. The similar expression can be developed for unit weld line moment equations. Then, the structural stress at each nodal point ($\sigma_{s,i}$) is calculated with the obtained unit nodal line forces and moments as follows:

$$\sigma_{s,i} = \sigma_{m,i} + \sigma_{b,i} = \frac{f_i}{t} \pm \frac{6m_i}{t^2} \tag{10.15}$$

where

$\sigma_{m,i} = $ the membrane stress at nodal i
$\sigma_{b,i} = $ the bending stress at nodal i
$f_i = $ the unit weld line force at node i
$m_i = $ the unit weld line moment at node i

For local notch stress calculation, Dong et al. (2003) introduced a small notch radius at the weld root to avoid any stress singularity and they assumed the self-equilibrium condition holds for the local notch stress distribution, which can be estimated by a bilinear distribution with a characteristic depth of $t_c = 0.1t$.

For illustration, the bilinear notch stress distribution induced by a weld root radius under unit-nodal weld-line forces can be considered as two linear stress distributions in regions 1 and 2 along the member thickness direction, as illustrated in Figure 10.6(a), where $\sigma_1^{(1)}$ and $\sigma_2^{(1)}$ are the local stresses at points 1 and 2 in region 1; and $\sigma_2^{(2)}$ and $\sigma_3^{(2)}$ are the local stresses at points 2 and 3 in region 2.

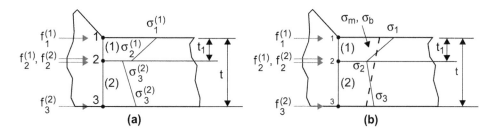

Figure 10.6
(a) Static equilibrium condition to the balanced weld line forces and the two linear stress distributions in regions 1 and 2; (b) static equilibrium condition to the balanced nodal forces and modified bilinear stress distributions.

The static equilibrium state leads to the following relations between the nodal weld line forces and the local stresses:

$$f_1^{(1)} = \frac{t_c}{6}\left(2\sigma_1^{(1)} + \sigma_2^{(1)}\right)$$

(10.16)

$$f_2^{(1)} = \frac{t_c}{6}\left(\sigma_1^{(1)} + 2\sigma_2^{(1)}\right)$$

(10.17)

$$f_2^{(2)} = \frac{t - t_c}{6}\left(2\sigma_2^{(2)} + \sigma_3^{(2)}\right)$$

(10.18)

$$f_3^{(2)} = \frac{t - t_c}{6}\left(\sigma_2^{(2)} + 2\sigma_3^{(2)}\right).$$

(10.19)

Figure 10.6(b) shows the bilinear stress distribution with the stress compatibility condition at point 2, where σ_1, σ_2, and σ_3 are the local stresses points 1, 2, and 3, respectively. The static equilibrium state will result in the earlier similar equations in terms of σ_1, σ_2, and σ_3.

Thus, by enforcing the static equilibrium condition and continuity condition at point 2, the following equations are obtained:

$$2\sigma_1^{(1)} + \sigma_2^{(1)} = 2\sigma_1 + \sigma_2$$

(10.20)

$$t_c\left(2\sigma_2^{(1)} + \sigma_1^{(1)}\right) + (t - t_c)\left(2\sigma_2^{(2)} + \sigma_3^{(2)}\right) = t_c(2\sigma_2 + \sigma_1)$$
$$+ (t - t_c)(2\sigma_2 + \sigma_3)$$

(10.21)

$$2\sigma_3^{(2)} + \sigma_2^{(2)} = 2\sigma_3 + \sigma_2.$$

(10.22)

The three unknowns (σ_1, σ_2, and σ_3) can be solved with the three equations as

$$\sigma_1 = \frac{1}{2}\left(2\sigma_1^{(1)} + \sigma_2^{(1)} - \sigma_2^{(2)}\right) - \frac{t_c}{2t}\left(\sigma_2^{(1)} - \sigma_2^{(2)}\right)$$

(10.23)

$$\sigma_2 = \sigma_2^{(2)} + \frac{t_c}{t}\left(\sigma_2^{(1)} - \sigma_2^{(2)}\right)$$

(10.24)

$$\sigma_3 = \sigma_3^{(2)} + \frac{t_c}{2t}\left(\sigma_2^{(2)} - \sigma_2^{(1)}\right).$$

(10.25)

In regions 1 and 2, the local notch membrane and bending stresses due to the introduction of a weld root radius are

$$\sigma_m^{(1)} = \frac{\sigma_1 + \sigma_2}{2} \tag{10.26}$$

$$\sigma_b^{(1)} = \frac{\sigma_1 - \sigma_2}{2} \tag{10.27}$$

$$\sigma_m^{(2)} = \frac{\sigma_2 + \sigma_3}{2} \tag{10.28}$$

$$\sigma_b^{(2)} = \frac{\sigma_2 - \sigma_3}{2}. \tag{10.29}$$

In contrast to the local notch stress calculation in this particular weld modeling example, the structural stresses can be directly calculated by the following static equilibrium condition:

$$\sigma_m = \frac{1}{t} \left(f_1^{(1)} + f_2^{(1)} + f_2^{(2)} + f_3^{(2)} \right) \tag{10.30}$$

$$\sigma_b = \frac{3}{t^2} \left(f_1^{(1)} \cdot t + \left(f_2^{(1)} + f_2^{(2)} \right) \cdot (t - 2t_c) + f_3^{(2)} \cdot t \right). \tag{10.31}$$

Since crack propagation dominates fatigue lives of welded joints, the stress intensity factor in the linear elastic fracture mechanics approach is a preferable damage parameter to describe the complex stress state at a crack tip and to relate to fatigue lives by the Paris crack growth law. The stress intensity factor is a function of crack size, far-field stress state, and geometry. For most of the welded joints, a close-form solution for the stress intensity factor is not available.

Fortunately, the structural stress definition is consistent with the far-field stress definition in the linear elastic fracture mechanics. Thus, the structural stress calculated from a complicated welded joint under arbitrary loading is analogous to the far-field stress in a simple fracture specimen, where the loading and geometric effects are captured in the form of membrane and bending stresses. As a result, the stress intensity factor for any welded joint can be estimated by using the existing stress intensity factor solution for a simple fracture mechanics specimen under membrane tension and bending.

Consider a two-dimensional single-edge-notched (SEN) fracture specimen under combined membrane tension and bending, as shown in Figure 10.7. The stress

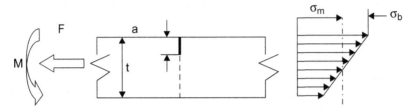

Figure 10.7
Two-dimensional single-edge-notched fracture specimen under combined
membrane tension and bending.

intensity factor (K_n) for an edge crack under structural (far-field) stresses, as
presented by Tada et al. (2000), is

$$K_n = \sqrt{t}[(\sigma_m + \sigma_b)f_m - \sigma_b(f_m - f_b)] = \sqrt{t}\sigma_s[f_m - r(f_m - f_b)]. \qquad (10.32)$$

Here r is the bending ratio $= \sigma_b/\sigma_s$; f_m and f_b are the membrane compliance
function and bending compliance function, depending on the ratio of crack propagation length to the member thickness (a/t), defined as follows:

$$f_m\left(\frac{a}{t}\right) = \left[0.752 + 2.02\frac{a}{t} + 0.37\left(1 - \sin\frac{\pi a}{2t}\right)^3\right]\frac{\sqrt{2\tan\dfrac{\pi a}{2t}}}{\cos\dfrac{\pi a}{2t}} \qquad (10.33)$$

$$f_b\left(\frac{a}{t}\right) = \left[0.923 + 0.199\left(1 - \sin\frac{\pi a}{2t}\right)^4\right]\frac{\sqrt{2\tan\dfrac{\pi a}{2t}}}{\cos\dfrac{\pi a}{2t}}. \qquad (10.34)$$

Equation (10.32), which is the stress intensity factor for a long edge crack as a
result of structural (far-field) stresses, has been extended by Dong and coworkers
(2003) to determine the stress intensity factor for a short edge crack due to local
notch membrane and bending stresses in region 1 as

$$K_{a<t_c} = \sqrt{t}\left[\left(\sigma_m^{(1)} + \sigma_b^{(1)}\right)f_m - \sigma_b^{(1)}(f_m - f_b)\right]. \qquad (10.35)$$

Here the characteristic depth, t_c, is a transition crack length from a small crack
regime to a long crack regime. Assuming there is a relation between the stress

intensities due to structural stresses and local notch stresses, a stress intensity magnification factor (M_{kn}) is then introduced as

$$M_{kn} = \frac{K_{a<t_c}}{K_n}.$$ (10.36)

Dong and his coworkers (2003) reported that M_{kn} approaches to unity as a crack size is close to 0.1t for all of the cases studied. Consequently, $t_c = 0.1t$ can be considered as a characteristic depth beyond which the notch stress effect becomes negligible. They also found that, for an edge crack in a T fillet weld, the stress intensity factor solution is due to the structural stresses; Equation (10.32) provides an accurate estimation for a crack size larger than 0.1t. As the crack size becomes smaller than 0.1t, the local notch stresses at the weld toe or root should be considered and will introduce an elevated stress intensity factor.

Figure 10.8 schematically illustrates the two distinguished patterns of the stress intensity factors for a crack size ratio a/t varying from a/t < 0.1 to a/t > 0.1. Thus, the so-called "two-state growth model" was proposed, where the stress intensity factor ($K_{a/t\leq0.1}$) dominated by the local notch stresses and stress intensity factor ($K_{a/t>0.1}$) controlled by the structural stresses can be used to characterize the small crack and the long crack regimes, respectively.

Based on the two-stage growth model, the Paris crack growth law is then modified accordingly as follows:

$$\frac{da}{dN} = C[f_1(\Delta K_{a/t\leq0.1}) \times f_2(\Delta K_{a/t>0.1})].$$ (10.37)

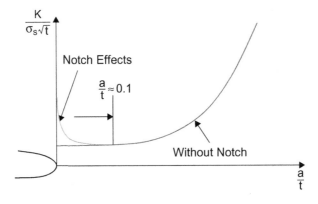

Figure 10.8
Schematics of the stress intensity factor versus the crack size ratio relation.

Here C is a fitting material coefficient; $\Delta K_{a/t \leq 0.1}$ and $\Delta K_{a/t > 0.1}$ are the stress intensity factor ranges in the short and long crack growth regimes; and $f_1()$ and $f_2()$ are the functional expressions. For the long crack growth regime, $f_2()$ follows the Paris crack growth model as

$$f_2(\Delta K_{a/t>0.1}) = (\Delta K_n)^m \tag{10.38}$$

where

m = the crack growth rate exponent for the long crack growth regime

By the introduction of M_{kn} in the short crack growth regime, $f_1()$ is assumed to have the similar power law express as $f_2()$, and can be expressed by

$$f_1(\Delta K_{a/t \leq 0.1}) = (M_{kn}\Delta K_n)^n \tag{10.39}$$

where

n = the crack growth rate exponent for the first stage of the crack growth

Finally, by substituting Equations (10.38) and (10.39), Equation (10.37) can be written

$$\frac{da}{dN} = C[(M_{kn}\Delta K_n)^n \times (\Delta K_n)^m]. \tag{10.40}$$

To determine the fatigue life N, the two-stage crack growth equation (Equation 10.40) can be integrated as

$$N = \int_{a_i}^{a=a_f} \frac{da}{C(M_{kn})^n (\Delta K_n)^m} = \int_{a_i/t=0.01}^{a_f/t=1} \frac{td \cdot \left(\frac{a}{t}\right)}{C(M_{kn})^n (\Delta K_n)^m}$$

$$= \frac{1}{C} \cdot t^{1-\frac{m}{2}} \cdot (\Delta \sigma_S)^{-m} I(r) \tag{10.41}$$

where

$$I(r) = \int_{a_i/t=0.01}^{a_f/t=1} \frac{d \cdot \left(\frac{a}{t}\right)}{(M_{kn})^n \left[f_m\left(\frac{a}{t}\right) - r\left(f_m\left(\frac{a}{t}\right) - f_b\left(\frac{a}{t}\right)\right)\right]^m}. \tag{10.42}$$

$I(r)$ is a dimensionless function of r, which takes into account the loading mode effect. If the nominal weld quality is not well defined, it is recommended

(Dong, 2005) that the initial crack-life defect size ratio ($a_i/t = 0.01$ or less) be used to calculate I(r) for S-N data correlations.

The I(r) function depends on the test loading types such as load control test and displacement control test. For the load control application, the estimated numerical solution of I(r) is given (Dong et al., 2004) as

$$I(r) = 0.294r^2 + 0.846r + 24.815. \tag{10.43}$$

Then, Equation (10.41) can be rewritten in terms of N as

$$\Delta\sigma_S = C^{-\frac{1}{m}} \cdot t^{\frac{2-m}{2m}} \cdot I(r)^{\frac{1}{m}} \cdot N^{-\frac{1}{m}}. \tag{10.44}$$

or, a typical S-N curve is expressed as

$$\Delta S_S = C^{-\frac{1}{m}} \cdot N^{-\frac{1}{m}} \tag{10.45}$$

where

$$\Delta S_s = \frac{\Delta\sigma_s}{t^{\frac{2-m}{2m}} I(r)^{\frac{1}{m}}}. \tag{10.46}$$

ΔS_s is the so-called equivalent structural stress range used for life predictions of welded joints. Figure 10.9 (Kong, 2011) shows that when the equivalent structural stress range is used, all of the S-N data for various joint types made of aluminum alloys, loading modes, and thickness are collapsed into a narrow band.

Fermer's Approach

Fermer and coworkers (1998) proposed a fatigue damage parameter based on the concept of maximum structural stress amplitude to predict fatigue lives of welded joints from the baseline S-N curves generated by testing fabricated laboratory welded specimens.

The structural stress calculation in this approach is exactly identical to Dong's approach where the structural stresses are calculated from balanced nodal forces and moments in a linear elastic FEA, using the elementary structural mechanics theory. Thus, it is supposed to be another mesh-insensitive approach. But, in

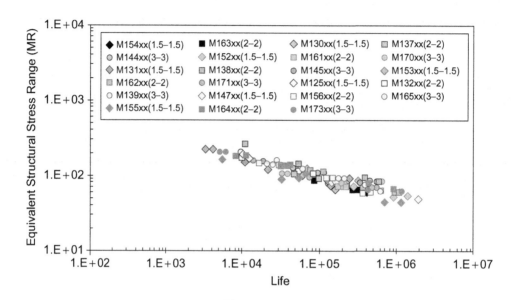

Figure 10.9

Correlation of existing S-N data for various joint types, loading modes, and plate thickness. *Source: Personal communication from Dr. J. K. Kong, February 7, 2011.*

this approach, welded joints are required to be molded by plate or shell elements with the following mesh rules:

- The 4-node shell elements are used for all the thin sheet structures and the welds.

- The shell elements present the mean surfaces of the thin sheet structures.

- There is an offset by $t/2$ at the nodes of the shell elements for weld along the weld line.

- The thickness of shell elements for a weld is the effective throat size.

- The proper element size is about 10 mm.

- The local weld toe and root radii need not be modeled.

These rules are enforced to ensure local stiffness of a welded joint is captured with the same modeling techniques used to generate the baseline S-N data in laboratory testing. This approach has been successfully employed in welded thin sheet structures that can be effectively modeled by plate or shell elements. Moreover,

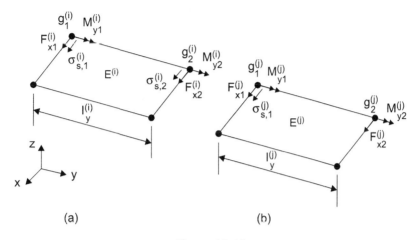

Figure 10.10
Forces (a) and moments (b) at each nodal point on shell elements.
Source: Adapted from Fermer et al. (1998).

the chosen maximum structural stress amplitude, as the fatigue damage parameter in Fermer's approach, is different from the equivalent structural stress range derived in Dong's approach.

Structural stresses can be defined by the following procedures. First, a local coordinate is defined (see Figure 10.10), where the weld line is located along the local y-axis. The unit line force, $f_x^{(i)}(y)$, and unit line moment, $m_y^{(i)}(y)$, in an element $(E^{(i)})$ can be converted from nodal forces, $F_{x1}^{(i)}(y)$ and $F_{x2}^{(i)}(y)$, and moments $M_{y1}^{(i)}(y)$ and $M_{y2}^{(i)}(y)$ that are obtained from a linear elastic FEA. The variation of weld line nodal force and moment is assumed to be linear along the weld line.

Then the following static equilibrium conditions hold:

$$F_{x1}^{(i)} + F_{x2}^{(i)} = \int_0^{l_y^{(i)}} f_x^{(i)}(y)dy = \frac{l_y^{(i)}}{2}\left(f_{x1}^{(i)} + f_{x2}^{(i)}\right) \tag{10.47}$$

$$l_y^{(i)}F_{x2}^{(i)} = \int_0^{l_y^{(i)}} yf_x^{(i)}(y)dy = \frac{\left(l_y^{(i)}\right)^2}{6}\left(f_{x1}^{(i)} + 2f_{x2}^{(i)}\right). \tag{10.48}$$

Here $l_y^{(i)}$ is the element edge length between the two nodal points ($g_1^{(i)}$ and $g_2^{(i)}$).

Solving Equations (10.47) and (10.48) leads to

$$f_{x1}^{(i)} = \frac{2}{l_y^{(i)}} \left(2F_{x1}^{(1)} - F_{x2}^{(i)} \right) \tag{10.49}$$

$$f_{x2}^{(i)} = \frac{2}{l_y^{(i)}} \left(2F_{x2}^{(1)} - F_{x1}^{(i)} \right). \tag{10.50}$$

Following the similar fashion, $m_{y1}^{(i)}$ and $m_{y2}^{(i)}$ can be derived in the following relations:

$$m_{y1}^{(i)} = \frac{2}{l_y^{(i)}} \left(2M_{y1}^{(i)} - M_{y2}^{(i)} \right) \tag{10.51}$$

$$m_{y2}^{(i)} = \frac{2}{l_y^{(i)}} \left(2M_{y2}^{(i)} - M_{y1}^{(i)} \right). \tag{10.52}$$

With Equations (10.49) through (10.52), the weld line force distribution can be easily obtained as follows:

$$f_x^{(i)}(y) = \frac{2}{l_y^{(i)}} \left(F_{x1}^{(i)} \left(2 - \frac{3y}{l_y^{(i)}} \right) + F_{x2}^{(i)} \left(\frac{3y}{l_y^{(i)}} - 1 \right) \right) \tag{10.53}$$

$$m_y^{(i)}(y) = \frac{2}{l_y^{(i)}} \left(M_{y1}^{(i)} \left(2 - \frac{3y}{l_y^{(i)}} \right) + M_{y2}^{(i)} \left(\frac{3y}{l_y^{(i)}} - 1 \right) \right). \tag{10.54}$$

The structural stress, $\sigma_s^{(i)}(y)$, which is normal on the crack surface along a weld line direction y in an element $(E^{(i)})$ can be expressed in terms of membrane stress, $\sigma_m^{(i)}(y)$, and bending stress, $\sigma_b^{(i)}(y)$. The structural stress is calculated by the elementary structural mechanics theory as:

$$\sigma_s^{(i)}(y) = \sigma_m^{(i)}(y) + \sigma_b^{(i)}(y) = \frac{f_x^{(i)}(y)}{t} \pm \frac{6m_y^{(i)}(y)}{t^2}. \tag{10.55}$$

For a continuous weld, using the nodal forces and moments, the grid structural stresses on the top surface in an element $(E^{(i)})$ can be calculated as follows:

$$\sigma_s^{(i)}(0) = \sigma_{s,1} = \frac{2 \left(2F_{x1}^{(1)} - F_{x2}^{(1)} \right)}{l_y^{(1)} t} + \frac{12 \left(2M_{y1}^{(1)} - M_{y2}^{(1)} \right)}{l_y^{(1)} t^2} \tag{10.56}$$

$$\sigma_s^{(i)}(l_y) = \sigma_{s,2} = \frac{2\left(2F_{x2}^{(1)} - F_{x1}^{(1)}\right)}{l_y^{(1)}t} + \frac{12\left(2M_{y2}^{(1)} - M_{y1}^{(1)}\right)}{l_y^{(1)}t^2}. \qquad (10.57)$$

This means that each nodal point inside a continuous weld will have two structural stresses. However, if a nodal point is located at a weld start, stop, or weld corner, as shown in Figure 10.11, it will have four structural stresses as calculated according to the nodal forces and nodal moments in the three surrounding elements. The maximum value of the structural stress amplitudes at each nodal point will be used for fatigue-life assessments of the weld.

Next, information required to evaluate fatigue life of welded joints is the maximum structural stress amplitude versus number of cycles to failure ($\sigma_{s,a} - N$) curve, which can be determined by fatigue-testing fabricated specimens at various

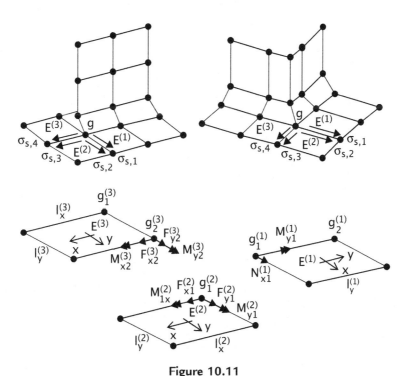

Figure 10.11
Moments and forces for the structural stress calculation at a grid point
located at weld start/stop and sharp corner.
Source: Adapted from Fermer et al. (1998).

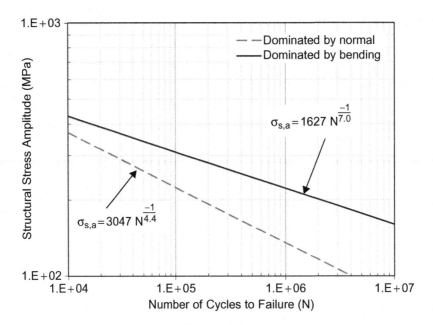

Figure 10.12
Structural stress amplitude versus number of cycles to failure.
Source: Adapted from Fermer et al. (1998).

load amplitude levels with the load ratio of $R = -1$. Applied load amplitudes can be converted to maximum structural stress amplitudes by using the procedures explained in the previous section.

Figure 10.12 shows the $\sigma_{s,a}$-N curves with the best-fit equations against experimental data. It is found that the slope of the $\sigma_{s,a}$-N curve is sensitive to the bending ratio (r) defined as the ratio of bending to structural stresses. The flatter $\sigma_{\perp,a}$-N curve (the upper one) was generated based on the data where the bending stress is dominant ($0.5 < r \leq 1.0$). On the other hand, the stiffer curve (the lower one) was obtained from the data where the membrane stress is dominant ($0 \leq r \leq 0.5$).

When a mean structural stress ($\sigma_{s,m}$) exists, the equivalent fully reversed stress amplitude ($\sigma_{s,a,R=-1}$) at the load ratio of $R = -1$ could be modified using the following equations:

- For $-\infty \leq R \leq 0$,

$$\sigma_{s,a,R=-1} = \sigma_{s,a} + M_{ms,1}\sigma_{s,m}. \tag{10.58}$$

- For $0 < R < 0.5$,

$$\sigma_{s,a,R=-1} = (1 + M_{ms,1}) \frac{\sigma_{s,a} + M_{ms,2}\sigma_{s,m}}{1 + M_2}. \tag{10.59}$$

Here $M_{ms,1}$ $(= 0.25)$ and $M_{ms,2}$ $(= 0.097)$ are the mean stress sensitivity factors defined in Haigh's diagram.

Notch Pseudo Stress or RXMS Approach

This approach considers the local pseudo stress in a stress concentration area as a fatigue damage parameter and requires fine meshes in the local weld toes and/or roots geometry to capture the accurate notch stresses from a linear elastic FEA. The pseudo stress analysis could be virtually impossible in most practical situations used to analyze large structures with all of the details required. However, this approach could be feasible if the substructure modeling technique is used to extract pseudo stresses in a desired location from a small but detailed local model with input and boundary conditions that are derived from the analysis results of its global model with coarse meshes.

The substructure modeling technique is used to model the nominal weld geometry with solid finite elements, where a specific notch radius is introduced at every weld/base material intersection. Depending on the member thickness of interest, the nominal notch radius ρ is recommended as follows:

$$\rho = 0.3\,\mathrm{mm}$$
$$0.5\,\mathrm{mm} \leq t \leq 6\,\mathrm{mm}$$

for

$$\rho = 0.05\,\mathrm{mm}$$

for

$$6.0\,\mathrm{mm} \leq t \leq 20\,\mathrm{mm}$$
$$\rho = 1.0\,\mathrm{mm}$$

for

$$20\,\mathrm{mm} \leq t \leq 100\,\mathrm{mm}.$$

This technique to assign various notch radii will allow better adoption of S-N curves for the size effect. This detailed substructure model can be created by using any automatic mesh generation software.

The substructure analysis technique to extract notch pseudo stresses involves two finite element analyses. First, a user will start to set up the cut boundary for the substructure model and analyze the global model for the cut boundary displacements. Second, the user will apply the cut boundary displacements and other relevant boundary conditions to the local detailed model for further stress analysis.

The pseudo stress life approach is employed for fatigue life predictions of welded joints. The synthetic pseudo stress life curve can be constructed by a given slope factor k and a median pseudo endurance limit (σ_a^e) at 2×10^6 cycles. Luckily, it has been found that the slope factor $(k = 3.0)$ is constant for seam-welded joints made of steels and aluminum alloys due to the fact that the crack propagation dominates fatigue lives of welded joints.

The pseudo endurance limit can be obtained by fatigue testing welded laboratory specimens with the staircase test method for the endurance load amplitude at 2×10^6 cycles. A linear elastic FEA is required to calculate the median pseudo endurance limit by applying the median endurance load. This is the way to covert the experimentally determined endurance load amplitude to the pseudo endurance limit. It is worth mentioning that residual stresses due to weld solidifications and distortions have been implicitly taken into account in the calculated pseudo endurance limit.

For some simple fabricated welded specimens used in laboratory testing, where nominal stresses can be easily defined, the pseudo endurance limit can be simply calculated by the product of the component endurance limit and the elastic stress concentration factor. Fortunately, many elastic stress concentration formulas in welds have been derived and can be found in the literature (Lida & Uemura, 1996; Monahan, 1995).

Therefore, the detailed FE substructure modeling and analysis procedures could be replaced by using these formulas for pseudo stress calculation. For example, the elastic stress concentration factors (K_{ta} and K_{tb}) of a fillet weld subjected to axial load and bending load, respectively, derived by Monahan (1995), are expressed in the following equations:

$$K_{ta} = 0.388 \left(\frac{\pi\theta_t}{180}\right)^{0.37} \left(\frac{t}{\rho}\right)^{0.454} \tag{10.60}$$

$$K_{tb} = 0.512 \left(\frac{\pi\theta_t}{180} \right)^{0.572} \left(\frac{t}{\rho} \right)^{0.469}. \tag{10.61}$$

Finally, the notch pseudo stress approach is also known as the RXMS approach; RX refers to the nominal notch <u>R</u>adius of <u>X</u> mm introduced in the substructure modeling technique and MS refers to the back-calculated <u>M</u>ean and <u>S</u>tandard deviation or <u>S</u>catter of the pseudo endurance limit.

Examples

Correlation studies between analytical solutions and experimental fatigue data of a seam-weld structure are discussed in this section. Two commercial fatigue analysis modules based on structural stress approaches were used to predict fatigue lives of the welded joints subjected to prescribed loading. Dong's structural stress approach has been implemented into a module called VERITY® (Battelle Memorial Institute) or FE-Safe® (Safe Technology Inc.), while Fermer's structural stress approach has been coded into the other module called DesignLife® (HBM United Kingdom Limited).

Fatigue testing on welded perch-mount specimens made of DP600 and HSLA steels has been conducted by Bonnen et al. (2009). As shown in Figure 10.13, the geometry and dimensions of perch-mount specimens are expressed in millimeters, and the specimen thickness (t) is 3.4 mm. The specimen is subjected to either cyclic normal or cyclic shear loading with respect to the weld line, as shown in Figure 10.14.

The fatigue test results in terms of load range in Newtons versus number of cycles to failure are shown in Figure 10.15, where two load life curves with the slope factors $k = 3.4 \left(\approx \frac{1}{0.292} \right)$ and $5.1 \left(\approx \frac{1}{0.198} \right)$ for cyclic normal and cyclic shear loads, respectively, can be seen.

To obtain nodal forces and moments at critical locations along the weld toe line, any linear elastic FE analysis with quasistatic loading conditions can be employed for the welded structure subjected to cyclic loads. For example, MSC-NASTRAN® (MSC-Software Corporation) is chosen here for illustration. As shown in Figure 10.14, the single FE welded structure for both the normal and shear load cases can be analyzed by using the SUBCAE option in MSC-NASTRAN.

After this analysis is done, the nodal forces and moments along the weld line can be extracted for structural stress calculations, and then used as an input to the two

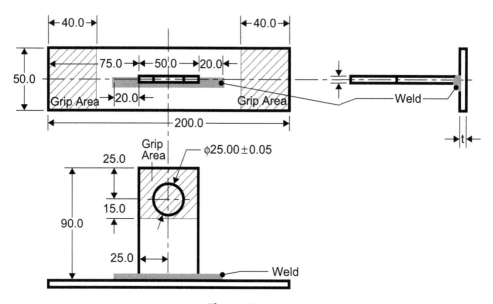

Figure 10.13
Geometry and dimension of perch-mount specimens.
Source: Adapted from Bonnen et al. (2009).

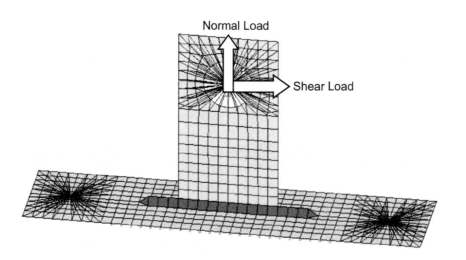

Figure 10.14
FE model for perch-mount specimen and loading directions.

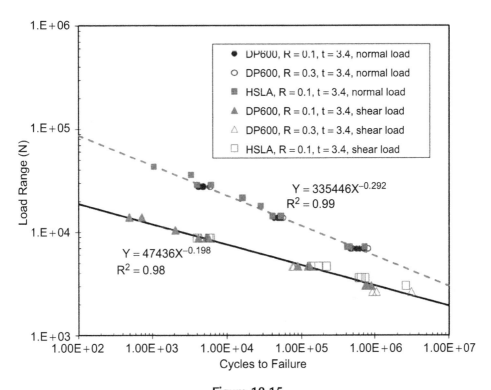

Figure 10.15

Fatigue test results of the perch-mount specimen subjected to normal and shear load.
Source: Adapted from Kang et al. (2011b).

commercial fatigue analysis modules for fatigue life predictions based on the fatigue properties of welded joints provided in the software's material database.

To examine validity of the two structural stress approaches in life predictions, the structural stress ranges calculated from VERITY and DesignLife due to physically applied load ranges in testing are plotted against the experimental fatigue lives in Figures 10.16 and 10.17, respectively (Kang et al., 2011b). Both the structural stress approaches significantly improve the relationship between number of cycles to failure and structural stress range because all the data points are now consolidated into one single S-N curve.

As shown in this example, both the structural stress approaches are very effective for the simple experimental test results. However, a user should be cautioned when

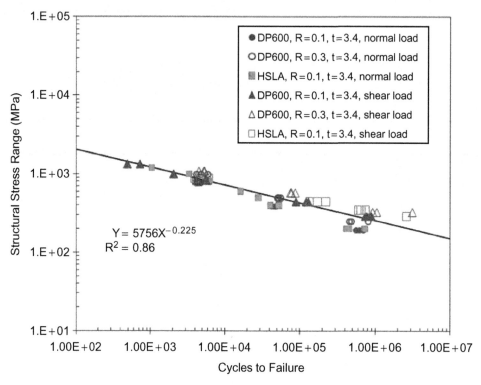

Figure 10.16
Structural stress range obtained from Dong's approach versus cycles
to failure of the perch-mount specimens.
Source: Adapted from Kang et al. (2011b).

using the material database provided in the commercial analysis module for life predictions because the welding and manufacturing processes included in the supplied fatigue properties could be different from those in actual welded joints of interest, resulting in different residual stresses.

Residual stresses from the welding process can have a significant effect on the fatigue behavior of a welded structure. A local tensile residual stress is often induced at the weld toe when the weld solidifies. Additional residual stresses as a result of distortion may also be present. It is a common assumption that the tensile residual stress could reach the yield strength of the parent material, resulting in a detrimental effect on the fatigue strength of the welded joint.

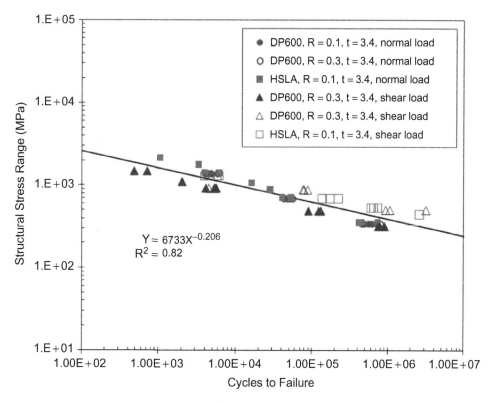

Figure 10.17

Structural stress range obtained from Fermer's approach versus cycles
to failure of the perch-mount specimens.
Source: Adapted from Kang et al. (2011b).

This postulation is valid for most of the welded construction specimens because the residual stress due to distortion could be negligible for these stiff specimens. Thus, welding-induced tensile residual stresses at the weld toe should be relieved by heat treatment to improve the fatigue performance of welded joints (Cheng et al., 2003; Webster & Ezeilo, 2001). All the structural-stress–based or pseudo-stress–based fatigue properties available in the commercial software are calibrated from the experimental fatigue data on these welded construction specimens, where tensile residual stresses have been implicitly taken into account.

However, welding-induced residual stresses are not always in tension as reported by Ohta et al. (1990) and Kang et al. (2008). They observed compressive residual stresses on some thin-walled tubular specimens after a welding process. The existence of compressive residual stress at the weld toe has been proven by the X-ray diffraction method and by fatigue-testing as-welded and heat-treated specimens. Figure 10.18 shows that the heat treatment process to relieve welding residual stresses reduces the fatigue performance of the seam-welded joints.

Fatigue life prediction results for the thin-walled tubular specimens are presented in Figure 10.18 using the two structural stress approaches coded by FAT1 and FAT2. The data indicate that both approaches are comparable to each other, and could overestimate and underestimate the fatigue performance of the welded joints in the low cycle and high cycle fatigue regimes, respectively.

It is thus concluded that the analytical solutions show no correlation with the experimental data. Before the two structural stress approaches are used, you

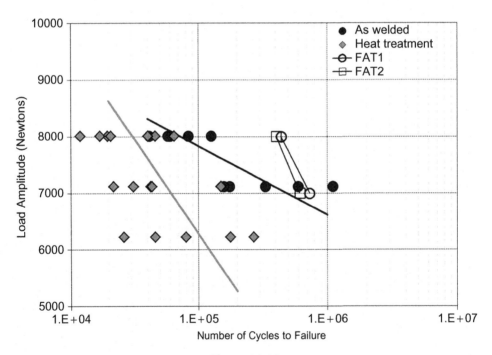

Figure 10.18
Comparison of predicted and experimental fatigue data.
Source: Adapted from Kang et al. (2008).

should understand the local residual stress states on the welded joints of interest and should generate in-house fatigue properties by fatigue-testing some welded samples with the same welding process as real products.

Variable Amplitude Multiaxial Fatigue Analyses

Welding is a common joining technique that has been widely used in many engineering components. These welded components are usually subjected to a loading history, which may be nonproportional and of variable amplitude. Thus, development of a convenient and accurate life prediction technique for welded components under multiaxial random loading is essential to an engineering design and analysis.

Nonproportional cyclic loading on welded components causes equal (Archer, 1987) or more damage (Siljander et al., 1992; Sonsino, 1995) than proportional loading. In nonproportional loading tests for components made of the same material, fatigue lives are decreased as compared to those in proportional loading tests with the same von Mises stress range. This phenomenon can be explained by the slip behavior of the material (Itoh et al., 1995). The change of the principal stress–strain axes due to nonproportional loading increases the interaction between slip systems, which is responsible for additional hardening.

In addition, a typical material such as Type 304 stainless steel (Itoh et al., 1995; Doong et al., 1990) shows more damaging effects than aluminum alloy (Krempl & Lu, 1983; Itoh et al., 1997) due to identical nonproportional loading tests. The degree of damage in a material is highly dependent on the ease on which slip systems interact.

Strong interaction occurs in Type 304 stainless steel and weak interaction in aluminum alloy. Therefore, the fatigue life reduction is strongly connected to additional nonproportional hardening due to both loading history and material, and a robust and reliable life prediction model for welded joints under nonproportional loading should take into account the nonproportional hardening effect.

Life prediction models for welded joints under nonproportional loading have been developed, among which Dong's structural stress-based model is the most popular. It is well known that Dong's mesh-insensitive model is a noble approach derived on the linear elastic fracture mechanics approach for Mode I loading only.

More recently, two separate fatigue damage models (Dong & Hong, 2006; Dong et al., 2010) were proposed for welded joints under nonproportional loading. The Mode II loading effect was accounted for by the definition of in-plan shear structural stress τ_s as

$$\tau_s = \frac{f_y}{t} + \frac{3m_x}{t^2} \tag{10.62}$$

and by a von Mises-like equivalent stress formula in Dong et al. (2010):

$$S_e = \frac{1}{D(\Phi)} \sqrt{\sigma_s^2 + 3\tau_s^2}. \tag{10.63}$$

Here f_y and m_x are the unit weld line in-plane force and torsion on the crack propagation plane. $D(\Phi)$ is the damage parameter for nonproportional hardening due to out-of-phase sinusoidal loads with a phase angle Φ. In this approach, a multiaxial rainflow reversal counting technique as described in Chapter 3 has been used for fatigue life predictions of seam-welded joints under variable amplitude multiaxial loading.

The local stress-based fatigue life prediction models, such as critical plane approaches (Siljander et al., 1992; Backstrom & Marquis, 2001), and the effective equivalent stress amplitude method (Sonsino, 1995) are commonly adopted for use in assessing nonproportional fatigue damage on welded joints, but their uses are very limited. The critical plane approaches using Findley's shear-stress amplitude (Findley, 1959) provide a physical interpretation of the damage process by identifying crack orientations; however, it fails to account for nonproportional hardening.

The effective equivalent stress amplitude method (Sonsino, 1995) has received much attention because not only is it easy to use, but it also gives reasonable physical meaning to account for nonproportional hardening due to material and loading history. However, Sonsino's approach is only valid for out-of-phase sinusoidal loading histories and cannot be applied to general nonproportional loads.

More recently, a variant of the local stress-based fatigue life prediction model was proposed, the equivalent nonproportional stress amplitude approach (Lee et al., 2007). This model is a modified version of Sonsino's effective equivalent stress amplitude model, where the nonproportional hardening effect is taken into account by a nonproportional coefficient for the material dependence of additional hardening as well as a nonproportional factor for the severity of non-proportional loading.

This nonproportional hardening model, which was originally developed by Itoh et al. (1995) in the *equivalent nonproportional strain amplitude* approach, was correlated well with nonproportional fatigue life data under complex nonproportional cyclic strain paths for Type 304 stainless steel. The applicability of this equivalent nonproportional stress amplitude approach has been validated with experimental results for various welded joint configurations due to nonproportional constant amplitude loading. What follows is a brief introduction of this model.

The equivalent nonproportional stress amplitude ($\sigma_{VM,a,NP}$) is defined as follows:

$$\sigma_{VM,a,NP} = \sigma_{VM,a}(1 + \alpha_{NP}f_{NP}). \tag{10.64}$$

Here $\sigma_{VM,a}$ is the equivalent proportional stress amplitude. The term of $(1 + \alpha_{NP}f_{NP})$ accounts for the additional strain hardening observed during nonproportional cyclic loading. α_{NP} is the nonproportional hardening coefficient for the material dependence, and f_{NP} is the nonproportional loading path factor for the severity of loading paths.

In Equation (10.64), $\sigma_{VM,a}$ is derived according to the von Mises hypothesis, but employs the maximum stress amplitude between two arbitrary stress points among all multiple points in a cycle. For the example of a plane stress condition with three stress amplitude components ($\sigma_{x,a}$, $\sigma_{y,a}$, $\tau_{xy,a}$), $\sigma_{VM,a}$ is maximized with respect to time, and defined here to account for the mean stress effect:

$$\sigma_{VM,a} = \max\left\{\sqrt{\sigma_{x,a}^2 + \sigma_{y,a}^2 - \sigma_{x,a}\sigma_{y,a} + 3\alpha_S^2\tau_{xy,a}^2} \times \left(\frac{\sigma_f'}{\sigma_f' - \sigma_{eq,m}}\right)\right\}. \tag{10.65}$$

Here α_S is the sensitivity shear-to-normal stress parameter, σ_f' is the fatigue strength coefficient determined from the best fit of the proportional loading data with a stress ratio $R = -1$, and $\sigma_{eq,m}$ is the equivalent mean stress. Ignoring the effect of torsional mean stress on fatigue lives, $\sigma_{eq,m}$ is calculated in the following equation:

$$\sigma_{eq,m} = \sigma_{x,m} + \sigma_{y,m} \tag{10.66}$$

where

$\sigma_{x,m}$ and $\sigma_{y,m}$ = the mean stress values in x- and y-axes, respectively

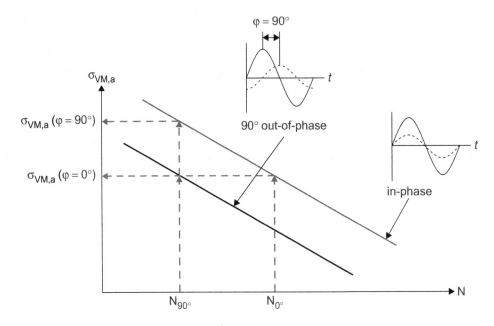

Figure 10.19
Graphical representation of nonproportional hardening due to 90°
out-of-phase loading in S-N curves.

The nonproportional material coefficient (α_{NP}) is related to the additional harden-ing of the materials under 90° out-of-phase loading. Alternatively, this coefficient can be obtained from the von Mises stress amplitude versus life curves of the same material under in-phase and 90° out-of-phase fatigue testing. As shown in Figure 10.19, at a same $\sigma_{VM,a}(\Phi = 0°)$ value, the life ($N_{90°}$) for 90° out-of-phase loading is shorter or more damaging than that ($N_{0°}$) for in-phase loading.

Since the in-phase $\sigma_{VM,a}$-N curve is the baseline S-N curve for life predictions, the higher stress amplitude $\sigma_{VM,a}(\Phi = 90°)$ than $\sigma_{VM,a}(\Phi = 0°)$ is found to produce an equivalent damage or $N_{90°}$ life to the 90° out-of-phase loading. This is assumed to be attributable to the nonproportional strain-hardening phenomenon. Thus, ($\alpha_{NP} + 1$) can be determined by the ratio of $\sigma_{VM,a}(\Phi = 90°)$ to $\sigma_{VM,a}(\Phi = 0°)$.

The nonproportional loading factor (f_{NP}), varying from zero to one, represents the effect of a loading path on nonproportional hardening. In-phase loading generates the value of f_{NP} equal to zero, whereas 90° out-of-phase loading produces the value of f_{NP} equal to one, indicating the most damaging loading condition. As reported by Itoh et al. (1995), this factor is calculated by integrating the contributions of all maximum principal stresses $\left(\vec{\sigma}_{1,max}(t)\right)$ on the plane being perpendicular to the plane of the largest absolute principal stress $(\vec{\sigma}^{\,ref}_{1,max})$.

This factor is mathematically represented as follows:

$$f_{NP} = \frac{C_{NP}}{T\left|\vec{\sigma}^{\,ref}_{1,max}\right|} \int_0^T \left(\left|\sin\xi(t) \times \left|\vec{\sigma}_{1,max}(t)\right|\right|\right) dt \qquad (10.67)$$

where

$\left|\vec{\sigma}_{1,max}(t)\right|$ = the maximum absolute value of the principal stress at time t, depending on the larger magnitude of the maximum and the minimum principal stresses at time t (maximum of $\left|\vec{\sigma}_1(t)\right|$ and $\left|\vec{\sigma}_3(t)\right|$)

$\xi(t)$ = the angle between $\vec{\sigma}^{\,ref}_{1,max}$ and $\vec{\sigma}_{1,max}(t)$, as shown in Figure 10.20

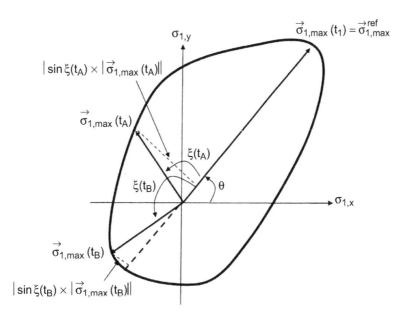

Figure 10.20
Schematic plot of principal stress vectors at various time steps.

Also $\vec{\sigma}_{1,max}(t)$ is oriented at an angle of θ with respect to the x-axis. f_{NP} is normalized to $|\vec{\sigma}_{1,max}^{ref}|$ and T (the time for a cycle). The constant C_{NP} is chosen to make f_{NP} unity under $90°$ out-of-phase loading.

A flowchart on how to apply the equivalent nonproportional stress amplitude approach for the cumulative damage assessment of welded joints under multiaxial variable amplitude loading is presented in Figure 10.21. After the material

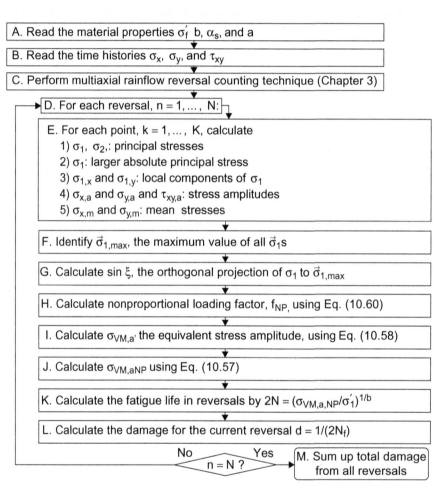

Figure 10.21
Flowchart to apply the equivalent nonproportional stress amplitude model for cumulative damage assessment under variable amplitude multiaxial loading.

properties (i.e., σ'_f, fatigue strength coefficient; b, fatigue strength exponent, α_S, sensitivity shear-to-normal stress parameter; and α_S, nonproportional hardening factor) and the stress time histories are read (steps A and B), the multiaxial rain-flow reversal counting technique, as outlined in Chapter 3, is invoked (step C) to count the reversals of the input data.

For each reversal (step D), the following procedures are then performed to calculate the damage.

1. Point-by-point calculations (step E) are performed.

2. Identify (step F) the maximum absolute principal stress in the reversal.

3. The sin ξ can be calculated for all points (step G).

4. The nonproportional loading factor f_{NP} is then calculated (step H) using the discrete form of Equation (10.60).

5. The maximum equivalent proportional stress amplitude, $\sigma_{VM,a}$, equivalent nonproportional stress amplitude, $\sigma_{VM,a,NP}$, fatigue life in reversals, and damage for the reversal are calculated as shown in steps I, J, K, and L.

6. After the damage is calculated for all reversals, the total damage is linearly summed (step M) using the linear damage rule.

Summary

This chapter has described various methods to predict fatigue life of seam-welded joints, including nominal stress approaches, structural stress approaches, and notch pseudo stress approaches. Nominal stress approaches are relatively simple but have limitations to disclose stresses and strains at the critical regions at the welded joint. It also appears difficult to determine weld class for complex weld shapes and loadings.

The structural stress approaches reviewed are finite element mesh insensitive because the structural stress of a welded joint is calculated by the elementary structural mechanics theory, which requires the input of nodal forces and moments obtained from a linear elastic FEA. It is fairly easy to use, thus these approaches have been commonly adopted in the automotive industry.

The notch pseudo stress approach considers the local pseudo stresses in stress concentration areas such as fatigue damage parameters and requires fine meshes in weld toes and roots to capture the notch stresses in a linear elastic finite element analysis. A substructure modeling technique is required to model the nominal local weld geometry with solid finite elements, where a constant notch radius is introduced at every weld-base material intersection.

There are no universal fatigue properties of welded joints because of the complexity of this subject. All the fatigue properties generated for use in the three approaches are derived from a large empirical fatigue database for welded construction steels. The manufacturing effects have been directly included in the large empirical material database for welded construction steels, but the residual stress effect as a result of different manufacturing processes is not taken into account in any of these methods. Thus, we should understand the local residual stress states on welded joints and generate in-house fatigue properties by fatigue-testing of welded laboratory samples with the same manufacturing processes as real products.

Present fatigue life prediction models for seam-welded joints under variable amplitude nonproportional loading histories, such as Dong's structural stress approaches, stress-based critical plane approaches, Sonsino's effective equivalent stress amplitude method, and the equivalent nonproportional stress amplitude approach, have been reviewed. None of them is perfect to use and there is always room for improvement. However, the equivalent nonproportional stress amplitude approach has been discussed in depth to address the important factors that are needed for future model development.

This model is an extension of Sonsino's effective equivalent stress amplitude method and it consists of four material parameters accounting for the severity of nonproportional loading paths, the material's susceptibility to nonproportional hardening, the material's fatigue life under shear-versus-normal stresses, and the mean stress effect. The four parameters can be obtained by fatigue testing for S-N curves under bending only, under torsion only, and with 90° out-of-phase loading. In addition, a procedure has been described to predict fatigue lives of welded joints under nonproportional variable amplitude loading histories, based on the linear cumulative damage rule, the equivalent nonproportional stress amplitude, and the multiaxial rainflow reversal counting technique.

References

Archer, R. (1987). Fatigue of welded steel attachments under combined direct stress and shear stress. In *International conference on fatigue of welded constructions*. Paper No. 50, April 7–9, 1987. Brighton, UK: The Welding Institute.

ASME. (1989). *Boiler and pressure vessel code, section III, rule for construction of nuclear power plant components, division 1, subsection NB, class 1 components.* New York: American Society of Mechanical Engineers.

ASME. (2007). *ASME Boiler and Pressure Vessel Code,* Section VIII-Division 2. New York: American Society of Mechanical Engineers.

Backstrom, M., & Marquis, G. (2001). A review of multiaxial fatigue of weldments: Experimental results, design code and critical plane approaches. *Journal of Fatigue & Fracture of Engineering Materials & Structures, 24,* 279–291.

Berge, S. (1985). On the effect of plate thickness in fatigue of welds. *Engineering Fracture Mechanics, 21*(2), 423–435.

Bonnen, J. F., Mandapati, R., Kang, H., Mohan, R., Khosrovaneh, A. K., Amaya, M. A., et al. (2009). *Durability of advanced high strength steel gas metal arc welds* (SAE Technical Paper 2009-01-0257). Warrendale, PA: SAE International.

Branco, C. M., Ferreira, J. M., & Radon, J. C. (1985). Fatigue of fillet welded joints. *Theoretical and Applied Fracture Mechanics, 3,* 13–22.

BS5400, Pt10. (1980). *Specification for steel, concrete, and composite bridges. Code of practice for fatigue design.* London: British Standards Institution.

BS5500. (1988). *Specification for unfired, fusion-welded pressure vessels.* London: British Standards Institution.

BS7608. (1993). *Code of practice for fatigue design and assessment of steel structures.* London: British Standards Institution.

BS7910. (2004). *Guide on methods for assessing the acceptability of flaws in metallic structures.* London: British Standards Institution.

BS8118-1. (1991). *Structural use of aluminum. Code of practice for design.* London: British Standards Institution.

Caccese, V., Blomquist, P. A., Berube, K. A., Webber, S. R., & Orozco, N. J. (2006). Effect of weld geometric profile on fatigue life of cruciform welds made by laser/GMAW processes. *Marine Structures, 19,* 1–22.

Chandel, R. S. (1988). Mathematical modeling of gas arc metal welding features. In *Proceedings of the fourth international conference on modeling of casting and welding processes* (pp. 109–120). Palm Coast, FL, April 17–22. Warrendale, PA: The Metals, Materials, and Minerals Society.

Chandel, R. S., Seow, H. P., & Cheong, F. L. (1997). Effect of increasing deposition rate on the bead geometry of submerged arc welds. *Journal of Materials Processing Technology, 72,* 124–128.

Cheng, X., Fisher, J. W., Prask, H. J., Gnaupel-Herold, T., Yen, B. T., & Roy, S. (2003). Residual stress modification by post-weld treatment and its beneficial effect on fatigue strength of welded structures. *International Journal of Fatigue,* *25,* 1259–1269.

Deutsches Institut Fur Normung. (1984). *DIN15018—cranes, steel structures, principles of design and construction: Dimensioning.* Berlin: Beuth-Verlag.

Doherty, J., Shinoda, T., & Weston, J. (1978). *The relationships between arc welding parameters and fillet weld geometry for MIG welding with flux cored wires* (Welding Institute Report 82/1978/PE). Cambridge, UK: The Welding Institute.

Dong, P. (2001a). *A robust structural stress procedure for characterizing fatigue behavior of welded joints* (SAE Technical Paper No. 2001-01-0086). Warrendale, PA: SAE International.

Dong, P. (2001b). A structural stress definition and numerical implementation for fatigue analysis of welded joints. *International Journal of Fatigue, 23,* 865–876.

Dong, P. (2005). *Mesh-insensitive structural stress method for fatigue evaluation of welded structures, Battelle SS JIP training course material.* Columbus, OH: Center for Welded Structures Research, Battelle Memorial Institute.

Dong, P., & Hong, J. K. (2002). *CAE weld durability prediction: A robust single damage parameter approach* (SAE Technical Paper No. 2002-01-0151). Warrendale, PA: SAE International.

Dong, P., & Hong, J. K. (2006). A robust structural stress parameter for evaluation of multiaxial fatigue of weldments. *Journal of ASTM International, 3,* 1–17.

Dong, P., Hong, J. K., & Cao, Z. (2003). Stresses and stress intensities at notches: 'Anomalous crack growth' revisited. *International Journal of Fatigue, 25,* 811–825.

Dong, P., Hong, J. K., Potukutchi, R., & Agrawal, H. (2004). *Master S-N curve method for fatigue evaluation of aluminum MIG and laser welds* (SAE Technical Paper No. 2004-01-1235). Warrendale, PA: SAE International.

Dong, P., Wei, Z., & Hong, J. K. (2010). A path-dependent cycle counting method for variable-amplitude multi-axial loading. *International Journal of Fatigue, 32,* 720–734.

Doong, S. H., Socie, D. F., & Robertson, I. M. (1990). Dislocation substructures and nonproportional hardening. *ASME Journal of Engineering Materials and Technology, 112*(4), 456–465.

ECCS/CECM/EKS. (1985). *Recommendations for the fatigue design of steel structures.* Brussels: ECCS

Eurocode 3. (2005). *Design of steel structures, Part 1-9, Fatigue,* EN 1993-1-9. Brussels: CEN.

Fermer, M., Andreasson, M., & Frodin, B. (1998). *Fatigue life prediction of MAG-welded thin-sheet structures* (SAE Technical Paper No. 982311). Warrendale, PA: SAE International.

Ferreira, J. A. M., & Branco, A. A. M. (1989). Influence of the radius of curvature at the weld toe in the fatigue strength of fillet welded joints. *Internal Journal of Fatigue, 11,* 29–36.

Ferreira, J. M., & Branco, C. M. (1991). Influence of fillet weld joint geometry on fatigue crack growth. *Theoretical and Applied Fracture Mechanics, 15,* 131–142.

Findley, W. N. (1959). A theory for the effect of mean stress on fatigue of metals under combined torsion and axial loading or bending. *Journal of Engineering for Industry, 81*, 301–306.

Foth, J., Marissen, R., Trautmann, K. H., & Nowack, H. (1986). Short crack phenomena in high strength aluminum alloy and some analytical tools for their prediction. In *The behavior of short fatigue cracks*, EGF Pub. 1 (pp. 353–368). London: Mechanical Engineering Publication.

Gurney, T. R. (1979). *Fatigue of welded structures*. Cambridge: Cambridge University Press.

Huissoon, J. P., Strauss, D. L., Rempel, J. N., Bedi, S., & Kerr, H. W. (1994). Multi-variable control of robotic gas metal arc welding. *Journal of Materials Processing Technology, 43*, 1–12.

International Institute of Welding (IIW). (1982). Design recommendations for cyclic loaded welded steel structures. *Welding in the World, 20*(7/8), 153–165.

IIW. (1990). *Guidance on assessment of fitness for purpose of welded structures* (IIW/IIS-SST-1157-90). France: The International Institute of Welding.

Itoh, T., Sakane, M., Ohnami, M., & Socie, D. F. (1995). Nonproportional low cycle fatigue criterion for type 304 stainless steel. *ASME Journal of Engineering Materials and Technology, 117*, 285–292.

Itoh, T., Nakata, T., Sakane, M., & Ohnami, M. (1999). Nonproportional low cycle fatigue of 6061 aluminum alloy under 14 strain paths. *European Structural Integrity Society, 25*, 41–54.

Jiang, C. (2006). AET integration, Presentation to the Advanced High Strength Sheet Steel Fatigue Committee of A-SP.

JSSC. (1995). *Fatigue design recommendations for steel structures* (JSSC Technical Report 32). Tokyo: Japanese Society of Steel Construction.

Kang, H., Khosrovaneh, A., Todd, L., Amaya, M., Bonnen, J., & Shih, M. (2011a). *The effect of welding dimensional variability on the fatigue life of gas metal arc welded joints* (SAE Technical Paper No. 2011-01-0196).

Kang, H., Khosrovaneh, A., Amaya, M., Bonnen, J., Todd, L., Mane, S., et al. (2011b). *Application of fatigue life prediction methods for GMAW joints in vehicle structures and frames* (SAE Technical Paper No. 2011-01-0192). Warrendale, PA: SAE International.

Kang, H., Lee, Y. -L., & Sun, X. (2008). Effects of residual stress and heat treatment on fatigue strength of weldments. *Materials Science and Engineering, A, 497*(1–2), 37–43.

Kim, I. S., Son, J. S., Kim, I. G., Kim, J. Y., & Kim, O. S. (2003). A study on relationship between process variables and bead penetration for robotic $CO2$ arc welding. *Journal of Materials Processing Technology, 136*, 139–145.

Krempl, E., & Lu, H. (1983). Comparison of the stress responses of an aluminum alloy tube to proportional and alternate axial and shear strain paths at room temperature. *Mechanics of Materials, 2*, 183–192.

Lawrence, F. V. (1973). Estimation of fatigue crack propagation life in butt welds. *Welding Journal, 52*, 213–220.

Lawrence, F. V., Ho, N. J., & Mazumdar, P. K. (1981). Predicting the fatigue resistance of welds. *Annual Review of Mater Science, 11*, 401–425.

Lawrence, F. V., & Munse, W. H. (1973). Fatigue crack propagation in butt weld containing joint penetration defects. *Welding Journal, 52*, 221–225.

Lee, Y. L., Tjhung, T., & Jordan, A. (2007). A life prediction model for welded joints under multiaxial variable amplitude loading histories. *International Journal of Fatigue, 29*, 1162–1173.

Lida, K., & Uemura, T. (1996). Stress concentration factor formulae widely used in Japan. *Fatigue & Fracture of Engineering Materials & Structures, 19*(6), 779–786.

Maddox, S. J. (1991). *Fatigue strength of welded structures*. Cambridge: Abington Publishing.

Maddox, S. J. (2003). Review of fatigue assessment procedures for welded aluminum alloy structures. *Internal Journal of Fatigue, 25*(12), 1359–1378.

Maul, G. P., Richardson, R., & Jones, B. (1996). Statistical process control applied to gas metal arc welding. *Computers Industrial Engineering, 31*(1/2), 253–256.

McGlone, J. C., & Chadwick, D. B. (1978). *The submerged arc butt welding of mild steel. Part 2. The prediction of weld bead geometry from the procedure parameters* (Welding Institute Report 80/1978/PE). Cambridge, UK: The Welding Institute.

Miner, M. A. (1945). Cumulative damage in fatigue. *Journal of Applied Mechanics, 67*, A159–A164.

Monahan, C. C. (1995). *Early fatigue cracks growth at welds*. Southampton, UK: Computational Mechanics Publications.

Nelson, D. V. (1982). Effect of residual stress on fatigue crack propagation. In *Residual stress effects in fatigue* (pp. 172–194). *ASTM STP 776*. West Conshohocken, PA: ASTM International.

Neuber, H. (1946). *Theory of notch stresses*. Ann Arbor, MI: Edwards.

Newman, J. C., & Raju, I. S. (1981). An empirical stress intensity factor equation for surface crack. *Engineering Fracture Mechanics, 15*, 185–192.

Ninh Nguyen, T., & Wahab, M. A. (1995). A theoretical study of the effect of weld geometry parameters on fatigue crack propagation life. *Engineering Fracture Mechanics, 51*(1), 1–18.

Niu, X., & Glinka, G. (1989). Stress intensity factors for semi-elliptical surface cracks in welded joints. *International Journal of Fracture, 40*, 255–270.

Norris, C. H. (1945). Photoelastic investigation of stress distribution in transverse fillet welds. *Welding Journal Research Supplement, 24*(10), 557s ff.

Ohta, A., Mawari, T., & Suzuki, N. (1990). Evaluation of effect of plate thickness on fatigue strength of butt welded joints by a test maintaining maximum stress at yield strength. *Engineering Fracture Mechanics, 37*, 987–993.

Palmgren, A. (1924). Die lebensdauer von kugellagern. *Zeitschrift des Vereinesdeutscher Ingenierure, 68*(14), 339–341.

Parker, A. P. (1981). *The mechanics of fracture and fatigue: An introduction*. London: E. & F. N. Spon Ltd.

Peterson, R. E. (1959). *Analytical approach to stress concentration effects in aircraft materials* (Technical Report 59-507). Dayton, OH: U.S. Air Force-WADC Symp. Fatigue Materials.

Peterson, R. E. (1975). *Stress concentration factors*. New York: John Wiley.

Potukutchi, R., Agrawal, H., Perumalswami, P., & Dong, P. (2004). *Fatigue analysis of steel MIG welds in automotive structures* (SAE Technical Paper No. 2004-01-0627). Warrendale, PA: SAE International.

Poutiainen, I., & Marquis, G. (2006). A fatigue assessment method based on weld stress. *International Journal of Fatigue, 28*, 1037–1046.

Radaj, D. (1990). *Design and analysis of fatigue resistant welded structures*. Cambridge: Abington Publication.

Radaj, D. (1996). Review of fatigue-strength assessment of non-welded and welded structures based on local parameters. *Internal Journal of Fatigue, 18*(3), 153–170.

Radaj, D. (1997). Fatigue notch factor of gaps in welded joints reconsidered. *Engineering Fracture Mechanics, 57*(4), 405–407.

Radaj, D., Sonsino, C. M., & Fricke, W. (2006). *Fatigue assessment of welded joints by local approaches* (2nd ed.). Boca Raton, FL: CRC Press.

Seto, A., Yoshida, Y., & Galtier, A. (2004). Fatigue properties of arc-welded lap joints with weld start and end points. *Fatigue Fracture Engineering Materials and Structures, 27*, 1147–1155.

Siljander, A., Kurath, P., & Lawrence, F. V., Jr. (1992). Nonproportional fatigue of welded structures. In M. R. Mitchell, & R. W. Langraph (Eds.), *Advanced in fatigue lifetime predictive technique, ASTM STP 1122* (pp. 319–338). Philadelphia: American Society for Testing and Materials.

Solakian, A. G. (1934). Stresses in transverse fillet welds by photoelastic methods. *Welding Journal, 13*(2), 22ff.

Sonsino, C. M. (1995). Multiaxial fatigue of welded joints under in-phase and out-of-phase local strains and stresses. *Internal Journal of Fatigue, 17*(1), 55–70.

Sonsino, C. M. (2009). Multiaxial fatigue assessment of welded joints recommendations for design codes. *Internal Journal of Fatigue, 31*, 173–187.

Tada, H., Paris, P. C., & Irwin, G. R. (2000). *The stress analysis of crack handbook* (3rd ed.). New York: ASME.

Webster, G. A., & Ezeilo, A. N. (2001). Residual stress distributions and their influence on fatigue lifetimes. *International Journal of Fatigue, 23*, S375–S383.

Yang, L. J., Chandel, R. S., & Bibby, M. J. (1993). The effects of process variables on the weld deposit area of submerged arc welds. *Welding Journal, 72*(1), 11–18.

Yee, R., & Burns, D. J. (1988). Thickness effect and fatigue crack development in welded. *Proceedings of 7th international conference on offshore mechanics and arctic engineering* (pp. 447–457), Houston. New York: ASME.

CHAPTER 11

Fatigue Life Prediction Methods of Resistance Spot-Welded Joints

Hong-Tae Kang
The University of Michigan–Dearborn

Yung-Li Lee
Chrysler Group LLC

Introduction

Some parameters, and their relation to resistance spot-welded fatigue, are as follows:

- *Manufacturing*: Properly formed spot welds are the result of a combination of the appropriate current, pressure, and hold time for a particular sheet thickness and material property combination. Residual stresses are normally inevitable.

- *Metallurgy*: Base sheet metal properties will change in the weld nugget and heat affected zone (HAZ) of the spot weld.

- *Statistics*: Electrode wear through the manufacturing cycle, the nonuniform nature of the joining surface, and missed or incomplete welds creates variability at both the local spot-weld level and the structural level.

- *Stress Analysis*: Large-scale automotive structural modeling of resistance spot welds relies on finite element analysis techniques based on crudely simplified spot-weld models. And single-weld analysis typically relies on mechanics of materials, elasticity, and fracture mechanics approaches in which detail models are often employed.

- *Fatigue Analysis*: Fatigue critical weld locations must be determined and then appropriate damage models and damage accumulation techniques must be chosen to model resistance spot-welded fatigue behavior.

Many researchers have been working on developing simple, reliable fatigue life prediction methods for resistance spot-welded joints, some of which have been successfully correlated with experimental results. This chapter starts with a brief summary of the primary factors affecting fatigue life of spot-welded joints and then focuses on some particular spot-welded fatigue analysis techniques and the relation of basic material specimen test to design.

Parameters Affecting Spot-Welded Joints

The primary mechanical parameters affecting fatigue life of resistance spot-welded joints include weld nugget diameter, sheet metal thickness, specimen width, specimen types, base metal strength, and multiaxial loading. Their effects are described in the following subsections.

Nugget Diameter

The weld nugget is formed from the molten material by the use of electrical resistance spot welding. The resistance spot welds are created by bringing electrodes in contact with sheet metal. Electrical current flows through the electrodes and encounters high resistance at the interface, or faying surface, between the sheets. This resistance creates a large amount of heat, which locally melts the sheet materials. The current flowing through the electrodes is then stopped, but the electrodes remain in place as the weld nugget forms. The weld nugget depends on the combination of current, electrode tip force, and the timing of these parameters, called the weld schedule.

The weld nugget diameter is defined as the average diameter of the major and minor button diameter determined by peeling one of the sheets back over the spot weld, leaving a hole in one sheet and a button of material attached to the other sheet. The weld nugget is responsible for transferring loads and is known to affect fatigue life and failure mode of resistance spot-welded joints.

The fatigue life increases as the weld nugget diameter increases (Wilson & Fine, 1981) for a concentric tube specimen subjected to fully reversed axial loading. Some researchers (Abe, Kataoka, & Satoh, 1986; Pollard, 1982) also observed the beneficial effect as the nugget diameter increased on the fatigue life of single resistance spot-welded joints subjected to cyclic loading.

Interestingly enough, Davidson (1983) found that the larger weld diameter showed longer fatigue life than the smaller one in the low cycle fatigue (LCF) regime and that the effect of nugget diameter on the fatigue life is insignificant for the high cycle fatigue (HCF) regime ($N_f > 10^6$).

Sheet Metal Thickness

The sheet metal thickness is another important parameter. Wilson and Fine (1981) studied the effect of sheet metal thickness ranging from 0.5 mm to 1.4 mm on fatigue performance. In their study, as compared to fatigue strength of the resistance spot welds with a baseline sheet thickness, the welds with thicker sheet metals could have higher fatigue strength in the LCF regime but lower fatigue strength in the HCF fatigue regime.

Other researchers (Davidson & Imhof, 1984; Pollard, 1982) also observed that the fatigue life of resistance spot-welded joints increased as sheet metal thickness increased. Similar reports on this subject could be found elsewhere (Bonnen et al., 2006; Jung, Jang, & Kang, 1996; Kitagawa, Satoh, & Fujimoto, 1985; Zhang & Taylor, 2000).

Specimen Width

The specimen width is also a factor affecting the fatigue life of spot welds. However, a few researchers studied the effect of sheet width on fatigue performance of a spot-welded joint. Rivett (1983) studied the shear fatigue properties of spot welds with three different widths of carbon steel and hot rolled HSLA steel. The fatigue life increased as the specimen width increased, but the incremental rate decreased as the specimen width increased. Sheppard (1993) included the effect of specimen width in the structural stress calculation and correlated the structural stress and fatigue life of spot-welded joints.

Base Metal Strength

Base metal strength is generally referred to as the ultimate tensile strength of the sheet metals spot-welded together. Abe et al. (1986) showed that the base metal strength did not affect the crack propagation life, but it was a very influential factor for the crack initiation life. Moreover, all the studies (Bonnen et al., 2006; Davidson & Imhof, 1984; Gentilcore, 2004; Kitagawa et al., 1985; Nordberg, 2005; Pollard, 1982; Wilson & Fine, 1981) showed the beneficial effect of higher base metal strength on fatigue performance in the LCF regime and the diminishing effect in the HCF regime.

Specimen Type

The effects of different material specimen types such as the tensile-shear, coach-peel, and cross-tension specimens were also widely studied. Figure 11.1 shows

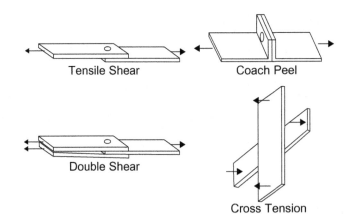

Figure 11.1
Various types of resistance spot-welded specimens under uniaxial loading. *Source: Adapted from Swellam (1991).*

the frequently employed material specimen types (Swellam, 1991). Generally a tensile-shear specimen shows better fatigue performance than that of coach-peel and cross-tension specimens at the same applied load range.

Fatigue properties of spot welds are most commonly evaluated by using the preceding single-welded specimens. But the interpretation of these material specimen data and application of the data to automotive structural analysis must be done with caution, since the deformation characteristics of these material specimens do not reflect those of spot welds in typical automotive structures. This means that the deformation of an automotive spot weld is constrained by that of the surrounding welds and structure.

In addition, the failure definition for the laboratory test may not reflect an appropriate failure definition for a weld in an automotive structure. Thus, other specimens with multiple welds, such as box beams, have been proposed for more direct application to automotive structures. However, at present no particular standard for a multiple spot-welded specimen has been developed.

Multiaxial Loading

For multiaxial spot-weld testing, two basic types of multiaxial spot-welded specimens as shown in Figure 11.2 recently have been proposed (Barkey & Kang, 1999;

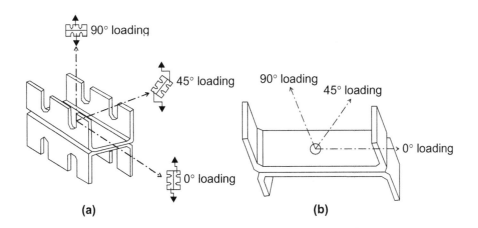

Figure 11.2
Two types of resistance spot-welded specimens under combined tension and shear. *Source: Adapted from Lee et al. (1998).*

Barkey & Han, 2001; Barkey, Kang, & Lee, 2001; Gieske & Hahn, 1994; Hahn et al., 2000; Lee, Wehner, Lu, Morrissett, & Pakalnins, 1998). The test fixture is designed to apply the combined tension and shear loads on the spot-welded specimens by changing the loading direction, as shown in Figure 11.3. This test set-up has been employed to validate the present fatigue damage models of spot-welded joints under multiaxial loading.

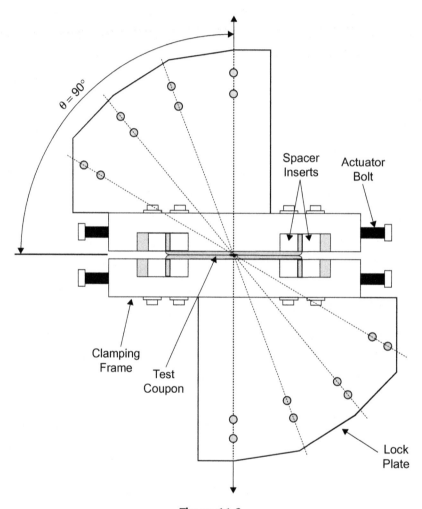

Figure 11.3
The test fixture to apply the combined tension and shear loads
on the spot-welded specimens.
Source: Adapted from Lee et al. (1998).

Fatigue Life Prediction Methods

If enough experimental data are available, a multivariable load-life type of approach (Kang & Barkey, 1999) can be employed. However, it can be advantageous to employ a fatigue damage parameter for resistance spot welds. In damage parameter approaches, an analytical model of the joint is developed to determine how the stress, deformation, or stress intensity depends on the applied load. These quantities are then related to a fatigue damage parameter and are calibrated by specimen tests.

The popular fatigue damage parameters for resistance spot welds were developed analytically or experimentally, and can be classified into three groups that are briefly reviewed in the following sections.

1. Load life approach

2. Linear elastic fracture mechanics approach

3. Structural stress approach

Load Life Approach

The applied loads, load ranges, or load amplitudes can be used to correlate the fatigue test results in the load life approaches. Pollard (1982) proposed empirical equations to calculate the fatigue life of resistance spot-welded joints from the relationship among the sheet thickness, nugget diameter, load range, and load ratio.

For in-plane shear tests and cross-tension tests of high strength low alloy steel (HSLA), the empirical equations are expressed in Equations (11.1) and (11.2), respectively:

$$\Delta F_{S,N_f} = 124{,}500 N_f^{-0.247} e^{-\left(\frac{0.139}{r} + \frac{0.0218}{0.5t}\right)} \tag{11.1}$$

$$\Delta F_{N,N_f} = 8.143 \times 10^6 N_f^{-0.288} r^{0.382} (0.5t)^{1.80} \tag{11.2}$$

where

$\Delta F_{S,N_f}$ and $\Delta F_{N,N_f}$ = the fatigue strength ranges in pounds in in-plane shear and out-of-plane normal loading, respectively

N_f = the fatigue life in cycles
r = the nugget radius
t = the sheet thickness

For fatigue lives between 5×10^3 and 5×10^6 cycles, these empirical equations showed good agreement with experimental results for both in-plane shear and cross-tension.

For a resistance spot weld subjected to combined in-plane shear ΔF_S and out-of-plane normal load ΔF_N ranges, the following constant-life criterion can be employed to determine the fatigue life of the weld:

$$\left(\frac{\Delta F_S}{\Delta F_{S,N_f}}\right)^{\alpha} + \left(\frac{\Delta F_N}{\Delta F_{N,N_f}}\right)^{\alpha} = 1.0 \tag{11.3}$$

where α is the exponent for the shape of the failure surface, usually setting $\alpha = 1.0$ for conservatism, if the test data are not available.

Equation (11.3) is the most direct and least sophisticated approach for determining fatigue life because it involves the assumptions that the weld nugget, free of rotation, does not resist any moment or torque; the mean stress effect is negligible; and that the nugget is subjected to proportional in-plane and out-of-plane loading histories.

For a variable amplitude proportional loading history on the weld nugget, the uniaxial rainflow cycle counting technique can be used to count the number of cycles on either the in-plane shear or out-of-plane normal loading history, and to calculate the resulting fatigue damage based on each load range and the number of extracted cycles.

Linear Elastic Fracture Mechanics Approach

The linear elastic fracture mechanics approach for spot welds was first developed by Pook (1975a, 1975b), in which fatigue strength of tensile-shear resistance spot weld was assessed in terms of the stress intensity factor at the spot weld. The work by Yuuki, Ohira, Nakatsukasa, and Yi (1986) first showed the feasibility of the stress intensity factor in correlating fatigue strength of spot welds for different specimens.

The finite element method has been widely used to determine the stress intensity factors at spot welds. Various modeling techniques for spot welds and sheet metals include the detailed three-dimensional solid elements (Cooper & Smith, 1986; Smith & Copper, 1988; Swellam, Ahmad, Dodds, & Lawrence, 1992) and the boundary elements (Yuuki & Ohira, 1989), and the combined three-dimensional solid element for spot welds and shell elements for sheets (Radaj, 1989; Zhang, 1997, 1999a,b,c, 2001, 2003).

All these models require refined finite element meshes at and near the spot welds and result in accurate calculation of the stress intensity factors. But the computational CPU time consumption is the drawback of modeling the spot-welded joints with the detailed three-dimensional elements, which prohibits engineers from using the analysis of any structure containing a large number of spot welds.

Two commonly used solutions to stress intensity factors at spot welds were therefore developed, without using three-dimensional solid finite elements. One is the structural-stress-based solution using the stresses in the shell elements (plate theory stresses) around the spot weld, and the other is the force-based solution using the interface forces and moments in the beam element for a spot weld.

Structural-Stress-Based Stress Intensity Factors

The classic work in fracture mechanics by Tada, Paris, and Irwin (1985) provided the fundamental solution to the stress intensity factors at a crack between two sheets under edge loads in terms of the structural stresses at the crack tip multiplying a square root of the sheet metal thickness. Pook (1979) later validated the solution by Tada et al. (1985) and concluded that the crack tip stresses can be directly calculated from the simple plate theory. This is an innovational work to obtain the stress intensity factors of spot welds based on the structural stress concept.

Radaj (1989), Zhang (1997, 1999a,b,c, 2001), and Lin, Wang, and Pan (Lin & Pan, 2008a,b; Lin, Pan, Tyan, & Prasad, 2003; Lin, Pan, Wu, Tyan, & Wung, 2002; Lin, Wang, & Pan, 2007; Wang, Lin, & Pan, 2005; Wang & Pan, 2005) further refined the structural stress concept and provided a number of formulas for stress intensity factors at spot welds based on the input of structural stresses.

For example, the formulas (Zhang, 1999b) for the spot-welded specimen type as shown in Figure 11.4 are as follows:

$$K_I = \frac{1}{6}\left[\frac{\sqrt{3}}{2}(\sigma_{ui} - \sigma_{uo} + \sigma_{li} - \sigma_{lo}) + 5\sqrt{2}(\tau_{qu} - \tau_{ql})\right]\sqrt{t} \qquad (11.4)$$

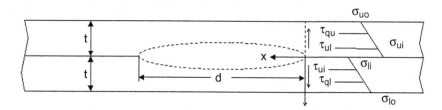

Figure 11.4
Structural stresses (plate theory stresses) around a spot weld
according to Zhang (1999b).

$$K_{II} = \left[\frac{1}{4}(\sigma_{ui} - \sigma_{li}) + \frac{2}{3\sqrt{5}}(\tau_{qu} + \tau_{ql}) \right] \sqrt{t} \qquad (11.5)$$

$$K_{III} = \frac{\sqrt{2}}{2}(\tau_{ui} - \tau_{li})\sqrt{t} \qquad (11.6)$$

where

σ_{ui}, σ_{uo}, σ_{li}, and σ_{lo} = the normal stresses
τ_{ui} and τ_{li} = the circumferential stresses
τ_{qu} and τ_{ql} = the transverse shear stresses on the verge of the spot weld

Note that these structural stresses are the plate theory stresses without any issue of stress singularity, and generally represent external loads.

An appropriate finite element model is needed to extract plate theory stresses at spot welds. Zhang (1999b) proposed that a special spoke pattern as illustrated in Figure 11.5 is considered as a simplified model without three-dimensional solid elements. The central beam in the spoke pattern is actually a cylindrical elastic beam element with a diameter of that of the nugget. The base material properties should be used for the beam element.

The spoke pattern consists of rigid bar elements transferring all the six transla-tional and rotational degrees of freedom between the master nodes of the beam element and those (slave nodes) of the shell elements. The diameter of the pat-tern is equal to the nugget diameter. It is suggested to constrain only the three translational and the rotational (in the radial direction) degrees of freedom, with the other two rotational degrees of freedom being set free. The structural stresses

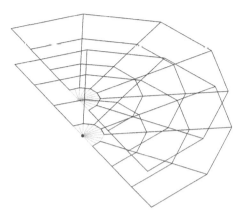

Figure 11.5
A special spoke pattern to represent a spot weld in a finite element analysis
model. *Source: Adapted from Zhang (1999b).*

obtained from this simplified finite element analysis is mesh dependent and
adequate mesh refinements should be introduced around the spoke pattern for
the accurate stress outputs.

Interface Force-Based Stress Intensity Factors

There is a need for a mesh insensitive solution to stress intensity factors at a
spot weld, as applied to automotive structures that have a large number of spot
welds. A common practice in the finite element analysis (FEA) is to model
spot welds with beam elements, which connect two sheet metals modeled by
shell elements without mesh refinements. The interface forces and moments in
the beam elements are employed to calculate the structural stresses around a
spot weld.

Many methodologies (Maddox, 1992; Rupp, Grubisic, & Buxbaum, 1994; Swellam,
1991; Swellam et al., 1992) for estimating the structural stresses based on the interface forces and moments have been developed. For example, Swellam et al. (1992)
proposed a fatigue damage parameter (K_i) based on the linear elastic fracture
mechanics concept. They assumed that a spot-welded joint consists of the two half-
spaces joined by a circular area under combined out-of-plane normal force, in-plane
shear force, and in-plane bending moment.

The resistance spot weld is subjected to the combinations of Mode I and Mode II loadings and the stress intensity factors at the edge of the spot-weld nugget are expressed (Tada et al., 1985) as:

$$K_I = \frac{\sqrt{2}F_z}{d\sqrt{\pi d}} + \frac{6\sqrt{2}\sqrt{M_x^2 + M_y^2}}{d^2\sqrt{\pi d}} \tag{11.7}$$

$$K_{II} = \frac{\sqrt{2}\sqrt{F_x^2 + F_y^2}}{d\sqrt{\pi d}} \tag{11.8}$$

where

F_x and F_y = the in-plane interface forces
M_x and M_y = the in-plane interface moments
d = the weld nugget diameter

The equivalent stress intensity factor is derived by linear superposition as shown in this equation:

$$K_{I_{eq}} = \frac{\sqrt{K_I^2 + \beta_1 K_{II}^2}}{G} . \tag{11.9}$$

Here $K_{I_{eq}}$ is an equivalent stress intensity factor of Mode I, and β_1 is a material constant that can be determined by collapsing the total fatigue life data of the only Mode I loading case and the combined Mode I and II loading case. The geometrical correction factor (G) is:

$$G = \sqrt{\frac{8Wt^2}{d^3}\left(\frac{9t^2}{d^2} + 1\right)} \tag{11.10}$$

where

W = the specimen width
t = the sheet metal thickness

A theoretical estimation of the stress intensity factor for the geometric effect was given by Zhang (1997, 1999a). For example, the stress intensity factor at a spot weld is as follows:

$$K_I = \frac{\sqrt{3}\sqrt{F_x^2 + F_y^2}}{2\pi \cdot d\sqrt{t}} + \frac{5\sqrt{2}F_z}{3\pi \cdot d\sqrt{t}} + \frac{2\sqrt{3}\sqrt{M_x^2 + M_y^2}}{\pi dt\sqrt{t}} \tag{11.11}$$

$$K_{II} = \frac{2\sqrt{F_x^2 + F_y^2}}{\pi d\sqrt{t}} \qquad (11.12)$$

$$K_{III} = \frac{\sqrt{2}\sqrt{F_x^2 + F_y^2}}{\pi d\sqrt{t}} + \frac{2\sqrt{2}M_z}{\pi d^2\sqrt{t}} \qquad (11.13)$$

where

M_z = the out-of-plane interface moment

The stress intensity factors given in Equations (11.11), (11.12), and (11.13) are the maximum values on the spot-weld edge; for a spot weld with unequal sheet thickness, the smaller sheet thickness is suggested as a crude approximation. The equivalent stress intensity (K_{eq}) can be obtained for combined actions of K_I, K_{II}, and K_{III}:

$$K_{I,eq} = \sqrt{K_I^2 + \beta_1 K_{II}^2 + \beta_2 K_{III}^2} \qquad (11.14)$$

where

β_2 = a material parameter to correlate K_{III} mode fatigue data to K_I mode fatigue data

In terms of correlating the equivalent stress intensity factor of Mode I to the fatigue life, Swellam et al. (1992) proposed a new fatigue damage parameter (K_i) to account for the load ratio effect as follows:

$$K_i = K_{I,eq,max} \times (1 - R)^{b_o}. \qquad (11.15)$$

Here $K_{I,eq,max}$ is the equivalent stress intensity factor of Mode I at the maximum applied load and R is the load ratio defined as the ratio of minimum to maximum loads. And b_o is a load ratio exponent to present a better correlation between the total fatigue life and K_i in log-log scale. If no test data is available, set a default value of $b_o = 0.85$.

Then a fatigue damage parameter and life relationship can be derived based on the plot using the least squares method as shown:

$$K_i = A(N_f)^h \qquad (11.16)$$

where

A and h = the constants from the curve-fitting for the fatigue test data

For a variable amplitude proportional loading history on the weld nugget, the uniaxial rainflow cycle counting technique can be used to count the number of cycles on either the in-plane shear or out-of-plane normal loading history that is responsible for the maximum equivalent stress intensity factor of Mode I, and to calculate the resulting fatigue damage based on each $K_{I,eq,max}$ and the number of extracted cycles.

Structural Stress Approach

The structural stress approach is to characterize some critical aspect of the stress state at the crack initiation location of the spot-welded joint and incorporate this stress state into a fatigue damage parameter that depends on the nugget forces, moments, and weld geometry such as the nugget diameter and sheet metal thickness. The structural stresses are not the true stresses at the weld, but are the local nominal stresses that are related to the loading mode and geometry in a linear elastic fashion.

The linear elastic finite element analysis showed the limitation to calculate the stresses at the edge of spot-welded joints even with a finely meshed model because of considerable plastic deformation at the edge of the weld nugget (Rupp, Storzel, & Grubisic, 1995). In structural stress approaches, a rigid beam element typically represents the weld nugget in finite element models to obtain forces and moments at the spot-welded joints (Kang et al., 2000; Kang, 2005). Based on the obtained forces and moments, structural stresses are calculated using the linear elastic equations of beam and plate theory.

Sheppard

Sheppard (1993, 1996) developed a structural stress approach to predict fatigue life of spot-welded joints. It was assumed that the fatigue life of spot-welded joints can be directly related to the maximum structural stress range at the critical region of the welded joints. In this approach, the fatigue crack initiation life is assumed to be negligible as compared to the crack propagation life. The weld nugget rotation during service loading was also assumed to be negligible in this approach. The structural stress range (Sheppard, 1996) is expressed as:

$$\Delta S_{ij} = \frac{\Delta Q_{ij}}{\omega t_i} + \frac{6 \cdot \Delta M_{ij}^*}{W t_i^2} + \frac{\Delta P_i}{t_i^2} \qquad (11.17)$$

where

ω = the effective specimen width (= $\pi d/3$)
W = the width of the specimen
t_i = the sheet thickness
ΔM_{ij}^* = the bending moment ranges
ΔQ_{ij} = the membrane load ranges
ΔP_i = the axial load range in the weld nugget

The subscript i refers to the number of sheet (i = 1, 2) and the subscript j refers to the number of elements in the particular sheet (j = 1, 2, 3, 4).

The forces around the weld nugget were obtained from linear elastic FEA. Plate elements simulated the behavior of sheet metals and a beam element simulated the weld nugget in the finite element model. All compressive membrane forces are set equal to zero since it was assumed that compressive forces do not contribute fatigue damage.

The maximum structural stress (ΔS_{max}) can be obtained from Equation (11.17) and correlated to the fatigue life of the spot weld. The relationship between ΔS_{max} and propagation life was derived from Forman's equation (Forman, Kearney, & Eagle, 1967):

$$\frac{N_{pt}}{(1-R)} = A_1 (\Delta S_{max})^{-m}. \qquad (11.18)$$

Here A_1 and m can be obtained from a curve fitting of maximum structural stress ranges (ΔS_{max}) versus measured fatigue life ($N_{pt}/(1-R)$) in log-log scale. R is the load ratio. It was assumed that the crack propagation life (N_{pt}) is equal to the total fatigue life of the spot weld.

Rupp, Storzel, and Grubisic

Rupp et al. (1995) used local structural stresses at the spot welds instead of using notch-root stresses or stress intensities to correlate with the fatigue life of the spot-welded joints. The local structural stresses were calculated based on the cross-sectional forces and moments using beam, sheet, and plate theory. A stiff beam element represented the spot weld in a finite element model to connect both sheet metals. The length of the beam element was recommended to be one-half of the summation of both sheet thicknesses (Hayes & Fermer, 1996).

When fatigue cracks are developed in the sheet metals, the local structural stresses are calculated from the formulas of the circular plate with central loading. In this case, a spot-welded specimen is considered as a circular plate with a rigid circular kernel at the center (Young, 1989), and the outer edges of the plate are treated as fixed, as shown in Figure 11.6. Then, the equivalent stresses for the damage parameter of the spot-welded joints can be derived by combination and superposition of the local structural stresses.

The solution of the radial stresses for this plate problem is presented in Roark's formulas for stress and strain (Young, 1989). The maximum radial stress ($\sigma_{r,max}$) resulting from lateral forces is determined as follows:

$$\sigma_{r,max} = \frac{F_{x,y}}{\pi dt} \tag{11.19}$$

where

$F_{x,y}$ = a lateral force in the local x or y direction, as shown in Figure 11.6

The radial stress due to a normal force F_z on the weld nugget is given by

$$\sigma_r = \frac{k_1 F_z}{t^2} \tag{11.20}$$

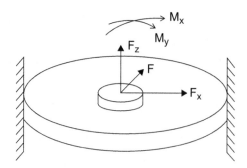

Figure 11.6
Circular plate model for a spot-welded specimen.

where

k_1 = a parameter that depends on the ratio of the nugget radius and specimen span

The maximum radial stress due to applied moments occurs at the edge of the nugget and is expressed as follows:

$$\sigma_{r,max} = \frac{k_2 M_{x,y}}{dt^2} \tag{11.21}$$

where

k_2 = a parameter that depends on the ratio of the nugget radius and specimen span

$M_{x,y}$ = an applied moment in the local x or y direction

The equivalent stresses for the damage parameter are calculated by the appropriate combination and superposition of the local radial structural stresses.

For nonproportional loading, the equivalent stresses can be determined as a function of the angle θ around the circumference of the spot weld. Here θ is the angle measured from a reference axis in the plane of the weld nugget. The equivalent stress of the fatigue damage parameter is:

$$\sigma_{eq}(\theta) = -\sigma_{max}(F_x)\cos\theta - \sigma_{max}(F_y)\sin\theta + \sigma(F_z) \\ + \sigma_{max}(M_x)\sin\theta - \sigma_{max}(M_y)\cos\theta \tag{11.22}$$

where

$$\sigma_{max}(F_{x,y}) = \frac{F_{x,y}}{\pi dt} \tag{11.23}$$

$$\sigma_{max}(M_{x,y}) = \kappa_3\left(\frac{k_2 M_{x,y}}{dt^2}\right) \tag{11.24}$$

$$\text{for } F_z > 0 \quad \sigma(F_z) = \kappa_3\left(\frac{k_1 F_z}{t^2}\right) \tag{11.25}$$

$$\text{for } F_z \leq 0 \quad \sigma(F_z) = 0. \tag{11.26}$$

Here $k_1 = 1.744$ and $k_2 = 1.872$ are determined based on an assumed ratio of radius to span of 0.1. Parameter $\kappa_3 (= 0.6\sqrt{t})$ is a material-dependent geometry factor applied to the stress terms calculated from the bending moment. It effectively reduces the sensitivity of these stress terms to the sheet thickness.

When fatigue cracks are developed through the weld nugget (nugget failure), the local structural stresses are calculated from the formulas of the beam subjected to tension, bending, and shear loads, as shown in Figure 11.7. This failure mode can occur when a spot weld is used to connect relatively thick sheets. In this case, the spot-weld nugget is modeled as a circular cross section of a beam.

The normal stress (σ_n), bending stress (σ_b), and maximum shear stress (τ_{max}) are given by the following formulas:

$$\sigma_n = \frac{4F_z}{\pi d^2} \tag{11.27}$$

$$\sigma_b = \frac{32M_{x,y}}{\pi d^3} \tag{11.28}$$

$$\tau_{max} = \frac{16F_{x,y}}{3\pi d^2}. \tag{11.29}$$

The nominal nugget stresses in this case can be calculated by superpositioning these formulas, and a failure orientation can be determined by using the

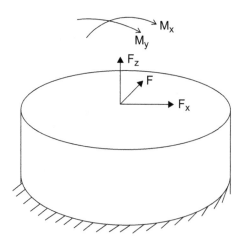

Figure 11.7
Beam model for the weld nugget subjected to tension,
bending, and shear.

stress-based critical plane approach. The equivalent stress (tensile normal stress) on the critical plane is taken as the damage parameter.

As a function of the angle θ along the circumference of the spot weld, these stresses are

$$\tau(\theta) = \tau_{max}(F_x)\sin\theta + \tau_{max}(F_y)\cos\theta \qquad (11.30)$$

$$\sigma(\theta) = \sigma(F_z) + \sigma_{max}(M_x)\sin\theta - \sigma_{max}(M_y)\cos\theta \qquad (11.31)$$

where

$$\tau_{max}(F_{x,y}) = \frac{16F_{x,y}}{3\pi d^2} \qquad (11.32)$$

$$\sigma_{max}(M_{x,y}) = \frac{32M_{x,y}}{\pi d^3} \qquad (11.33)$$

$$\text{for } F_z > 0 \quad \sigma(F_z) = \frac{4F_z}{\pi d^2} \qquad (11.34)$$

$$\text{for } F_z \leq 0 \quad \sigma(F_z) = 0. \qquad (11.35)$$

Stress histories of sheet or nugget stresses detailed previously are used to calculate the fatigue life of the spot-welded structures. Rainflow cycle counting of these histories can be used to determine the equivalent stress amplitude and mean stress associated with each cycle. If the structure is subject to proportional loading, a single crack initiation site near each nugget can be readily determined. If the structure is subjected to nonproportional loading, the many potential sites for crack initiation around the circumference of each weld nugget must be examined.

In either proportional or nonproportional loading, a correction may be made for mean stress sensitivity. Rupp et al. (1995) proposed the following equivalent stress amplitude at $R = 0$ ($\sigma_{eq,0}$) by modifying the equivalent stress amplitude ($\sigma_{eq,a}$) and equivalent mean stress ($\sigma_{eq,m}$) based on a Goodman-type mean stress correction procedure:

$$\sigma_{eq,0} = \frac{\sigma_{eq,a} + M\sigma_{eq,m}}{M + 1} \qquad (11.36)$$

where

$M =$ the mean stress sensitivity factor

Hence the total fatigue life can be correlated to the calculated equivalent stress amplitude at $R = 0$.

The failure mode of spot-welded joints was determined using a simple guideline that is generally accepted in industry (Rupp et al., 1995). When the weld nugget diameter versus sheet thickness is plotted, the boundary value of the cracking in the sheet and through the nugget is $3.5\sqrt{t}$, where t is the sheet thickness. When the weld nugget diameter is larger than the boundary value, the crack will be in the sheet. Alternatively, when the weld nugget diameter is smaller than the boundary value, the crack will be through the weld nugget.

Dong

Dong and his coworkers (Dong, 2001a,b, 2005; Dong & Hong, 2002; Potukutchi, Agrawal, Perumalswami, & Dong, 2004) applied the structural stress approach described in Chapter 10 to predict fatigue life of spot-welded joints (Bonnen et al., 2006; Dong, 2005; Kang, Dong, & Hong, 2007). This structural stress is calculated from the nodal forces and moments derived from the linear elastic FEA. At each nodal point along the weld line, the forces and moments have to be resolved into the local coordinate systems that define Mode I loading.

The nodal forces and moments in a local coordinate system are then converted to line forces and moments along the weld line. Application of this approach is also discussed further in the next section.

Applications

This section describes some examples for the applications of the fatigue life pre-diction methods of spot-welded joints discussed in previous sections. Prediction results were also compared with experimental results to assess the effectiveness of those methods. The test results that were used in this session were from Bonnen et al. (2006), and included seven advanced high strength steel (AHSS) grades, mild steels, and a conventional high strength low alloy (HSLA) grade. Tensile shear (TS) and coach peel (CP) specimen geometries were employed for the spot-welded specimens. The specimen dimensions are shown in Figures 11.8 and 11.9.

All specimens were tested at the single frequency in the range of 5 to 30 Hz with the load ratio of $R = 0.1$ or $R = 0.3$. Fatigue failure was defined as the specimens

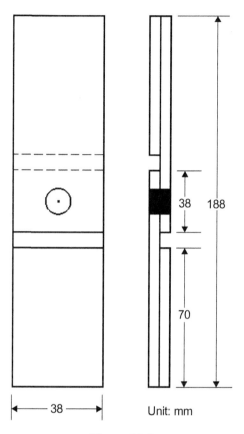

Figure 11.8
Schematic of tensile shear specimen.
Source: Adapted from Bonnen et al. (2006).

being completely separated in the two parts, or as the testing machine reaching the displacement limit due to extensive cracking of specimens.

The fatigue test results of TS and CP specimens are shown in Figure 11.10, in terms of load amplitude versus number of cycles to failure. The test results show that fatigue life of a tensile shear specimen is clearly separated from that of a coach peel specimen.

As reviewed in the previous section, many different approaches were proposed to estimate fatigue life of spot-welded joints. However, only two structural stress

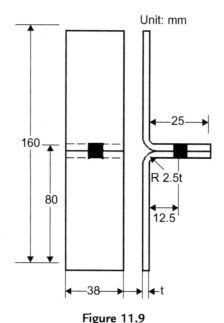

Figure 11.9
Schematic of coach peel specimen.
Source: Adapted from Bonnen et al. (2006).

approaches, those of Rupp and coworkers (1995) and Dong (2005), are demonstrated for the examples. Those approaches were well explained in the previous section.

The two approaches have differences in modeling of the spot-welded joint in the finite element model to obtain nodal forces and moments. For the approach of Rupp and coworkers, a simple rigid beam element represents the spot-welded joint as shown in Figure 11.11.

On the other hand, for Dong's approach, it requires more effort to represent the spot-welded joint, as shown in Figure 11.12. A beam element represents the spot-weld and rigid elements connect from the center node to periphery nodes of the weld nugget at each plate. The length of the beam element is equal to one half the summation of the two sheet thicknesses.

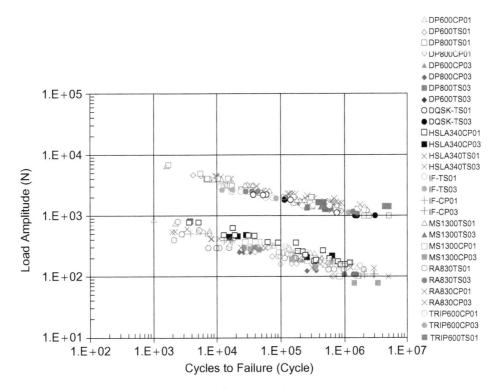

Figure 11.10
Fatigue results for TS and CP specimens.
Source: Adapted from Bonnen et al. (2006).

The structural stress ranges were calculated using the grid forces and moments obtained from FE analyses, and plotted with experimental fatigue life to obtain the best fit curve equation for the data as shown in Figures 11.13(a) and 11.13(b). The structural stresses were obtained for TS and CP specimens tested by Bonnen et al. (2006). Unlike applied load range as shown in Figure 11.10, most data points are within the factor of 5 lines. The predicted fatigue life is determined from the best-fit curve equation for the specific applied load range.

The predicted fatigue life versus experiment fatigue life is plotted in Figures 11.13 and 11.14 for Dong's approach and Rupp and coworkers' approach, respectively. The dotted and dashed line represents the perfect correlation between the prediction

Tensile Shear

Coach Peel

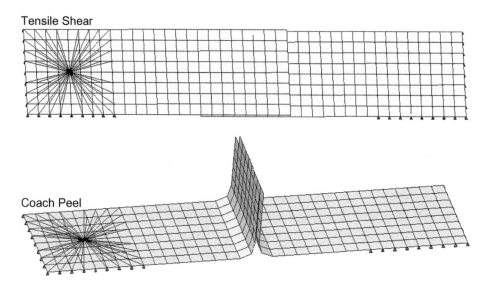

Figure 11.11
Finite element models for Rupp and coworkers' approach.

Tensile Shear

Coach Peel

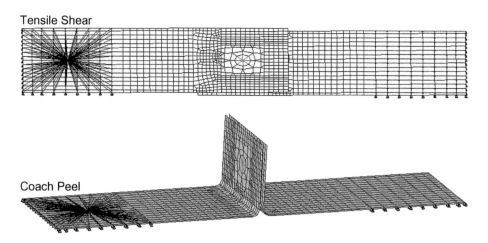

Figure 11.12
Finite element models for Dong's approach.

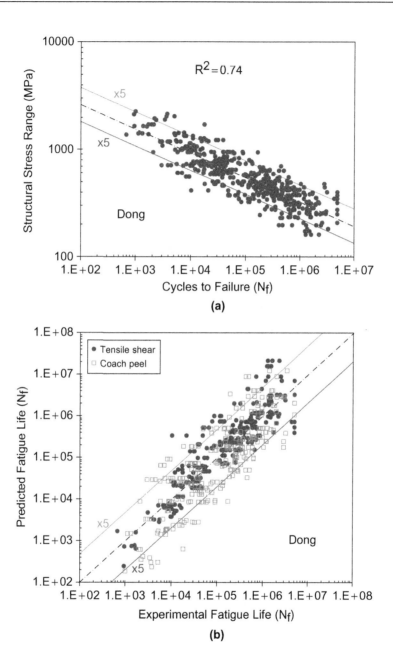

Figure 11.13

Structural stress range versus cycles to failure (a) and predicted fatigue
life versus experimental fatigue life (b) for Dong's approach.
Source: Adapted from Bonnen et al. (2006).

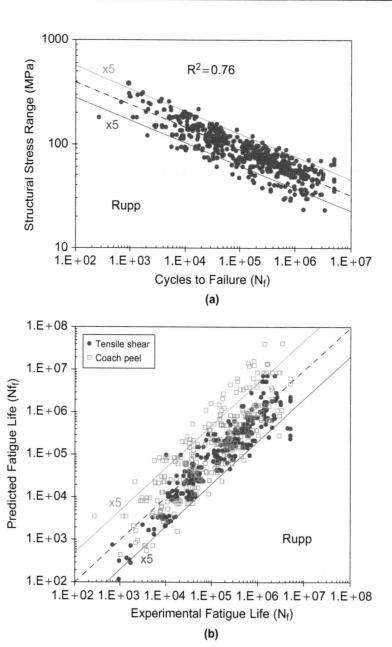

Figure 11.14
Structural stress range versus cycles to failure (a) and predicted fatigue life
versus experimental fatigue life (b) for Rupp and coworkers' approach.
Source: Adapted from Bonnen et al. (2006).

and experiment fatigue life. Factor of 5 lines are also presented above and below the perfect correlation line. Most of the data points are within the factor of 5 lines for both approaches.

Summary

In this chapter, the effects of nugget diameter, specimen thickness, base metal strength, and specimen width have been reviewed. Increasing nugget diameter and specimen thickness improves the fatigue lives of spot welds. The effects of base metal strength are negligible on relatively long fatigue lives, although significant at short fatigue lives. The fatigue lives increase with increasing specimen width, but with a diminishing rate for larger widths.

The primary methods for fatigue life prediction of resistance spot-welded joints are the load-life approach, linear elastic fracture mechanics approach, and the structural stress-life approach. These approaches have been discussed in detail in this chapter.

The load-life approach fails to correlate fatigue data from different geometries. Therefore, this approach requires testing for each type of geometry. Nevertheless, this approach provides understanding of the geometric factors that affect the total fatigue life of resistance spot-welded joints.

The linear elastic fracture mechanics approach requires accurate determination of the stress intensity factors at spot welds and the relationship of these stress intensity factors to a single fatigue damage parameter, which correlates to fatigue lives of spot welds. The stress intensity factors are expressed by the structural stresses around the spot weld or estimated by the interface forces and moments in the weld nugget. In general, force-based solutions are less accurate than stress-based ones. The equivalent Mode I stress intensity factor to account for the load ratio effect, developed by Swellam et al. (1992), is the one recommended for use.

The structural stress approach requires the calculated local structural stresses using plate, sheet, and beam theory based on the cross-sectional forces and moments. This method is suitable for application to large finite element models because mesh refinement is not necessary for spot welds. The applying theory depends on the failure mode. The failure mode determines which theory will be employed to calculate the local structural stress.

Both Dong's and Rupp's approaches show relatively accurate predictions for fatigue life of spot-welded joints. These methods require the following information:

- The estimated failure mode

- The forces and moments around nuggets

- Local structural stresses as given by the equations presented

- Mean stress sensitivity

- A relation between maximum equivalent stress amplitude and total fatigue life, such as a power law relation, obtained from a fit of experimental data

The linear elastic fracture mechanics and the structural stress approaches have the following limitations. First, the derived formulas for the stress intensity factors and local structural stresses are linear solutions, and are therefore bound to elastic and small-deformation behavior of the material. The results should be applicable to brittle fracture and high-cycle fatigue of spot welds where plasticity is contained by a large elastic stress field. For low-cycle fatigue and ultimate failure of spot welds where plasticity or large deformation may prevail, the results are not applicable. Second, the material heterogeneity, residual stress, welding imperfection, and other welding-related factors have not been considered here.

References

Abe, H., Kataoka, S., & Satoh, T. (1986). *Empirical formula for fatigue strength of single-spot-welded joint specimens under tensile-shear repeated load* (SAE Technical Report No. 860606). Warrendale, PA: SAE International.

Barkey, M. E., & Han, J. (2001). *Fatigue analysis of spot welds subjected to a variable amplitude loading history* (SAE Technical Paper 2001-01-0435). Warrendale, PA: SAE International.

Barkey, M. E., & Kang, H. (1999). Testing of spot welded coupons in combined tension and shear. *Experimental Techniques, 23*(5), 20–22.

Barkey, M. E., Kang, H., & Lee, Y. (2001). Failure modes of single resistance spot welded joints subjected to combined fatigue loading. *International Journal of Materials and Product Technology, 16*(6/7), 510–526.

Bonnen, J. F., Agrawal, H., Amaya, M., Mohan Iyengar, R., Kang, H., Khosrovaneh, A. K., Link, T. M., Shih, H-. C., Walp, M., & Yan, B. (2006). *Fatigue of advanced high strength steel spot-welds* (SAE Technical Paper No. 2006-01-0978). Warrendale, PA: SAE International.

Cooper, J. F., & Smith, R. A. (1986). Initial fatigue crack growth at spot welds. In *Fatigue of engineering materials and structures*. London: Institution of Mechanical Engineers.

Davidson, J. A. (1983). *A review of the fatigue properties of spot-welded sheet steels* (SAE Technical Report No. 830033). Warrendale, PA: SAE International.

Davidson, J. A., & Imhof, E. I. (1984). *The effect of tensile strength on the fatigue life of spot welded sheet steels* (SAE Technical Report No. 840110). Warrendale, PA: SAE International.

Dong, P. (2001a). *A robust structural stress procedure for characterizing fatigue behavior of welded joints* (SAE Technical Paper No. 2001-01-0086). Warrendale, PA: SAE International.

Dong, P. (2001b). A structural stress definition and numerical implementation for fatigue analysis of welded joints. *International Journal of Fatigue, 23*, 865–876.

Dong, P. (2005). *Mesh-insensitive structural stress method for fatigue evaluation of welded structures, Battelle SS JIP training course material*. Columbus, OH: Center for Welded Structures Research, Battelle Memorial Institute.

Dong, P., & Hong, J. K. (2002). *CAE weld durability prediction: A robust single damage parameter approach* (SAE Technical Paper No. 2002-01-0151). Warrendale, PA: SAE International.

Forman, R. G., Kearney, V. E., & Eagle, R. M. (1967). Numerical analysis of crack propagation in a cyclic-loaded structure. *Transaction of ASME, Journal of Basic Engineering, D89*(3), 459–464.

Gentilcore, M. (2004). *An assessment of the fatigue performance of automotive sheet steels* (SAE Technical Paper No. 2004-01-0629). Warrendale, PA: SAE International.

Gieske, D., & Hahn, O. (1994). Neue einelement-probe zum prüfen von punktschweiß verbindungen unter kombinierten belastungen. *Schweiß en und Schneiden, 46*, 9–12.

Hahn, O., Dölle, N., Jendrny, J., Koyro, M., Meschut, G., & Thesing, T. (2000). Prüfung und berechnung geklebter blech-profil-verbindungen aus aluminium. *Schweißen und Schneiden, 52*, 266–271.

Heyes, P., & Fermer, M. (1996). A spot-weld fatigue analysis module in the MSC/FATIGUE environment. *Symposium on International Automotive Technology*, Pune, India.

Jung, W. W., Jang, P. K., & Kang, S. S. (1996). *Fatigue failure and reinforcing method of spot welded area at the stage of vehicle development* (SAE Technical Paper No. 960553). Warrendale, PA: SAE International.

Kang, H. (2005). A fatigue damage parameter of spot welded joints under proportional loading. *International Journal of Automotive Technology, 6*(3), 285–291.

Kang, H., & Barkey, M. E. (1999). Fatigue life estimation of resistance spot-welded joints using an interpolation/extrapolation technique. *International Journal of Fatigue, 21*, 769–777.

Kang, H., Barkey, M. E., & Lee, Y. (2000). Evaluation of multiaxial spot weld fatigue parameters for proportional loading. *International Journal of Fatigue, 22*, 691–702.

Kang, H., Dong, P., & Hong, J. K. (2007). Fatigue analysis of spot welds using a mesh-insensitive structural stress approach. *International Journal of Fatigue, 29*, 1546–1553.

Kitagawa, H., Satoh, T., & Fujimoto, M. (1985). *Fatigue strength of single spot-welded Joints of rephosphorized high-strength and low-carbon steel sheets* (SAE Technical Report No. 850371). Warrendale, PA: SAE International.

Lee, Y., Wehner, T., Lu, M., Morrissett, T., & Pakalnins, E. (1998). Ultimate strength of resistance spot welds subjected to combined tension and shear. *Journal of Testing and Evaluation, 26*, 213–219.

Lin, P.-C., & Pan, J. (2008a). Closed-form structural stress and stress intensity factor solutions for spot welds under various types of loading conditions. *International Journal of Solids and Structures, 45*, 3996–4020.

Lin, P.-C., & Pan, J. (2008b). Closed-form structural stress and stress intensity factor solutions for spot welds in commonly used specimens. *Engineering Fracture Mechanics, 75*, 5187–5206.

Lin, S.-H., Pan, J., Tyan, T., & Prasad, P. (2003). A general failure criterion for spot welds under combined loading conditions. *International Journal of Solids and Structures, 40*, 5539–5564.

Lin, S.-H., Pan, J., Wu, S.-R., Tyan, T., & Wung, P. (2002). Failure loads of spot welds under combined opening and shear static loading conditions. *International Journal of Solids and Structures, 39*, 19–39.

Lin, P.-C., Wang, D.-A., & Pan, J. (2007). Mode I stress intensity factor solutions for spot welds in lap-shear specimens. *International Journal of Solids and Structures, 44*, 1013–1037.

Maddox, S. J. (1992). Fatigue design of welded structures. In *Engineering design in welded constructions* (pp. 31–56). Oxford, UK: Pergamon.

Nordberg, H. (2005). *Fatigue properties of stainless steel lap joints spot welded, adhesive bonded, weldbonded, laser welded and clinched joints of stainless steel sheets—a review of their fatigue properties* (SAE Technical Paper No. 2005-01-1324). Warrendale, PA: SAE International.

Pollard, B. (1982). *Fatigue strength of spot welds in titanium-bearing HSLA steels* (SAE Technical Report No. 820284). Warrendale, PA: SAE International.

Pook, L. P. (1975a). *Approximate stress intensity factors for spot and similar welds* (NEL Report, p. 588). Glasgow, Scotland: National Engineering Laboratory.

Pook, L. P. (1975b). Fracture mechanics analysis of the fatigue behavior of spot welds. *International Journal of Fracture, 11*, 173–176.

Potukutchi, R., Agrawal, H., Perumalswami, P., & Dong, P. (2004). *Fatigue analysis of steel MIG welds in automotive structures* (SAE Technical Paper No. 2004-01-0627). Warrendale, PA: SAE International.

Radaj, D. (1989). Stress singularity, notch stress and structural stress at spot welded joints. *Engineering Fracture Mechanics, 34*, 495–506.

Rivett, R. M. (1983). *Assessment of resistance spot welds in low carbon and high strength steel sheet* (SAE Technical Report No. 830126). Warrendale, PA: SAE International.

Rupp, A., Grubisic, V., & Buxbaum, O. (1994). Ermittlung ertragbarer beanspruchungen am schweißpunkt auf basis der übertragenen schnittgrößen. *FAT Schriftenreihe, 111*, Frankfurt.

Rupp, A., Storzel, K., & Grubisic, V. (1995). *Computer aided dimensioning of spot welded automotive structures* (SAE Technical Report No. 950711). Warrendale, PA: SAE International.

Sheppard, S. D. (1993). Estimation of fatigue propagation life in resistance spot welds. *ASTM STP, 1211*, 169–185.

Sheppard, S. D. (1996). Further refinement of a methodology for fatigue life estimation in resistance spot weld connections. *Advances in Fatigue Lifetime Prediction Techniques: Vol. 3*, ASTM STP 1292 (pp. 265–282). Philadelphia: American Society for Testing and Materials.

Smith, R. A., & Cooper, J. F. (1988). Theoretical predictions of the fatigue life of shear spot welds. In *Fatigue of Welded Constructions* (pp. 287–295). Abington, Cambridge: The Welding Institute.

Swellam, M. H. (1991). *A fatigue design parameter for spot welds*. Ph.D. thesis, The University of Illinois at Urbana-Champaign.

Swellam, M. H., Ahmad, M. F., Dodds, R. H., & Lawrence, F. V. (1992). The stress intensity factors of tensile-shear spot welds. *Computing Systems in Engineering, 3*, 487–500.

Tada, H., Paris, P., & Irwin, G. (1985). *The stress analysis of cracks handbook*. Hellertown, PA: Del Research Corporation.

Wang, D.-A., Lin, P.-C., & Pan, J. (2005). Geometric functions of stress intensity factor solutions for spot welds in lap-shear specimens. *International Journal of Solids and Structures, 42*, 6299–6318.

Wang, D.-A., & Pan, J. (2005). A computational study of local stress intensity factor solutions for kinked cracks near spot welds in lap-shear specimens. *International Journal of Solids and Structures, 42*, 6277–6298.

Wilson, R. B., & Fine, T. E. (1981). *Fatigue behavior of spot welded high strength steel joints* (SAE Technical Report No. 810354). Warrendale, PA: SAE International.

Young, W. C. (1989). *Roark's Formulas for Stress & Strain* (6th ed.). New York: McGraw-Hill.

Yuuki, R., & Ohira, T. (1989). *Development of the method to evaluate the fatigue life of spot-welded structures by fracture mechanics* (IIW Doc. III-928-89).

Yuuki, R., Ohira, T., Nakatsukasa, H., & Yi, W. (1986). Fracture mechanics analysis of the fatigue strength of various spot welded joints. *Symposium on resistance welding and related processes,* Osaka, Japan.

Zhang, S. (1997). Stress intensities at spot welds. *International Journal of Fracture, 88*, 167–185.

Zhang, S. (1999a). Approximate stress intensity factors and notch stresses for common spot-welded specimens. *Welding Journal, 78*, 173-s–179-s.

Zhang, S. (1999b). Recovery of notch stress and stress intensity factors in finite element modeling of spot welds. In *NAFEMS world congress on effective engineering analysis* (pp. 1103–1114). Newport, Rhode Island.

Zhang, S. (1999c). Stress intensities derived from stresses around a spot weld. *International Journal of Fracture, 99,* 239–257.

Zhang, S. (2001). Fracture mechanics solutions to spot welds. *International Journal of Fracture, 112,* 247–274.

Zhang, S. (2003). *A strain gauge method for spot weld testing* (SAE Technical Report No. 2003-01-0977). Warrendale, PA: SAE International.

Zhang, Y., & Taylor, D. (2000). Sheet thickness effect of spot welds based on crack propagation. *Engineering Fracture Mechanics, 67,* 55–63.

CHAPTER 12

Design and Analysis of Metric Bolted Joints

VDI Guideline and Finite Element Analysis

Yung-Li Lee
Chrysler Group LLC

Hsin-Chung Ho
Chrysler Group LLC

Chapter Outline

Introduction

The *Verein Deutscher Ingenieure* (VDI) 2230 guideline of the The Association of German Engineers can be a useful tool for designing bolted joints in many cases, but its use is limited due to the inherent assumptions of the analytical calculation models, such as the simplified cone shape deformation model for elastic compliance of clamped plates, the empirical parameters to account for the additional bolt compliances from the head and nut, the analysis technique to reduce multibolted joints to a single-bolted joint, the model to estimate the load plane factor, and the nominal stress approach for fatigue analysis. The finite element (FE) analysis is the alternative way to design and analyze the bolted structure.

This chapter introduces the basics of the VDI 2230 guideline for ISO metric bolted joints to prevent potential failure modes such as embedding, clamp load loss, clamped plates crushing and slipping, bolt yielding, thread stripping, and bolt fatigue failure. The assumptions and the calculation procedure to estimate the elastic compliances of the bolt and clamped plates are also described.

The last portion of this chapter also presents the commonly used FE bolt modeling techniques for forces and stress analysis, and recommends the appropriate fatigue damage assessment method for two different threaded bolts (such as rolled before and after heat treatment).

Overview of Bolted Joints and Mechanical Properties

The ISO metric bolted joints and their mechanical properties are presented in this section as background information. Figure 12.1 shows the nomenclature for ISO

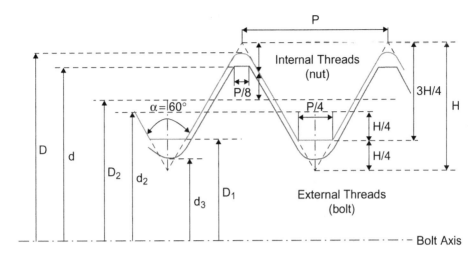

Figure 12.1
Nomenclature of ISO metric threads.

metric threads. The ISO threads can be described in the form of M $d \times P$ and a coarse pitch is assumed when P is omitted in the specification. The following parameters are used to define the ISO metric thread geometry:

H = height of fundamental triangle in mm
P = pitch in mm
D = major diameter of internal threads (nut) in mm
D_1 = minor diameter of internal threads in mm
D_2 = pitch diameter of internal threads in mm
d = major (nominal) diameter of external threads (bolt) in mm
d_2 = pitch diameter of external threads in mm = $d - 0.649519\,P$
d_3 = minor diameter of external threads in mm = $d - 1.226869\,P$
α = thread (flank) angle

A through-bolted joint and a tapped thread joint (also termed a screw joint) are the commonly used joint assemblies in the automotive industry. The through-bolted joint and the tapped thread joint are abbreviated to DSV and ESV in VDI 2230, respectively. Figures 12.2 and 12.3 depict the sections of these joints subjected to an external axial load and their theoretical deformation (pressure) cone shapes.

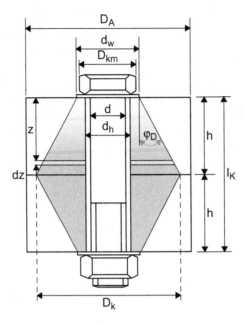

Figure 12.2
Sections of a through-bolted joint and its theoretical
deformation cone shape.

The following variables are used to define these bolted joints:

D_A = diameter of the interface surface of circular clamped plates

D_K = diameter of the deformation cone at the interface surface of circular clamped plates

D_{km} = effective diameter of bolt head or nut-bearing area

d_w = diameter of a washer head

d_h = clearance hole diameter

l_K = total clamping length

h = thickness of a clamped plate

φ_D = deformation cone angle for a through-bolted (DSV) joint

φ_E = deformation cone angle for a tapped thread (ESV) joint

z = distance from the washer (bearing) surface

dz = distance increment from the washer (bearing) surface

The induced internal forces and externally applied loads to a bolted joint are introduced here. The induced bolt preload, clamp loads, balanced thread torque,

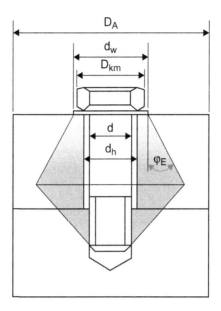

Figure 12.3
Sections of a tapped thread (screw) joint and its theoretical
deformation cone shape.

and under-head torque in a bolted joint due to an assembly tightening torque
are shown in Figure 12.4, and the applied loads to a through-bolted joint are
illustrated in Figure 12.5. The following symbols are used:

F_M = bolt preload load due to M_A
F_A = axial force on the bolted axis
F_Q = transverse force normal to the bolt axis
M_A = applied assembly torque
M_G = thread torque
M_K = under-head torque
M_B = bending moment at the bolting point
M_T = torque (twist moment) at the bolt position at the interface

Some design guidelines for hex cap screws, coefficients of friction on various
contact surfaces, and mechanical properties are presented. The recommended clearance hole diameter and the minimum bearing diameter of hex cap screws for a
specific ISO metric thread are shown in Tables 12.1 and 12.2.

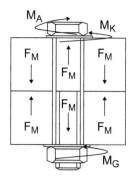

Figure 12.4
Induced bolt preload, clamp loads, balanced thread torque,
and under-head torque in a bolted joint due
to an assembly tightening torque.

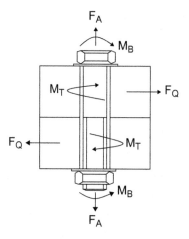

Figure 12.5
Applied loads to a through-bolted joint.

Moreover, Tables 12.3 through 12.5 summarize the coefficients of friction in the thread, in the bolt- or nut-bearing area, and between the clamp plate interfaces, where

μ_G = coefficient of friction in the thread
μ_K = coefficient of friction in the bolt or nut-bearing area
μ_T = coefficient of friction between the clamp plate interfaces

Table 12.1: Clearance Hole Diameter (d$_h$)

Thread Diameter	Clearance Hole d$_h$ (mm)
M3	3.4
M4	4.5
M5	5.5
M6	6.6
M8	9.0
M10	11.0
M12	13.5
M14	15.5
M16	17.5
M18	20.0
M20	22.0
M22	24.0
M24	26.0
M27	30.0
M30	33.0

Source: Adapted from Marbacher (1998).

Table 12.2: Minimum Bearing Diameter (d$_{w,min}$) of Hex Cap Screws

Thread Diameter	Width across the Flats (mm)	Bearing Diameter d$_{w,min}$ (mm)
M4	7	5.9
M5	8	6.9
M6	10	8.9
M8	13	11.6
M10	16	14.6
M10	17	15.9
M12	18	16.6
M12	19	17.4
M14	21	19.6
M14	22	20.5
M16	24	22.5
M18	27	25.3
M20	30	28.2
M22	32	30.0
M22	34	31.7
M24	36	33.6
M27	41	38.0
M30	46	42.7

Source: Adapted from Marbacher (1998).

Table 12.3: Coefficients of Friction in the Thread (μ_G)

Internal Threads (Nut) — Machined, Dry		External Thread (Bolt) — Steel								Adhesive
		Black Oxide or Phosphated				Zinc Plated		Cad. Plated (Machined or Rolled)		
		Rolled			Machined					
μ_G		Dry	Oiled	MoS$_2$	Oiled	Dry	Oiled	Dry	Oiled	Dry
Steel	Plain	0.12 to 0.18	0.10 to 0.16	0.08 to 0.12	0.10 to 0.16	—	0.10 to 0.18	—	0.08 to 0.14	0.16 to 0.25
Steel	Zinc Pl.	0.10 to 0.16	—	—	—	0.12 to 0.20	0.10 to 0.18	—	—	0.14 to 0.25
Steel	Cad. Pl.	0.08 to 0.14	—	—	—	—	—	0.12 to 0.18	0.12 to 0.14	—
Cast Iron	Plain	—	0.10 to 0.18	—	0.10 to 0.18	—	0.10 to 0.18	—	0.08 to 0.16	—
Al-Alloy	Plain	—	0.08 to 0.20	—	—	—	—	—	—	—

Note: MoS$_2$ = molybdenum disulfide (Moly-lub) lubricated.
Source: Adapted from Marbacher (1998).

Table 12.4: Coefficients of Friction in the Bolt or Nut-Bearing Area (μ_K)

Material of Joint Members (μ_K)				Bolt Head — Steel									
				Black Oxide or Phosphated					Ground	Zinc Plated		Cad. Plated	
				Cold Headed			Machined			Cold Headed			
Material	Finish	Surface	(Dry)	Dry	Oiled	MoS$_2$	Oiled	MoS$_2$	Oiled	Dry	Oiled	Dry	Oiled
Steel	Plain	Ground	Dry	—	0.16 to 0.22	—	0.1 to 0.18	—	0.16 to 0.22	0.10 to 0.18	—	0.08 to 0.16	—
Steel	Plain	Machined	Dry	0.12 to 0.18	0.10 to 0.18	0.08 to 0.12	0.10 to 0.18	0.08 to 0.12	—	0.10 to 0.18	0.10 to 0.18	0.08 to 0.18	0.08 to 0.14
Steel	Cad. Pl.		Dry	0.10 to 0.16	0.10 to 0.16	—	0.10 to 0.16	—	0.10 to 0.18	0.16 to 0.20	0.10 to 0.18	—	—
Steel	Zinc Pl.		Dry	0.08 to 0.16	0.08 to 0.16	0.08 to 0.16	0.08 to 0.16	0.08 to 0.16	0.08 to 0.16	—	—	0.12 to 0.20	0.12 to 0.14
Cast Iron	Plain	Ground	Dry	—	0.10 to 0.18	—	—	—	0.10 to 0.18	0.10 to 0.18	0.10 to 0.18	0.06 to 0.16	—
Cast Iron	Plain	Machined	Dry	—	0.14 to 0.20	—	0.10 to 0.18	—	0.14 to 0.22	0.10 to 0.18	0.10 to 0.16	0.08 to 0.16	—
Al. Alloy	Machined		Dry	—	0.08 to 0.20	0.08 to 0.20	0.08 to 0.20	0.08 to 0.20	—	—	—	—	—

Note: MoS$_2$ = molybdenum disulfide (Moly-lub) Lubricated.
Source: Adapted from Marbacher (1998).

Table 12.5: Coefficient of Friction μ_T between the Clamp Plate Interfaces

Material Combination	Static Friction Coefficient in the State	
	Dry	Lubricated
Steel-steel/cast steel	0.1 to 0.23	0.07 to 0.12
Steel-grey cast iron	0.12 to 0.24	0.06 to 0.1
Grey cast iron-grey cast iron	0.15 to 0.3	0.2
Bronze-steel	0.12 to 0.28	0.18
Grey cast iron-bronze	0.28	0.15 to 0.2
Steel-copper alloy	0.07	
Steel-aluminum alloy	0.1 to 0.28	0.05 to 0.18
Aluminum-aluminum	0.21	

Source: Adapted from VDI 2230, published by Beuth Verlag GmbH, Berlin (2003).

Table 12.6: Mechanical Properties for Metric Steel Bolts, Screws, and Studs

Strength Grade	$R_{P0.2min}$ (N/mm^2)	R_{mS} (N/mm^2)
4.6	240	400
4.8	340	420
5.8	420	520
8.8	660	830
9.8	720	900
10.9	940	1040
12.9	1100	1220

Source: Adapted from Shigley and Mischke (2001).

Table 12.6 shows the mechanical properties for metric steel bolts, screws, and studs, and Table 12.7 illustrates the ratio of ultimate shear strength to ultimate tensile strength for various materials, where

R_{mS} = ultimate tensile strength of the external threads (bolt) in N/mm^2
R_{mM} = ultimate tensile strength of the internal threads (nut) in N/mm^2
$R_{P0.2min}$ = minimum 0.2% yield strength of the external threads in N/mm^2
τ_{BM} = ultimate shear strength of the internal threads in N/mm^2
τ_{BS} = ultimate shear strength of the external threads in N/mm^2

Table 12.7: τ_{BM}/R_{mM} or τ_{BS}/R_{mS} Values for Various Materials

Materials	τ_{BM}/R_{mM} or τ_{BS}/R_{mS}
Annealing steel	0.60
Austenitic (solution heat treated)	0.80
Austenitic F60/90	0.65
Grey cast iron	0.90
Aluminum alloys	0.70
Titanium alloy (age-hardened)	0.60

Source: Adapted from VDI 2230, published by Beuth Verlag GmbH, Berlin (2003).

Estimate of a Bolt Diameter

The VDI 2230 guideline allows for the estimation of a bolt diameter and strength to withstand an applied work load F_A without risk of premature failure, using the data in Table 12.8. An operating temperature of 20°C (12.68°F) is assumed. Once a diameter is selected, double check the results by either calculating or testing the bolted joint.

The suggested procedure follows:

1. In column 1, select the next higher force to the work force F_A acting on the bolted joint.

2. Find the required minimum preload force (F_{Mmin}) that is higher than the work force F_A by proceeding the following step:
   a. four steps up for static or dynamic transverse (shear) force
   b. two steps up for dynamic and eccentric axial force
   c. one step up for either dynamic and concentric, or static and eccentric force
   d. no step for static and concentric axial force

3. Due to the preload scatter in various tightening techniques, obtain the required maximum bolt preload force (F_{Mmax}) by proceeding the following step from F_{Mmin}:
   a. two steps up for air driven torque tools
   b. one step up for tightening with a torque wrench or precision power screw driver
   c. no step up for the "turn of the nut" method or yield-controlled method

Table 12.8: Estimate of Bolt Diameter

Force (N)	Nominal Diameter (mm) Strength Grade		
	12.9	10.9	8.8
250			
400			
630			
1000	3	3	3
1600	3	3	3
2500	3	3	4
4000	4	4	5
6300	4	5	6
10,000	5	6	8
16,000	6	8	8
25,000	8	10	12
40,000	10	12	14
63,000	12	14	16
100,000	16	18	20
160,000	20	22	24
250,000	24	27	30
400,000	30	33	36
630,000	36	39	

Source: Adapted from VDI 2230, published by Beuth Verlag GmbH, Berlin (2003).

4. Determine the bolt size by the number next to this maximum preload force in columns 2 through 4 underneath the appropriate strength grade: 12.9, 10.9, 8.8.

These procedures provide a rough estimate of the bolt size as a starting point for more detailed VDI calculations to prevent potential failure modes of the joint. It is also recommended that the calculations be done in conjunction with finite element methods for a reliable design.

Estimate of Joint Elastic Compliance (Resilience)

Computation of the portion of external load transmitted to the bolt and the clamped plates requires the estimation of bolt and plate elastic compliances. The elastic compliance δ (i.e., resilience or flexibility) characterizes the capability of a component to deform elastically under a unit force, which can be defined as follows:

$$\delta = \frac{1}{k} = \frac{f}{F} = \frac{1}{EA} \tag{12.1}$$

where

k = elastic stiffness
F = applied force
f = deformation due to a force F
E = Young's modulus
A = cross-sectional area
l = component length

Even though there are many techniques to estimate elastic compliances of a bolt and clamped plates (Norton, 1996; Brickford & Nassar, 1998; Juvinall & Marshek, 2000; Shigley & Mischke, 2001; VDI, 2003), the VDI 2230 guideline is recommended and presented here.

Elastic Compliance of a Bolt

Experiments have shown that bolts exhibit more elastic compliances than the ones calculated based on the grip lengths. The VDI 2230 guideline specifies that additional compliance contributions from the bolt head, engaged threads, and the nuts or tapped threads should be included.

As shown in Figure 12.6, the compliance of a bolt δ_S can be obtained as follows:

$$\delta_S = \delta_K + \delta_1 + \delta_2 + \dots + \delta_G + \delta_M \tag{12.2}$$

where

δ_K = compliance of the bolt head
δ_1, δ_2 = compliance of the bolt sections
δ_3 = compliance of unengaged threads
δ_G = compliance of the engaged threads
δ_M = compliance of the nut or tapped hole

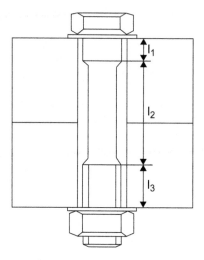

Figure 12.6
Schematic illustration of bolt dimensions used for calculation
of bolt elastic compliance.

The compliance of the bolt head δ_K is calculated by

- For hexagon head bolt

$$\delta_K = \frac{0.5d}{E_S A_N}. \tag{12.3}$$

- For socket head cap screw

$$\delta_K = \frac{0.4d}{E_S A_N}. \tag{12.4}$$

The compliance of the bolt section δ_i is

$$\delta_i = \frac{l_i}{E_S A_i}. \tag{12.5}$$

The compliance of unengaged threads is

$$\delta_3 = \frac{l_3}{E_S A_{d3}}. \tag{12.6}$$

The compliance of engaged threads is

$$\delta_G = \frac{0.5d}{E_S A_{d3}}. \qquad (12.7)$$

The compliance of the nut is

•   For a tapped thread joint (ESV)

$$\delta_M = \frac{0.33d}{E_M A_N}. \qquad (12.8)$$

•   For a through-bolted joint (DSV)

$$\delta_M = \frac{0.4d}{E_M A_N} \qquad (12.9)$$

where

A_N = nominal cross-sectional area of the bolt
A_i = cross-sectional area of an individual cylindrical bolt
A_{d3} = cross-sectional area of the minor diameter of the bolt threads
E_S = Young's modulus of the bolt
E_M = Young's modulus of the nut
l_i = length of an individual cylindrical bolt
l_3 = length of unengaged threads

Elastic Compliance of Clamped Plates for Concentrically Bolted Joints

When the bolt in a joint is tightened, the clamped plates deform. Linear elastic finite element analyses were performed to determine the axial compressive stress contour lines for a through-bolted joint and a tapped thread joint. As shown in Figure 12.7, the compressive stress distribution spreads out from the bolt contact areas and generates the conical volume that defines a clamping deformation cone (or pressure cone).

The slope of the deformation cone depends on the geometry of the clamped plates. It is assumed that elastic compliance of clamped plates can be calculated from this deformation cone. But this calculation could be complicated for such

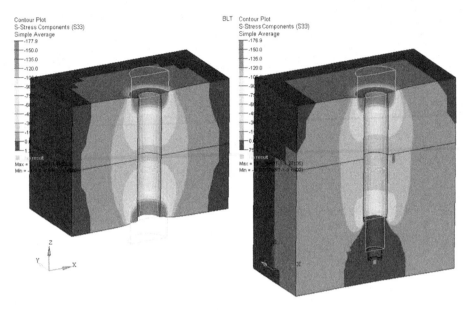

Figure 12.7
Axial compressive stress contour plots in the clamped plates for a through-bolted joint and a tapped thread joint.

a cone with continuous changing slope. With a simpler approach with a fixed cone angle φ, as introduced by VDI 2230, the elastic compliance of clamped plates can be easily estimated.

To illustrate how plate elastic compliance is calculated from the deformation cone, a concentrically through-bolted joint of a circular cross-section is used, where the center bolt line coincides with the center line of the clamped plates. As illustrated earlier in Figure 12.2, with a given Young's modulus of the clamped material E_p, the elastic compliance $d\delta_P$ of an infinite small plate thickness is calculated

$$d\delta_P = \frac{dz}{E_P A(z)} \qquad (12.10)$$

where

$$A(z) = \pi[(z\tan\varphi + d_W/2)^2 - (d_h/2)^2]$$
$$A(z) = \pi(z\tan\varphi + (d_W + d_h)/2)(z\tan\varphi + (d_W - d_h)/2)$$

So the elastic compliance of an upper plate is

$$\delta_P = \int_{z=0}^{z=h} \frac{dz}{E_P A(z)} = \frac{1}{\pi E} \int_0^h \frac{dx}{[x\tan\varphi + (d_w + d_h)/2][x\tan\varphi + (d_w - d_h)/2]}$$

$$= \frac{1}{\pi E_p d_h \tan\varphi} \ln\left[\frac{(d_w + 2h\tan\varphi - d_h)(d_w + d_h)}{(d_w + 2h\tan\varphi + d_h)(d_w - d_h)}\right].$$

$\qquad\qquad\qquad\qquad\qquad\qquad\qquad\qquad\qquad\qquad\qquad$ (12.11)

If the clamped plates of a through-bolted joint (DSV) have the same E_p with symmetrical frusta back to back, the two frusta would act as two identical springs in series. Setting the total clamped length as $l_K = 2h$ yields

$$\delta_P = \frac{2}{\pi E_p d_h \tan\phi} \ln\left[\frac{(d_w + d_h)(d_w + l_k\tan\phi - d_h)}{(d_w - d_h)(d_w + l_k\tan\phi + d_h)}\right].$$

$\qquad\qquad\qquad\qquad\qquad\qquad\qquad\qquad\qquad\qquad\qquad$ (12.12)

When the projected deformation cone exceeds the diameter of the interface surface of circular clamped plates of a through-bolted joint ($D_K > D_A$), the cone is then replaced by a deformation sleeve, as shown in Figure 12.8 where l_V and l_H are the one-side cone depth and the clamp sleeve length, respectively.

Similarly, Figure 12.9 depicts the same concept for calculation of the elastic compliance of the clamped plates of a tapped thread joint (ESV), except that one equivalent cone replaces the upper and the bottom cones with the same compliance.

The following are the procedures to estimate the elastic compliance of the clamped plates:

1. Calculate the limiting diameter of a deformation cone at an interface D_K.

$$D_K = d_w + w \times l_K \times \tan\phi \qquad\qquad (12.13)$$

   where $w = 1$ for a through-bolted joint (DSV); $w = 2$ for a tapped thread joint (ESV).

2. Determine the cone angle

   • For DSV joints

$$\tan\phi_D = 0.362 + 0.032\ln(\beta_L/2) + 0.153\ln y. \qquad (12.14)$$

   • For ESV joints

$$\tan\phi_E = 0.348 + 0.013\ln\beta_L + 0.193\ln y \qquad (12.15)$$

   where $\beta_L = l_K/d_W$ and $y = D_A/d_W$.

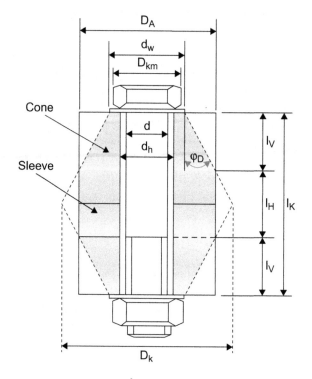

Figure 12.8
Through-bolted joint (DSV) with deformation
cones and a sleeve.

3. Estimate the plate elastic compliance
   - For $D_A \geq D_K$

$$\delta_P = \frac{2}{\pi w E_p d_h \tan \phi} \ln \left[\frac{(d_w + d_h)(d_w + w l_k \tan \phi - d_h)}{(d_w - d_h)(d_w + w l_k \tan \phi + d_h)} \right]. \qquad (12.16)$$

   - For $d_W < D_A < D_K$

$$\delta_P = \frac{1}{\pi E_p} \left\{ \frac{2}{w d_h \tan \phi} \ln \left[\frac{(d_w + d_h)(D_A - d_h)}{(d_w - d_h)(D_A + d_h)} \right] \right.$$

$$\left. + \frac{4}{D_A^2 - d_h^2} \left[l_K - \frac{(D_A - d_w)}{w \tan \phi} \right] \right\}. \qquad (12.17)$$

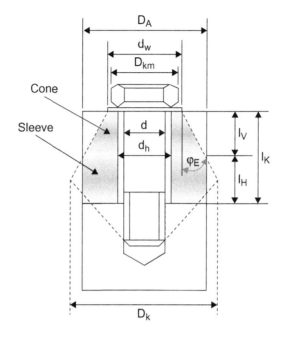

Figure 12.9
Tapped thread (screw) joint (ESV) with a deformation
cone and a sleeve.

External Loads—Load Plan Factor (n) and Load Factor (Φ)

An axial load to a bolted joint is not typically applied under the bolt head and the nut. Instead, the load is applied at some intermediate level of the clamped plates. Loading planes in a bolted joint are used to describe where the external load is applied. The location of the loading plane can be expressed by a percentage of the clamped plate length (l_k) with respect to the mating surface of the clamped plates.

For example, as depicted in Figure 12.10, the two loading planes of a joint under an external axial load F_A are located by $n_1 l_k$ and $n_2 l_k$. With the loading planes located, the load introduction factor n (= $n_1 + n_2$) can be determined. The n factor when multiplied by the clamped length (l_K), denotes the thickness of sections of clamped plates unloaded by F_A. Joint materials between the loading planes will be unloaded by F_A. Joint material outside of the loading planes will be further compressed.

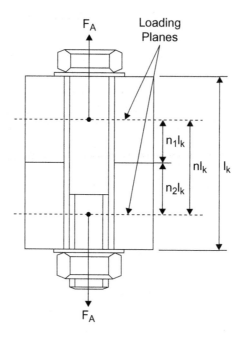

Figure 12.10
Loading plane factors.

Introducing the working load as close to the interface as possible is another method used to lessen the effect of the applied axial load on the bolt (F_{SA}). The closer the load introduction is to the interface the more the additional bolt load will be reduced. Given the theoretical case where the load is introduced directly at the interface ($n = 0$), $F_{SA} = 0$, the entire applied load is absorbed by the clamped plates.

The effect of a load plane factor n on the portion of an external load transmitted to the bolt can be analyzed by a spring model. Figure 12.11 shows that the plate elastic compliance can be decomposed into two parts: one $n\delta_P$ in series (a) and the other $(1-n)\delta_P$ in parallel (b) with the bolt compliance δ_S. Figure 12.11(b) presents the equivalent spring model to that in Figure 12.11(a).

Based on the theory of springs in series and in parallel, the following effective parallel springs δ'_P and δ'_S are defined as

$$\delta'_P = n\delta_P \quad \text{and} \quad \delta'_S = (1-n)\delta_P + \delta_S . \tag{12.18}$$

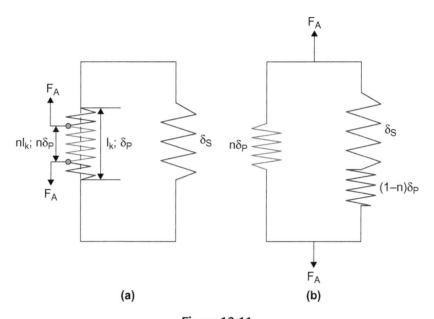

Figure 12.11
(a) A load plane factor on an actual spring model; (b) a combined
effective spring model.

The same elongation f as a result of the force transmitted to the clamped
plates F_{PA} and the bolt F_{SA} (the deformation compatibility condition) can be
calculated as

$$f = \delta'_P F_{PA} = \delta'_S F_{SA} \tag{12.19}$$

or

$$F_{PA} = \frac{\delta'_S}{\delta'_P} F_{SA}. \tag{12.20}$$

Since the force equilibrium ($F_A = F_{SA} + F_{PA}$) applies, we have the following load
distribution to the bolt:

$$F_{SA} = \frac{\delta'_P}{\delta'_S + \delta'_P} F_A = \frac{n\delta_P}{\delta_S + \delta_P} F_A = \Phi F_A \tag{12.21}$$

where

Φ = the load factor representing the percentage of the externally applied load to the bolt

For the load to the clamped plate, it follows,

$$F_{PA} = (1 - \Phi)F_A. \tag{12.22}$$

The n factor is not well determined and often is used as an adjustment coefficient. In practice, either an artificial number 1.0 or 0.5 is taken, or it can be obtained from finite element analysis or lab experiment.

Assembly Loads on Bolted Joints

When a bolted joint is tightened to an assembly tightening torque (M_A), the bolt is stretched and the clamped plates are compressed by the same amount of bolt preload (F_M). Also the applied torque can be balanced by the thread torque (M_G) and the under-head torque (M_K), which is expressed by

$$M_A = M_G + M_K. \tag{12.23}$$

Separate force-versus-deformation curves can be plotted on the same axes for the bolt and the clamped plates. Figure 12.12(a) shows plots of a preload joint, where the y-axis is the absolute value of the force and the x-axis represents the elongation of the bolt (f_{SM}) or the compressive deformation of the clamped plates (f_{PM}).

If the x-axis is the absolute value of the deformation, the two elastic curves can be combined into what is known as a joint diagram, as illustrated in Figure 12.12(b). The base-to-height ratio of the line represents the elastic compliance of the bolt or of the clamped plates (δ_S or δ_P).

When the tighten bolted joint is subjected to an applied axial load F_A, both the bolt and the clamped plates experience the same deformation increment ($f_{SA} = f_{PA}$) due to deformation compatibility, and the force equilibrium leads to $F_A = F_{SA} + F_{PA}$. From Figure 12.12(b), the bolt load will increase to $F_M + F_{SA}$ and the compressive clamped force will reduce to a residual clamp force F_{KR} (= $F_M - F_{PA}$).

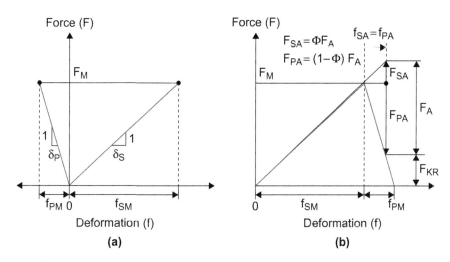

Figure 12.12
Elastic force versus deformation curves for the bolt
and clamped plates.

Embedding Effect

After the bolt assembly process, surface flattening occurs as a result of local plastic deformation of rough surfaces at contact areas, such as threads, head, and nut-bearing areas, and interfaces of clamped plates This phenomenon is referred to as *embedding*, which will significantly reduce the magnitude of preload.

It is assumed the same amount of preload loss F_Z in the bolt and clamped plates is due to the total plastic deformation (f_Z), which can be found in the joint diagram shown in Figure 12.13. The trigonometric relation confirms the following equation:

$$F_Z = \frac{f_Z}{f_{SM} + f_{PM}} F_M = \frac{f_Z}{\delta_S + \delta_P}. \qquad (12.24)$$

The amount of plastic deformation depends on the loading mode, the type of material, the average surface roughness value (R_Z) in μm, and the number of interfaces. According to VDI 2230, the empirical f_Z values for bolts, nuts, and clamped plates made of steels can be found in Table 12.9.

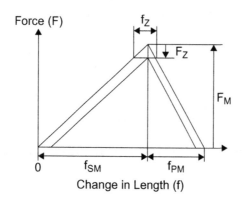

Figure 12.13
Effect of preload loss on the joint diagram.

Table 12.9: Embedding Deformation f_Z of Bolts, Nuts, and Compact Clamped Plates Made of Steel

Average Roughness Height (R_Z) in μm	Loading	Approximate Embedding Deformation in μm		
		Threads	Head or Nut-Bearing Area	Interfaces of Clamped Plates
$R_Z < 10$	axial	3	2.5	1.5
	shear	3	3	2
$10 \leq R_Z < 40$	axial	3	3	2
	shear	3	4.5	2.5
$40 \leq R_Z < 160$	axial	3	4	3
	shear	3	6.5	3.5

Source: Adapted from VDI 2230, published by Beuth Verlag GmbH, Berlin (2003).

Tightening Torque

The thread torque (M_G) can be calculated from the product of the tangential force on the thread contact surface and the thread pitch radius. Derived from the force equilibrium conditions for a bolt preload, normal and friction forces on a thread (flank) angle α on an inclined plane with an thread helix angle (lead angle) ϕ, the tangential force can be expressed as follows (Juvinall & Marshek, 2000; Shigley & Mischke, 2001):

$$F_t = F_M \frac{\mu_G \cos\phi + \cos(\alpha/2)\sin\phi}{\cos(\alpha/2)\cos\phi - \mu_G \sin\phi}. \qquad (12.25)$$

Since the value of $\mu_G \sin \phi$ is small and can be negligible, Equation (12.25) can be reduced to

$$F_t = F_M \left(\frac{\mu_G}{\cos(\alpha/2)} + \tan \phi \right). \tag{12.26}$$

If a full turn were developed for a thread advancement of a pitch, a triangle would be formed and would have the following relationship:

$$\tan \phi = \frac{P}{\pi d_2}. \tag{12.27}$$

Thus, the thread torque M_G is expressed by

$$M_G = \frac{d_2}{2}(F_t) = F_M \left(\frac{P}{2\pi} + \frac{d_2 \mu_G}{2\cos(\alpha/2)} \right). \tag{12.28}$$

Also, the under-head torque (M_K) is calculated by

$$M_K = F_M \frac{D_{km}}{2} \mu_K \tag{12.29}$$

where

D_{km} = the effective diameter of bolt head or nut-bearing area = $0.5(d_w + d_h)$

Finally the relation between the assembly tightening torque and the preload force is given:

$$M_A = M_G + M_K = F_M \left(\frac{P}{2\pi} + \frac{d_2 \mu_G}{2\cos(\alpha/2)} + \frac{D_{km}\mu_K}{2} \right). \tag{12.30}$$

For any metric bolt with $\alpha = 60°$, Equation (12.30) can be simplified to

$$M_A = M_G + M_K = F_M[(0.16P + 0.58d_2\mu_G) + 0.5D_{km}\mu_K]. \tag{12.31}$$

Assembly Preload

During tightening, an equivalent von Mises stress for a bolt under tension (σ_M) and torsional (τ_M) stresses is obtained:

$$\sigma_{redM} = \sqrt{\sigma_M^2 + 3\tau_M^2} \tag{12.32}$$

where

$$\sigma_M = \frac{F_M}{A_S} = \frac{4F}{\pi d_S^2}$$

$$\tau_M = \frac{M_G(d_S/2)}{l_P} = \frac{M_G}{W_P} = \frac{16M_G}{\pi d_S^3}.$$

d_S = stress diameter = $0.5(d_3 + d_2) = d - 0.9382P$
A_S = effective tensile stress area
l_P = polar moment of inertia = $\frac{1}{32}\pi d_S^4$
W_P = polar moment of resistance = $\frac{l_P}{d_S/2}$

Equation (12.32) can be reduced to

$$\sigma_{redM} = \sigma_M \sqrt{1 + 3\left(\frac{\tau_M}{\sigma_M}\right)^2} = \sigma_M \sqrt{1 + 3\left(\frac{4M_G}{d_o F_M}\right)^2}. \qquad (12.33)$$

It is also known from Equation (12.31) that

$$\frac{M_G}{F_M} = 0.16P + 0.58d_2\mu_G.$$

Thus,

$$\sigma_{redM} = \sigma_M \sqrt{1 + 3\left(\frac{2d_2}{d_S}\left(\frac{0.32P}{d_2} + 1.16\mu_G\right)\right)^2}. \qquad (12.34)$$

For an initial yielding occurring in the outer surface of the shank, the following criterion must be met:

$$\sigma_{redM} = \sigma_M \sqrt{1 + 3\left(\frac{2d_2}{d_S}\left(\frac{0.32P}{d_2} + 1.16\mu_G\right)\right)^2} = vR_{P0.2min} \qquad (12.35)$$

or

$$\sigma_M = \frac{vR_{P0.2min}}{\sqrt{1 + 3\left(\frac{2d_2}{d_S}\left(\frac{0.32P}{d_2} + 1.16\mu_G\right)\right)^2}} \qquad (12.36)$$

where

v = utilization of the initial or the gross yield stress during tightening (about 90%)

But VDI 2230 guideline specifies that the nominal stress at the onset of global yielding over the relevant cross-section should not exceed 90% of the minimum yield strength. In this case of fully plastic state, the following shear stress must be corrected accordingly as

$$\tau_M = \frac{12 M_G}{\pi d_S^3}. \tag{12.37}$$

Therefore, the following two acceptance criteria for permissible assembly stress (σ_{Mzul}) and preload (F_{Mzul}) apply:

$$\sigma_{Mzul} = \frac{v R_{P0.2min}}{\sqrt{1 + 3\left(\frac{3}{2}\frac{d_2}{d_S}\left(\frac{0.32P}{d_2} + 1.16\mu_G\right)\right)^2}} \tag{12.38}$$

$$F_{Mzul} = \sigma_M A_S = \frac{v R_{P0.2min} A_S}{\sqrt{1 + 3\left(\frac{3}{2}\frac{d_2}{d_o}\left(\frac{0.32P}{d_2} + 1.16\mu_G\right)\right)^2}}. \tag{12.39}$$

The smallest value of the thread friction coefficient must be used to calculate F_{Mzul}. For a 90% utilization of the minimum yield stress and a constant value for the coefficients of friction $\mu_G = \mu_K$, the assembly preloads F_{Mzul} and the tightening torque M_A for various bolts are tabulated in Table 12.10.

Bolt failure due to overtensioning occurs when the equivalent stress achieved during installation exceeds the tensile yield strength of the bolt. VDI 2230 specifies that to prevent the global yielding of a bolt, the following equation needs to be satisfied:

$$F_{Mmax} \leq F_{Mzul}. \tag{12.40}$$

Due to tool repeatability, M_A has a high and low end, resulting in maximum (F_{Mmax}) and minimum (F_{Mmin}) initial bolt preloads, immediately after tightening. Per VDI 2230, the maximum bolt preload F_{Mmax} is calculated as

$$F_{Mmax} = \alpha_A F_{Mmin}. \tag{12.41}$$

Table 12.10: Calculated Assembly Preload F_{Mzul} and Tightening Torque M_A

Size	Strength Grade	F_{Mzul} (kN) with μ_G =				M_A (Nm) with μ_G =			
		0.08	0.1	0.12	0.14	0.08	0.1	0.12	0.14
M4	8.8	4.6	4.5	4.4	4.3	2.3	2.6	3.0	3.3
	10.9	6.8	6.7	6.5	6.3	3.3	3.9	4.6	4.8
	12.9	8.0	7.8	7.6	7.4	3.9	4.5	5.1	5.6
M5	8.8	7.6	7.4	7.2	7.0	4.4	5.2	5.9	6.5
	10.9	11.1	10.8	10.6	10.3	6.5	7.6	8.6	9.5
	12.9	13.0	12.7	12.4	12.0	7.6	8.9	10.0	11.2
M6	8.8	10.7	10.4	10.2	9.9	7.7	9.0	10.1	11.3
	10.9	15.7	15.3	14.9	14.5	11.3	13.2	14.9	16.5
	12.9	18.4	17.9	17.5	17.0	13.2	15.4	17.4	19.3
M7	8.8	15.5	15.1	14.8	14.4	12.6	14.8	16.8	18.7
	10.9	22.7	22.5	21.7	21.1	18.5	21.7	24.7	27.5
	12.9	26.6	26.0	25.4	24.7	21.6	25.4	28.9	32.2
M8	8.8	19.5	19.1	18.6	18.1	18.5	21.6	24.6	27.3
	10.9	28.7	28.0	27.3	26.6	27.2	31.8	36.1	40.1
	12.9	33.6	32.8	32.0	31.1	31.8	37.2	42.2	46.9
M10	8.8	31.0	30.3	29.6	28.8	36	43	48	54
	10.9	45.6	44.5	43.4	42.2	53	63	71	79
	12.9	53.3	52.1	50.8	49.4	62	73	83	93
M12	8.8	45.2	44.1	43.0	41.9	63	73	84	93
	10.9	66.3	64.8	63.2	61.5	92	108	123	137
	12.9	77.6	75.9	74.0	72.0	108	126	144	160
M14	8.8	62.0	60.6	59.1	57.5	100	117	133	148
	10.9	91.0	88.9	86.7	84.4	146	172	195	218
	12.9	106.5	104.1	101.5	98.8	171	201	229	255
M16	8.8	84.7	82.9	80.9	78.8	153	180	206	230
	10.9	124.4	121.7	118.8	115.7	224	264	302	338
	12.9	145.5	142.4	139.0	135.4	262	309	354	395
M18	8.8	107	104	102	99	220	259	295	329
	10.9	152	149	145	141	314	369	421	469
	12.9	178	174	170	165	367	432	492	549
M20	8.8	136	134	130	127	308	363	415	464
	10.9	194	190	186	181	438	517	592	661
	12.9	227	223	217	212	513	605	692	773
M22	8.8	170	166	162	158	417	495	567	634
	10.9	242	237	231	225	595	704	807	904
	12.9	283	277	271	264	696	824	945	1057

Table 12.10: cont'd

Size	Strength Grade	F_{Mzul} (kN) with μ_G =				M_A (Nm) with μ_G =			
		0.08	0.1	0.12	0.14	0.08	0.1	0.12	0.14
M24	8.8	196	192	188	183	529	625	714	798
	10.9	280	274	267	260	754	890	1017	1136
	12.9	327	320	313	305	882	1041	1190	1329
M27	8.8	257	252	246	240	772	915	1050	1176
	10.9	367	359	351	342	1100	1304	1496	1674
	12.9	429	420	410	400	1287	1526	1750	1959
M30	8.8	313	307	300	292	1053	1246	1428	1597
	10.9	446	437	427	416	1500	1775	2033	2274
	12.9	522	511	499	487	1755	2077	2380	2662

Source: Adapted from VDI 2230, published by Beuth Verlag GmbH, Berlin (2003).

and

$$F_{Mmin} = F_{Kerf} + (1 - \Phi)F_A + F_Z \qquad (12.42)$$

where

α_A = tightening factor listed in Table 12.11
F_{Kerf} = required minimum clamp force to prevent plate slippage

It is important to have a minimum clamp force in the joint at all times to ensure sealing and to maintain a frictional grip in the interfaces. Per VDI 2230, the formula to calculate the required minimum clamp force in the interfaces is expressed as

$$F_{Kerf} = \max(F_{KQ}, F_{KP}) \qquad (12.43)$$

$$F_{KQ} = \frac{F_Q}{q_F \mu_T} + \frac{M_T}{q_F r_a \mu_T} \qquad (12.44)$$

$$F_{KP} = A_D p_{imax} \qquad (12.45)$$

where

A_D = sealing area
F_{KQ} = frictional grip to transmit a transverse load (F_Q) and a torque about the bolt axis (M_T)

Table 12.11: Approximate Tightening Factors (α_A)

Tightening Factor $\alpha_A = \dfrac{F_{Mmax}}{F_{Mmin}}$	Scatter (%) $\dfrac{\Delta F_M/2}{F_{Mmean}} = \dfrac{1-\alpha_A}{1+\alpha_A}$	Tightening Technique	Setting Technique
1.05 to 1.2	±2 to ±10	Elongation-controlled tightening with ultrasound	
1.1 to 1.5	±5 to ±20	Mechanical elongation measurement	Adjustment via longitudinal measurement
1.2 to 1.4	±9 to ±17	Yield-controlled tightening, motor or manually operated	Input of the relative torque and rotation angle coefficient
1.2 to 1.4	±9 to ±17	Angle-controlled tightening, motor or manually operated	Experimental determination of prelightening torque and angle rotation
1.2 to 1.6	±9 to ±23	Hydraulic tightening	Adjustment via length or pressure measurement
1.4 to 1.6	±17 to ±23	Torque-controlled tightening with torque wrench, indicating wrench, or precision tightening spindle with dynamic torque measurement	Experimental determination of required tightening torques on the original bolting part, e.g., by measuring bolt elongation
1.7 to 2.5	±26 to ±43	Torque-controlled tightening with torque wrench, indicating wrench, or precision tightening spindle with dynamic torque measurement	Determination of the required tightening torque by estimating the friction coefficient (surface and lubricating conditions)
2.5 to 4	±43 to ±60	Tightening with impact wrench or impact wrench with momentum control	Calibration of the bolt by means of retightening torque, made up of the required tightening torque (for the estimated friction coefficient) and an additional factor

Source: Adapted from VDI 2230, published by Beuth Verlag GmbH, Berlin (2003).

F_{KP} = special force required for sealing

q_F = number of slippage planes

r_a = torque radius from the bolt axis due to M_T

p_{imax} = maximum internal pressure to be sealed

Clamped Plate Pressure

Clamped plates crushing under excessive pressure occurs when the interfacial pressure achieved during installation exceeds the permissible pressure of one of the clamped plates. VDI 2230 specifies that the maximum surface contact pressure should not exceed the permissible pressure of the clamped materials, which is expressed by

$$p_{Mmax} = \frac{F_{Mzul}}{A_{Pmin}} \leq p_G \qquad (12.46)$$

where

p_{Mmax} = maximum surface pressure due to bolted joint assembly

p_G = permissible surface pressure of the clamped material

A_{Pmin} = minimum bolt head or nut-bearing area

Even though VDI 2230 lists the permissible surface pressure values for various clamped materials, Brickford and Nassar (1998) suggest that if the permissible pressure is not known for a material, use the minimum tensile yield strength of the material $R_{P0.2min}$.

Minimum Thread Engagement Length

The required minimum thread engagement length for a bolted joint is calculated based on the design criterion that the bolt shank should fail prior to the thread stripping, which requires that the maximum tensile force of the bolt must be lower than the maximum shear force of the nut thread during engagement. The following strength calculations can be made to evaluate the required thread engagement length.

Tensile Strength of a Bolt

When threaded bolts are tensile tested, these parts exhibit greater strengths than that predicted by the material strength and the root area, and behave as if they had a larger cross-sectional area. An empirical formula was developed to account for this phenomenon by defining the so-called tensile stress area, determined as follows:

$$A_S = \frac{\pi}{4}\left(\frac{d_2 + d_3}{2}\right)^2 = \frac{\pi}{4}(d - 0.9382P)^2. \tag{12.47}$$

The ultimate tensile strength of the threaded bolt can be computed

$$F_{mS} = R_{mS}A_S. \tag{12.48}$$

Shear Strength of External Threads

Stripping of external threads will occur as the material strength of the internal threads exceeds that of the external threads. As shown in Figure 12.14, stripping of the external threads would occur along the critical shear planes at the tips of the internal mating threads.

The critical shear area (A_{SGS}) for the length of external thread engagement (m_{eff}) is expressed as follows:

$$A_{SGS} = \frac{\pi D_1}{P}t_e m_{eff} = \frac{\pi D_1}{P}\left[\frac{P}{2} + (d_2 - D_1)\tan 30°\right]m_{eff} \tag{12.49}$$

where t_e = thickness of an external thread at the critical shear plane.

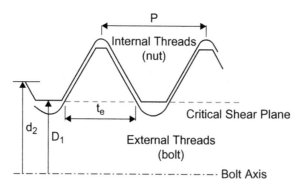

Figure 12.14
Shear stress area of external threads (bolt).

The ultimate shear force (F_{mGS}) that will fracture the external (bolt) threads is obtained:

$$F_{mGS} = \tau_{BS} A_{SGS} C_2 \tag{12.50}$$

where

C_2 = shear stress area reduction factor due to external thread bending

Shear Strength of Internal Threads

On the other hand, stripping of the internal threads will occur when the material strength of the external threads exceeds that of the internal threads. As shown in Figure 12.15, stripping of the internal threads will develop on the critical shear plane at the tips of the external mating threads.

The critical shear area (A_{SGM}) for the length of internal thread engagement (m_{eff}) is expressed as follows:

$$A_{SGM} = \frac{\pi d}{P} t_i m_{eff} = \frac{\pi d}{P} \left[\frac{P}{2} + (d - D_2) \tan 30° \right] m_{eff} \tag{12.51}$$

where

t_i = thickness of an internal thread at the critical shear plane

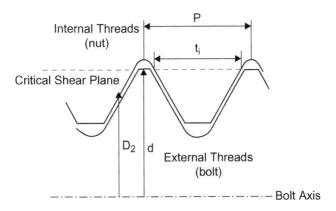

Figure 12.15
Shear stress area of internal threads (nut).

Thus, the ultimate shear force (F_{mGM}) that will fracture the internal (nut) threads is obtained:

$$F_{mGM} = \tau_{BM} A_{SGM} C_1 C_3 \tag{12.52}$$

where the correction factors (C_1 and C_3) are introduced to account for the reduction of the shear stress area due to the internal thread dilation (C_1) and thread bending (C_3) in the thread turns.

According to Barron (1998) and VDI 2230, the following empirical expressions for C_1, C_2, and C_3 can be determined:

- C_1, internal thread dilation strength reduction factor
  For $1.4 \leq s/d \leq 1.9$

$$C_1 = -2.61 + 3.8(s/d) - (s/d)^2. \tag{12.53}$$

  For $s/d > 1.9$

$$C_1 = 1.0. \tag{12.54}$$

- C_2, external thread strength reduction factor
  For $1 < R_S < 2.2$

$$C_2 = 5.594 - 13682 R_S + 14.107 R_S^2 - 6.057 R_S^3 + 0.9353 R_S^4. \tag{12.55}$$

  For $R_S \leq 1.0$

$$C_2 = 0.897. \tag{12.56}$$

- C_3, internal thread strength reduction factor
  For $0.4 < R_S < 1$

$$C_3 = 0.728 + 1.769 R_S - 2.896 R_S^2 + 1.296 R_S^3. \tag{12.57}$$

  For $R_S \geq 1$

$$C_3 = 0.897 \tag{12.58}$$

where s represents the width across flats for nuts and

$$R_S = \frac{\tau_{BM} A_{SGM}}{\tau_{BS} A_{SGS}} = \frac{\tau_{BM} d[0.5P + (d - D_2)\tan 30°]}{\tau_{BS} D_1 [0.5P + (d_2 - D_1)\tan 30°]}. \tag{12.59}$$

Design Rules

The basic design rules for axial loaded threaded joints are:

1. Any thread-stripping failure should be avoided

2. The bolt shank should fail in tension prior to thread stripping

The two rules require that the ultimate shear strength values of the external and internal threads should exceed the tensile strength of the bolted shank. And for serviceability (easy replacement), it is preferable that stripping of nut threads occurs prior to the failure of bolt threads, which implies

$$F_{mGS} \geq F_{mGM}. \tag{12.60}$$

Therefore, the overall design criteria can be expressed in the following form:

$$F_{mGS} \geq F_{mGM} \geq F_{mS}. \tag{12.61}$$

This inequality equation results in the minimum length of engagement as shown here:

$$m_{eff} = \frac{R_{mS} A_S P}{C_1 C_3 \tau_{BM} \left(\frac{P}{2} + (d - D_2)\tan 30°\right)\pi d}. \tag{12.62}$$

To account for the fact that a length of engagement of about 0.8P in the nut thread remains unloaded, the min length of engagement is

$$m_{eff} = \frac{R_{mS} A_S P}{C_1 C_3 \tau_{BM} \left(\frac{P}{2} + (d - D_2)\tan 30°\right)\pi d} + 0.8P. \tag{12.63}$$

Service Loads on Bolted Joints

The following failure modes as a result of excessive external loading should be prevented.

Bolt Yielding Due to Overstressing

Yielding in a bolt is due to overstressing in tension, bending, or torsion. It occurs when the equivalent stress achieved during service exceeds the tensile yield strength of the bolt, which can cause loss of preload. To prevent bolt yielding under service loads, the following equation applies:

$$S_F = \frac{R_{p0.2min}}{\sigma_{redB}} \geq 1.0 \qquad (12.64)$$

where

S_F = safety factor against bolt yielding

$\sigma_{redB} = \sqrt{\sigma_z^2 + 3(k_\tau \tau)^2}$ = von Mises equivalent stress

$\sigma_z = \frac{1}{A_S}(F_{Mzul} + F_{SAmax})$

$\tau = \frac{16M_G}{\pi d_S^3}$

k_τ = shear stress reduction factor for any working stress state = 0.5 recommended by VDI 2230

Clamp Load Loss Due to Gapping

Gapping occurs when the service load causes the clamp force to decrease to zero, resulting in gaps between the bolt and the clamped plates. To prevent the clamp load loss due to gapping, the following minimum residual clamp load F_{KRmin} at the interface is required:

$$F_{KRmin} = \frac{F_{Mzul}}{\alpha_A} - (1-\varphi)F_{Amax} - F_Z \geq 0. \qquad (12.65)$$

Thread Stripping

Thread stripping occurs when the shear stress at the threads exceeds the shear stress of the threaded material. Typically having the bolt break rather than the thread strip is preferred due to repair costs. The minimum engagement length in Equation (12.63) is required.

Clamped Plates Crushing Due to Excessive Pressure

Plate crushing occurs when the interfacial pressure achieved during service loading exceeds the permissible pressure of one of the clamped plates. As a result, it will crush the plates and cause loss of preload. The safety factor against clamped plates crushing, Sp, is defined as

$$S_P = \frac{p_G}{p_{Bmax}} \geq 1.0 \qquad (12.66)$$

where

$$p_{Bmax} = \frac{F_{Mmax} + F_{PAmax}}{A_{pmin}} = \text{induced surface pressure}$$

F_{Mmax} = maximum bolt preload immediately after tightening
F_{PAmax} = portion of the working axial load, which unloads the clamped plates
A_{pmin} = bearing or contact area

Slipping Due to Transverse Loads

Slipping occurs when a transverse force or torsional moment exceeds the reaction force of the normal force multiplied by the coefficient between clamped plates. A slip in a single direction may be acceptable. If slip occurs in opposite directions and repeats, fretting may occur causing loss of clamping load or the nut may back off. The following equation is required to prevent slipping due to transverse loads:

$$\frac{F_{KRmin}\mu_T}{S_G} \geq \frac{F_Q}{q_F} + \frac{M_T}{q_F r_a} \qquad (12.67)$$

where

F_{KRmin} = minimum residual clamp load as defined in Equation (12.65)
S_G = safety factor against slipping (= 1.2 for static loading; 1.8 for cyclic loading, recommended by VDI 2230)

Fatigue Failure

Fatigue failure occurs when alternating service loading amplitude exceeds the endurance limit of a bolt. Fatigue life may reduce significantly if pitting or corrosion occurs. VDI 2230 adopts the infinite-life design concept for bolted joints by using the nominal stress-life (S-N) approach. So the following failure criterion applies:

$$\frac{\sigma_{AS}}{S_D} \geq \sigma_a = \frac{F_{SAmax} - F_{SAmin}}{2A_S} = \frac{\Phi(F_{Amax} - F_{Amin})}{2A_S} \qquad (12.68)$$

where

σ_{AS} = fatigue limit = amplitude at 2×10^6 cycles (in N/mm^2 = MPa)
S_D = safety factor against fatigue (= 1.2 recommended by VDI 2230)
σ_a = nominal stress amplitude, in MPa, calculated on the tensile stress area

Experiments show that fatigue strengths of rolled threads made by rolling after heat treatment of the bolts, resulting in compressive residual stresses at the thread root, are better than those rolled before heat treatment. VDI 2230 recommends the fatigue limit of a bolted joint can be estimated as follows:

- For rolled threads before heat treatment (SV) is

$$\sigma_{ASV} = 0.85 \left(\frac{150}{d} + 45 \right). \tag{12.69}$$

- For rolled threads after heat treatment (SG) is

$$\sigma_{ASG} = \left(2 - \frac{F_{Sm}}{F_{0.2min}} \right) \sigma_{ASV} \tag{12.70}$$

where

F_{Sm} = mean nominal stress level = $\frac{F_{SAmax} + F_{SAmin}}{2} + F_{Mzul}$
$F_{0.2min}$ = minimum yield force = $A_S R_{P0.2min}$

The fatigue strength in the finite life regime where 10^4cycles $\leq N_f \leq 2 \times 10^6$ cycles can be calculated

- For rolled threads before heat treatment,

$$\sigma_{AZSV} = \sigma_{ASV} \left(\frac{2 \times 10^6}{N_f} \right)^{1/3}. \tag{12.71}$$

- For rolled threads after heat treatment,

$$\sigma_{AZSG} = \sigma_{ASG} \left(\frac{2 \times 10^6}{N_f} \right)^{1/6}. \tag{12.72}$$

Application

As shown in Figure 12.16, the tapped thread (screw) joint (ESV) is calculated as a concentrically loaded and concentrically clamped joint, where the clamped plates are made of plain structural steel that has the ultimate tensile strength value of

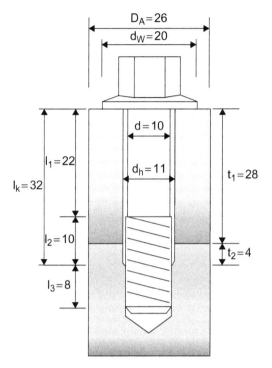

Figure 12.16
Dimensions of a tapped thread (screw) joint
(all the units in mm).

$R_{mM} = 470$ N/mm^2. This bolted joint is subjected to dynamic axial loads cycling from 15,000 N to 0 N. This joint is also to be tightened using an indicating torque wrench.

Both the fastener and the clamped plates are made of steel with $E_S = E_P = 206,850$ N/mm^2. The relevant surface roughness value is $R_Z = 16$ μm.

Solution

Preliminary Estimate of the Bolt Diameter

Using the information in Table 12.8 for the concentrically applied dynamic working load of $F_{Amax} = 15,000$ N, the bolt nominal diameter can be determined as follows:

1. Choose $F_A = 16,000$ N as the next comparable load to F_{Amax} listed earlier in Table 12.1.

Table 12.12: Bolt Dimensions

Name	Variable	Dimensions
Pitch	p (mm)	1.5
Nominal diameter	d (mm)	10
Pitch diameter	d_2 (mm)	9.03
Minor diameter	d_3 (mm)	8.16
Stress diameter	d_S (mm)	8.60
Shank diameter	d_{l_1} (mm)	9
Shank length	l_1 (mm)	22
Exposed thread length	l_2 (mm)	10
Engaged thread length	l_3 (mm)	8
Bearing diameter	d_W (mm)	20
Clamping length	l_K (mm)	32
Bore diameter	d_h (mm)	11
Maximum diameter of the solid circulate plates	d_A (mm)	26

2. Increase load level one step up for dynamically and concentrically applied axial load of $F_M = 16{,}000$ N, resulting $F_{Mmin} = 25{,}000$ N.

3. Increase the load level F_{Mmin} one step up for the preload variability of the torque wrench, having $F_{Mmax} = 40{,}000$ N.

4. Determine a bolt size M10 from column 2 of Table 12.1, provided the strength grade 12.9 is selected.

After a bolt is selected, $M10 \times 1.5 - 12.9$, the dimensions listed in Table 12.12 are obtained.

Tightening Factor

Based on the indicating torque wrench and the determination of the required tightening torque by estimating the friction coefficient, the tightening factor is obtained from Table 12.11:

$$\alpha_A = 1.7$$

Required Minimum Clamped Load

Since there is no minimum clamp force required in the joint at all times to ensure sealing due to any internal pressure $p_{i,max}$ and to maintain a frictional grip in the interfaces due to a transverse load (F_Q) or a torque (M_Y),

$$F_{Kerf} = \max(F_{KQ}, F_{KP}) = 0 \, kN$$

Calculation of Elastic Compliances and the Load Factor

The elastic compliance of a bolt δ_S can be obtained as follows:

$$\delta_S = \delta_K + \delta_1 + \delta_2 + \delta_G + \delta_M.$$

The compliance of the hexagon head bolt δ_K is calculated as follows:

$$\delta_K = \frac{0.5d}{E_S A_N} = \frac{0.5 \times 10}{206,850 \times 0.25 \times \pi \times 10^2} = 3.078 \times 10^{-7} \text{ mm/N}.$$

The compliance of the bolt shank δ_1 is

$$\delta_1 = \frac{l_1}{E_S A_1} = \frac{22}{206,850 \times 0.25 \times \pi \times 9^2} = 1.672 \times 10^{-6} \text{ mm/N}.$$

The compliance of unengaged threads is

$$\delta_2 = \frac{l_2}{E_S A_{d3}} = \frac{10}{206,850 \times 0.25 \times \pi \times 8.16^2} = 9.245 \times 10^{-7} \text{ mm/N}.$$

The compliance of engaged threads is

$$\delta_G = \frac{0.5d}{E_S A_{d3}} = \frac{0.5 \times 10}{206,850 \times 0.25 \times \pi \times 8.16^2} = 4.622 \times 10^{-7} \text{ mm/N}.$$

The compliance of the tapped thread plates is

$$\delta_M = \frac{0.33d}{E_M A_N} = \frac{0.33 \times 10}{206,850 \times 0.25 \times \pi \times 10^2} = 2.031 \times 10^{-7} \text{ mm/N}.$$

Thus, $\delta_S = 3.569 \times 10^{-6}$ mm/N.

For a tapped thread joint (ESV), the limiting diameter of a deformation cone at the interface D_K is calculated:

$$D_K = d_w + 2l_K \tan \phi$$

with

$$\beta_L = \frac{l_K}{d_W} = \frac{32}{20} = 1.6$$

$$y = \frac{D'_A}{d_W} = \frac{D_A}{d_W} = \frac{26}{20} = 1.3.$$

For ESV,

$$\tan\phi_E = 0.348 + 0.013\ln\beta_L + 0.193\ln y = 0.405$$

and

$$D_K = d_w + 2l_K\tan\phi = 20 + 2\times 32\times 0.405 = 45.9\ \text{mm}.$$

For $d_W = 20\ \text{mm} < D_A = 26\ \text{mm} < D_K = 45.9\ \text{mm}$, the elastic compliance of the clamped plates is calculated:

$$\delta_P = \dfrac{\dfrac{2}{wd_h\tan\phi}\ln\left[\dfrac{(d_w+d_h)(D_A-d_h)}{(d_w-d_h)(D_A+d_h)}\right] + \dfrac{4}{D_A^2-d_h^2}\left[l_K - \dfrac{(D_A-d_w)}{w\tan\phi}\right]}{\pi E_p}$$

$$\delta_P = \dfrac{\dfrac{2}{2\times 11\times 0.405}\ln\left[\dfrac{(20+11)(26-11)}{(20-11)(26+11)}\right] + \dfrac{4}{26^2-11^2}\left[32 - \dfrac{(26-20)}{2\times 0.405}\right]}{\pi\times 206{,}850}$$

$$= 3.881\times 10^{-7}\ \text{mm/N}.$$

Determination of the Loss of Preload Due to Embedding

Based on the axial load and the surface roughness value of $R_Z = 16\ \mu m$, the empirical f_Z values for clamped plates made of steel, which can be found earlier in Table 12.9, for the threads, the head-bearing area, and the inner interfaces are $3\ \mu m$, $3\ \mu m$, and $2\ \mu m$, respectively. So the total amount of embedment is

$$f_Z = 8\ \mu m = 8\times 10^{-3}\ \text{mm}.$$

The loss of the preload is calculated as

$$F_Z = \dfrac{f_Z}{\delta_S + \delta_P} = \dfrac{8\times 10^{-3}}{3.569\times 10^{-6} + 3.881\times 10^{-7}} = 2022\ \text{N}.$$

Determination of the Minimum Assembly Preload

The minimum assembly preload is calculated as

$$F_{Mmin} = F_{Kerf} + (1-\Phi)F_{Amax} + F_Z = 0 + (1-0.098)\times 15{,}000 + 2022 = 15{,}552\ \text{N}.$$

Determination of the Maximum Assembly Preload

The maximum bolt preload F_{Mmax} is calculated as

$$F_{Mmax} = \alpha_A F_{Mmin} = 1.7 \times 15,552 = 26,438 \, N.$$

Checking the Bolt Size and Strength

For a 90% utilization of the minimum yield stress and a coefficient of friction in the thread $\mu_G = 0.1$, the assembly preload capacity F_{Mzul} can be taken from Table 12.10 earlier in the chapter:

For strength grade 12.9: $F_{Mzul} = 52.1 \, kN$
For strength grade 10.9: $F_{Mzul} = 44.5 \, kN$
For strength grade 8.8: $F_{Mzul} = 30.3 \, kN$
For the cost effective bolt design, choose M10 × 1.5 – 8.8, since
  $F_{Mmax} = 26.4 \, kN < F_{Mzul} = 30.3 \, kN.$

Working Stress

The maximum bolt force is

$$F_{Smax} = F_{Mzul} + \varphi F_{Amax} = 30,300 + 0.098 \times 15,000 = 31,770 \, N.$$

The tensile stress area is calculated as

$$A_S = \frac{\pi}{4} d_S^2 = \frac{\pi}{4} (8.60)^2 = 58.09 \, mm^2.$$

So the maximum tensile stress is obtained:

$$\sigma_{zmax} = \frac{F_{Smax}}{A_S} = \frac{31,770}{58.09} = 546.91 \, N/mm^2.$$

Also the maximum torsional stress is calculated:

$$\tau_{max} = \frac{M_G}{W_P}$$

where

$$M_G = F_{Mzul}(0.16P + 0.58d_2\mu_G) = 303,00(0.16 \times 1.5 + 0.58 \times 9.03 \times 0.1)$$
$$= 23,141 \, N \, mm$$
$$M_G = 23,141 \, N \, mm$$
$$W_P = \frac{\pi}{16} d_s^3 = \frac{\pi}{16} \times 8.60^3 = 124.9 \, mm^3$$

So

$$\tau_{max} = \frac{23,141}{124.9} = 185.3 \, \text{N/mm}.$$

With $k_\tau = 0.5$. the von Mises equivalent stress is calculated as

$$\sigma_{redB} = \sqrt{\sigma_z^2 + 3(k_\tau \tau)^2} = \sqrt{546.91^2 + 3 \times (0.5 \times 185.3)^2} = 570 \, \text{N/mm}^2$$

Thus,

$$\sigma_{redB} = 570 \, \text{N/mm}^2 < R_{p0.2min} = 660 \, \text{N/mm}^2$$

and

$$S_F = \frac{R_{p0.2min}}{\sigma_{redB}} = 1.16 \geq 1.0.$$

The bolt of $M10 \times 1.5 - 8.8$ ($R_{mS} = 830 \, \text{N/mm}^2$; $R_{p0.2min} = 660 \, \text{N/mm}^2$) meets the design requirement.

Alternating Stress

The applied nominal stress amplitude on the bolt is determined:

$$\sigma_a = \frac{F_{SAo} - F_{SAu}}{2A_S} = \frac{\varphi(F_{Amax} - F_{Amin})}{2A_S} = \frac{0.098(15,000 - 0)}{2 \times 0.25 \times \pi \times 8.60^2} = 12.65 \, \text{N/mm}^2.$$

The endurance limit for bolts rolled before heat treatment (SV) is

$$\sigma_{ASV} = 0.85 \left(\frac{150}{d} + 45 \right) = 51 \, \text{N/mm}^2.$$

Therefore,

$$S_D = \frac{\sigma_{ASG}}{\sigma_a} = 4.0 \geq 1.2, \text{recommended by VDI 2230.}$$

The endurance limit for bolts rolled after heat treatment (SG),

$$\sigma_{ASG} = \left(2 - \frac{F_{Sm}}{F_{0.2min}} \right) \sigma_{ASV}$$

where

$$F_{Sm} = \frac{\varphi(F_{Amax} + F_{Amin})}{2} + F_{Mzul} = \frac{0.098(15,000 + 0)}{2} + 30,300 = 31,035 \, \text{N}$$

$$F_{0.2min} = A_S R_{0.2min} = 0.25 \times \pi \times 8.60^2 \times 660 = 38,338 \, \text{N}$$

So

$$\sigma_{ASG} = \left(2 - \frac{F_{Sm}}{F_{0.2min}}\right)\sigma_{ASV} = \left(2 - \frac{31,035}{38,338}\right) \times 51 = 60.7\,\text{N/mm}^2.$$

Therefore,

$$S_D = \frac{\sigma_{ASG}}{\sigma_a} = 4.8 \geq 1.2, \text{recommended by VDI 2230.}$$

Both bolts rolled before and after treatment will work for design for infinite life.

Determination of the Surface Pressure

The minimum bolt head-bearing area is

$$A_{pmin} = \frac{\pi}{4}(d_W^2 - d_h^2) = \frac{\pi}{4}(20^2 - 11^2) = 219.1\,\text{mm}^2.$$

For the assembly state:

$$p_{Mmax} = \frac{F_{Mzul}}{A_{pmin}} = \frac{30,300}{219.1} = 138.3\,\text{N/mm}^2 \leq p_G \approx R_{0.2min} = 660\,\text{N/mm}^2,$$

so the safety factor is

$$S_P = \frac{R_{p0.2min}}{p_{Mmax}} = 4.77 \geq 1.0.$$

For the service state:

$$p_{Bmax} = \frac{F_{Mmax} + F_{PAmax}}{A_{pmin}} = \frac{F_{Mzul} + (1 - \varphi) \times F_A}{A_{pmin}} = \frac{30,300 + (1 - 0.098) \times 15,000}{219.1}$$

$$p_{Bmax} = \frac{30,300 + (1 - 0.098) \times 15,000}{219.1} = 200\,\text{N/mm}^2$$

$$S_P = \frac{R_{p0.2min}}{p_{Bmax}} = \frac{660}{200} = 3.3 \geq 1.0.$$

Minimum Length of Engagement

With the choice of the bolt of M10 × 1.5 – 8.8, the ultimate tensile strength value of the steel bolt is $R_{mS} = 830\,\text{N/mm}^2$. The clamped plates are made of plain structural steel with $R_{mM} = 470$ N/mm$_2$. The ultimate shear strength values of the bolt and the clamped material can be estimated from what is shown earlier in Table 12.7 as

$$\tau_{BS} = 0.6 \times R_{mS} = 498 \, \text{N/mm}^2$$
$$\tau_{BM} = 0.6 \times R_{mM} = 282 \, \text{N/mm}^2.$$

The tensile stress area is determined:

$$A_S = 0.25 \times \pi \times 8.60^2 = 58.09 \, \text{mm}^2.$$

For the tapped thread joint (ESV), the dilation strength reduction C_1 is equal to 1.0, meaning little dilation of the internal thread parts due to the thicker walls. It is also assumed the dimensions of the internal threads follow

$$D_1 \approx d_3 \quad \text{and} \quad D_2 \approx d_2$$

thus,

$$R_S = \frac{\tau_{BM} A_{SGM}}{\tau_{BS} A_{SGS}} = \frac{\tau_{BM} d [0.5 \, P + (d - D_2) \tan 30°]}{\tau_{BS} D_1 [0.5 \, P + (d_2 - D_1) \tan 30°]}$$

$$R_S = \frac{282 \times 10 \times [0.5 \times 1.5 + (10 - 9.03) \tan 30°]}{498 \times 8.6 \times [0.5 \times 1.5 + (9.03 - 8.16) \tan 30°]} = 0.689.$$

The internal thread strength reduction factor C_3 is then calculated:

$$C_3 = 0.728 + 1.769 \times 0.689 - 2.896 \times 0.689^2 + 1.296 \times 0.689^3 = 0.996.$$

The minimum length of engagement is

$$m_{eff} = \frac{R_{mS} A_S P}{C_1 C_3 \tau_{BM} \left(\dfrac{P}{2} + (d - D_2) \tan 30° \right) \pi d} + 0.8 \, P$$

$$m_{eff} = \frac{830 \times 50.09 \times 1.5}{1.0 \times 0.996 \times 282 \left(\dfrac{1.5}{2} + (10 - 9.03) \tan 30° \right) \pi \times 10} + 0.8 \times 1.5 = 6.6 \, \text{mm}.$$

The actual length of engagement is 10 mm, which exceeds the minimum engagement length of 6.6 mm.

Determination of the Tightening Torque

For $\mu_G = \mu_K = 0.1$, the required tightening torque, according to Table 12.10, is $M_A(Nm) = 43$ Nm.

Finite Element-Based Fatigue Analysis of Thread Bolts

The direct use of the VDI 2230 guideline is rather difficult for most complex jointed structures. For example, an engine block structure with bolted joints, as shown in Figure 12.17, is too complicated to analyze with the VDI guide-line. The reasons are simple:

- The guideline is mainly valid for cylindrical or prismatic clamped plates.

- The guideline is also applicable for single-bolted joints, not for multi-bolted joints.

- The elastic compliance for clamped plates is calculated based on a simplified deformation cone/sleeve approach.

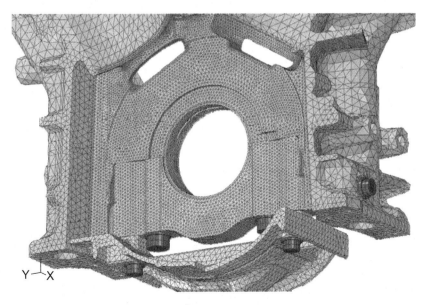

Figure 12.17
Detailed FE model for an engine block.

- The load introduction factor n that influences the load transmitting capability to a bolt is challenging to determine for a complex bolted joint structure.

These limitations can be overcome by an finite element method because the compliances of the bolt and clamped parts in a complex geometry and loaded structure can be accurately calculated. For a simple FE model with beam elements for bolts, the forces transmitted to the bolts can be easily captured and the nominal stress approach as recommended in the VDI guideline could be used for stress and fatigue assessment of the bolts.

For a detailed FE model including brick elements for bolt threads, the local stresses at the high stress concentration locations can be conveniently obtained and the local stress-based or strain-based multiaxial fatigue approach can be used for fatigue damage assessment of the bolts.

Two commonly used FE models for threaded bolts are introduced.

Solid Bolt Model

The solid bolt model assembly (Kim et al., 2007; Buhr et al., 2009) modeled by three-dimensional high-order tetrahedral elements, as shown in Figure 12.18, is the most accurate FE among all. This detailed model is created by an automatic meshing generation routine. In general, two element types such as bricks (also called hexahedrons) and tetrahedrons are available for the finite element analysis (FEA).

The linear brick elements have the advantage over the linear tetrahedral elements because it is well known that linear tetrahedral elements perform poorly in problems with plasticity, nearly incompressible materials, and acute bending. In addition, the use of high-order tetrahedral elements would overcome these issues and improve computational accuracy.

Moreover, there are quite robust meshing routines for many complex structural volumes using tetrahedral elements. Although there are meshing routines for volumes using brick elements, those routines are not as robust and may not always work for complex volumes. For a variety of reasons, an automatic meshing routine that will generate high-order tetrahedral elements for complex bolted joints is preferable.

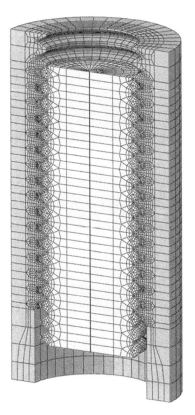

Figure 12.18
Portion of a solid brick model for a tapped thread bolt.

For finite element analysis, the finer the mesh size, the better the stress result at a stress concentration location and the more expensive the computational CPU times will be. The optimal mesh size can be determined by gradually reducing the mesh size until a stationary stress output is reached, or until the optimal linear element size is defined by one-tenth of the notch root radius, based on the rule of thumb.

For example, in the study of determining local stresses in a bolted joint using FEA, the authors (Fares et al., 2006; Lehnoff & Bunyard, 2000) use the first-order mesh size of about one-tenth of the notch radius. For a concentric bolted joint under concentric loads, the bolted-joint model can be simplified by an axi-symmetric model (Lehnhoff & Bunyard, 2000; DeJack et al., 2010).

The surface-to-surface contact elements need to be placed on all contacts between the mating surfaces. The thermal deformation model is employed in the bolt model to simulate the bolt pretension force. In this approach, the thermal expansion coefficient is assumed to be a unit and the temperature difference ΔT is calculated by

$$\Delta T = \frac{4F_M}{\pi d^2 E} \tag{12.73}$$

where

   E = elastic modulus of the material
   d = nominal diameter of the bolt
   F_M = bolt preload load after the assembly tightening torque

The three-dimensional solid bolt models are typically useful for thermo-mechanical fatigue analysis of a complex bolted structure if a visco-plastic-elastic constitutive model and its material parameters are given. The finite element stress outputs from high stress concentration locations can then be used to estimate the fatigue life of a bolted structure by using the stress-based or the strain-based multiaxial fatigue theories. However, this approach requires extensive computational CPU time and memory usage.

Spider Bolt Model

The spider bolt model (Kim et al., 2007; Buhr et al., 2009) is the simplest approach, which does not need to model the thread geometry, and it just ties the head–nut and mating surfaces together with beam elements. This model comprises one three-dimensional beam element for the bolt shank and two web-like beam elements for the head and nut contacts. The physical properties of the bolt such as the cross-sectional area, the moment of inertia, and the height can be used as an input to these beam elements.

All the tensile, bending, and thermal loads can be transferred through the spider elements to the bolt. The bolt pretension can be modeled by applying an initial strain ε_o to the beam element by

$$\varepsilon_o = \frac{4F_M}{\pi d^2 E}. \tag{12.74}$$

The spider bolt model does not require contact surface elements and is capable of calculating the bolt forces because the elastic compliances for bolts and clamped plates are well represented in these models. However, due to the lack of detailed solid elements for threads, the local stresses at high stress concentration areas cannot be determined.

For the fatigue damage assessment of a bolted structure, the nominal stress approach recommended by the VDI 2230 guideline can be employed without any additional FE bolt analysis. But if the local stress-based or strain-based multiaxial fatigue theories are chosen, an FE substructure technique, treating each bolt model with the detailed thread geometry as a super element for the global analysis, needs to be performed.

For example, a global FE analysis was first conducted on the complex bolted structure with an ABAQUS® (ABAQUS Inc.) thread bolt connection element as shown in Figure 12.19, and then an FE substructure analysis was performed on the detailed brick bolt model, as depicted in Figure 12.18, for local stresses and strains.

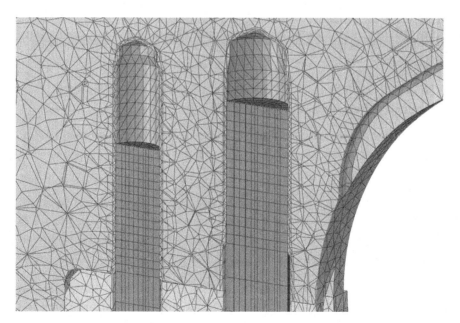

Figure 12.19
ABAQUS thread bolt connection model used in the global model.

Finally, for the fatigue damage assessment of rolled threads after heat treatment, the nominal stress approach in the VDI 2230 guideline is recommended and the spider bolt model should be the choice for an FE analysis because the presence of residual stress after a rolling process is empirically considered as part of the nominal S-N curve. The local stress-based or strain-based multiaxial fatigue analysis could be an option if these residual stresses can be analytically or experimentally quantified.

For the fatigue damage assessment of rolled threads before heat treatment where the beneficial residual stresses due to the rolling process are diminishing after heat treatment, either the nominal stress approach or the local stress-based or strain-based multiaxial fatigue analysis can be used and either FE bolt model can be employed for the stress analysis.

Summary

The VDI 2230 guideline to design and analyze ISO metric bolted joints has been presented, which has the following limitations:

- The guideline is mainly valid for cylindrical or prismatic clamped plates.

- The guideline is also applicable for single-bolted joints, not for multi-bolted joints.

- The elastic compliance for clamped plates is calculated based on a simplified deformation cone/sleeve approach.

- The load introduction factor n that influences the load transmitting capability to a bolt is challenging to determine for a complex bolted joint structure.

An FE analysis becomes necessary in a complex bolted structure because the compliances of the bolt and clamped parts in a complex geometry and loaded structure can be accurately calculated. Two different FE models for bolts (solid brick and spider bolt models) are reviewed.

For fatigue assessment of rolled threads after heat treatment, the nominal stress approach in the VDI 2230 guideline is recommended and the spider bolt model is the choice. For fatigue assessment of rolled threads before heat treatment, either the nominal stress approach or the local stress-based or the strain-based multiaxial fatigue analysis can be used due to the absence of residual stresses, and both FE bolt models can be employed for the stress analysis.

References

Brickford, J. H., & Nassar, S. (1998). *Handbook of bolts and bolted joints*. New York: Marcel Dekker.

Barron, J. (1998). Computing the strength of a fastener. In H. H. Brickford, & S. Nassar (Eds.), *Handbook of bolts and bolted joints* (pp. 163–185). New York: Marcel Dekker.

Brickford, J. H. (1998). Selecting preload for an existing joint. In H. H. Brickford, & S. Nassar (Eds.), *Handbook of bolts and bolted joints* (pp. 659–686). New York: Marcel Dekker.

Buhr, K., Haydn, W., Bacher-Hoechst, M., Wuttke, U., & Berger, C. (2009). Finite-element-based methods for the fatigue design of bolts and bolted joints (SAE Paper No. 2009-01-0041). Detroit, MI: SAE World Congress.

DeJack, M. A., Ma, Y., & Craig, R. D. (2010). Bolt load relaxation and fatigue prediction in threads with consideration of creep behavior for die cast aluminum (SAE Paper No. 2010-10-0965). Detroit, MI: SAE World Congress.

Fares, Y., Chaussumier, M., Daidie, A., & Guillot, J. (2006). Determining the life cycle of bolts using a local approach and the Dan Van criterion. *Fatigue & Fracture of Engineering Materials & Structures, 29,* 588–596.

Juvinall, R. C., & Marshek, K. M. (2000). *Fundamentals of machine component design* (3rd ed.). New York: John Wiley & Sons.

Kim, J., Yoon, J.-C, & Kang, B.-S. (2007). Finite element analysis and modeling of structure with bolted joints. *Applied Mathematical Modeling, 31,* 895–911.

Lehnhoff, T. F., & Bunyard, B. A. (2000). Bolt thread and head fillet stress concentration factors. *Journal of Pressure Vessel Technology, Transactions of the ASME, 122,* 180–185.

Marbacher, B. (1998). Metric fasteners. In H. H. Brickford, & S. Nassar (Eds.), *Handbook of bolts and bolted joints* (pp. 187–231). New York: Marcel Dekker.

Norton, R. L. (1996). *Machine design: An integration approach*. Englewood Cliffs, NJ: Prentice-Hall.

Shigley, J. E., & Mischke, C. R. (2001). *Mechanical engineering design* (6th ed.). New York: McGraw-Hill.

VDI 2230, Part 1. (2003). *Systematic calculation of high duty bolted joints—joints with one cylindrical bolt*. Berlin: Beuth Verlag GmbH.

Index

A

ABAQUS
    fixed reactive analysis problem 64
    inertia relief analysis 71
    spider bolt model 511
    thread bolt connection model *511*
ABS, *see* Antilock brake systems
Absolute acceleration, random vibration test,
    linear SDOF system 360
Absolute displacement, suspension
    component load analysis 34
Accelerated vibration test specification,
    examples 334
Acceleration
    central difference method 80
    finite element analysis 68–69
    inertia relief analysis 66
    linear SDOF system, vibration
      testing 339
    McPherson strut type suspension 33
    modal transient response analysis 71
    powertrain CG accelerations 27
    powertrain mount load analysis 24–25,
      25, 27
    random vibration test, linear SDOF
      system 356, 359–361
    spindle
      forces and velocity *51*
      vertical WFT force input method 37
    suspension component load analysis,
      inertia force 32
    vehicle performance simulation 49
    and velocity change 7

Accumulated plastic strain
    classical flow rules 261
    kinematic hardening 266
    yield surface function 259
Advanced cyclic plasticity models,
    nonproportional hardening *280–281*,
    280–283
Advanced Dynamic Analysis of Mechanical
    Systems (ADAMS)
    CDTire model validation 46
    multibody dynamics analysis 11
Advanced high strength steel (AHSS),
    resistance spot-welded joint fatigue
    life 448
Age-hardening aluminum alloys, temperature
    correction factor 125
AHSS, *see* Advanced high strength steel
All-wheel-drive (AWD) vehicle, powertrain
    mount load analysis 23, 25
Aluminum alloys
    constant amplitude nominal stress–life
      curve 147
    constant amplitude stress–life curve *148*
    equivalent strain approach 308
    fatigue damage models 178
    fatigue limit 119
    load correction factor 127
    medians method 223
    Mitchell's equation 120, 223
    nonproportional hardening 282
    nonproportional strain hardening 165
    relative stress gradient correction
      factor *238*

Note: Page numbers in *italics* indicate figures, tables, and footnotes

Printed in the United States
By Bookmasters